LAyer is inse

single layer in this book
is a 2 layer Bd.

module = ic

Controlling Radiated Emissions by Design

Second Edition

THE KLUWER INTERNATIONAL SERIES
IN ENGINEERING AND COMPUTER SCIENCE

Controlling Radiated Emissions by Design

Second Edition

Michel Mardiguian

Kluwer Academic Publishers
Boston/Dordrecht/London

Distributors for North, Central and South America:
Kluwer Academic Publishers
101 Philip Drive
Assinippi Park
Norwell, Massachusetts 02061 USA
Telephone (781) 871-6600
Fax (781) 871-6528
E-Mail <kluwer@wkap.com>

Distributors for all other countries:
Kluwer Academic Publishers Group
Distribution Centre
Post Office Box 322
3300 AH Dordrecht, THE NETHERLANDS
Telephone 31 78 6392 392
Fax 31 78 6546 474
E-Mail services@wkap.nl>

Electronic Services <http://www.wkap.nl>

Library of Congress Cataloging-in-Publication

Mardiguian, Michel.
Controlling radiated emissions by design / Michel Mardiguian.-- 2nd ed.
p. cm. -- (The Kluwer international series in engineering and computer science ; SECS 580)
Includes bibliographical references and index.
ISBN 0-7923-7978-0 (alk. paper)
1. Electronic circuits--Noise. 2. Electromagnetic interference. 3. Electronic circuit design. 4. Shielding (Electricity) I. Title. II. Series.

TK7867.5 .M37 2001
621.382'24--dc21

00-061081

Printed on acid-free paper.

Printed in the United States of America

Contents

Foreword

Somewhere around 1991, I was fortunate enough to preview another book, Controlling Conducted Emission by Design, by J.Fluke. My assessment at the time was that Mr. Fluke had written a very good and useful volume, its only flaw being that it focused on just half of the EMI emissions problem. This, of course, was also its strength; although radiated and conducted emissions do not exist in separate worlds, it can be useful, at least as a point of departure, to look at them as distinct phenomena. Expanding the scope of the conducted emissions book therefore would have served mostly to dilute its purpose and compromise its clarity.

After some discussions with the publisher it was decided to correct the omission by recruiting an author for a companion volume, to be titled Controlling Radiated Emissions by Design. I am gratified to have played a role in making that happen.

Many EMC engineers with whom I have worked are capable of writing a good radiated emissions book, but few can match Michel Mardiguian's combination of practical experience and proficiency with the pen. On the engineering side, he has mastered EMC on such diverse projects a weapon guidance systems, automotive electronics, jet engine test platforms and, most recently, the renowned tunnel that stretches beneath the English channel to join England with France. In terms of academic credentials, the author has written or coauthored six earlier books, including a highly doable and well respected one on electrostatic discharge. He has taught the principles of electromagnetic compatibility throughout the United States and Europe, and IEEE symposium records contain many of his short works.

Michel's books typically have several characteristics:

1. They are aimed at practical applications but provide enough theoretical material to support the suggested design and retrofit solutions.
2. They rely heavily on visual material, allowing the reader to "see" the EMI problems and solutions.
3. They omit everything that is not essential to the subject under scrutiny.

This third quality is less common that one might expect—many authors feel an irresistible urge to stuff everything they know into one volume. But Michel avoids egocentric on marginal subjects, preferring to approach each book simply as a job that needs to be done.

With this book the author follows his usual pattern but with perhaps a greater refinement. Mathematical material is largely confined to two chapters, allowing the book to look mostly at real-world applications. More than 175 tables and illustrations provide information in a highly accessible format. This type of book is difficult for the editors and expensive for the publisher, but the reader should benefit. Some subjects (e.g., EMC testing) are covered very briefly, but the reader is referred to other sources of pertinent information.

It would be premature and presumptuous to call this book a classic in EMC literature, but it does display some of the characteristics of enduring works. Controlling Radiated Emissions by Design constitutes a cohesive and clear examination of the subject and is perhaps the author's best work to date.

Jeffrey K. Eckert
J. K. Eckert & Company
Gainesville, Virginia

Preface

In 1999, after seven years and six reprints, this book, although quite successful, started to show its age. Interestingly, most of the EMI aspects of the future that we were anticipating in the first edition are actually happening. At the request of Kluwer Academic Publishing, I prepared this fully revised edition, taking into account the significant technology advance of year 2000. The quantum jump that we are witnessing at the end of the century, almost unprecedented in the history of electronic engineering, poses a severe challenge to the EMC discipline, which this new edition will hopefully help to win.

The speed and range of digital transmissions, exceptions accounted like fiber optics, is approaching the limits of what a pair of copper wires can handle, but this goes with an increasing radiated pollution. Clock frequencies, in the core of the most powerful processor chips are exceeding the Gigahertz, causing the printed circuits and the modules themselves to compete with, and sometimes override, cable RF radiation.

At the same time, a myriad of popular RF devices and services like portable cellular phones, laptop PCs with integrated wireless modems, wireless Internet for everyone etc .. are spreading over, such that the number of sensitive RF receivers operating within a square kilometer of urban area can be counted in tens of thousands.

The gathering of so many RF sources, turning into unauthorized transmitters, mixed with so many licit receivers is increasing more than ever the number of potential interference situations. Also, less obvious and still controversial is the possible health hazards associated with the long term exposure to the very near field of low power RF transmitters, especially in the upper VHF and UHF regions. This will probably result in a continuous search for lower radiations. This book is meant to help the design engineers, technicians and EMC specialists for more comprehensive, efficient and economical methods of RF emissions containment.

I am especially grateful to Alex Greene, chief editor, for his timely support and constant interest in the achievement of this project. My thanks to Don Sweeney and Maxine Martin, from DLS Electronic Systems Inc. and Ulf Nilsson from EMC Services in Sweden, who provided valuable suggestions for this second edition. My gratitude goes also to Fabienne Baudron who did the final editing with an impeccable and rigorous dedication. Last, but not least, million thanks to Corinne, who under the strain of being the wife of an engineer-who-has-a-book-to-write, supported me with her love and patience.

Michel Mardiguian
St. Remy les Chevreuse, France

Chapter 1

Generalities on Radiated Interference

1.1 THE PROBLEM OF RADIATED EMI

During the first part of the twentieth century, electromagnetic interference (EMI) was primarily a concern for telecommunications via wire and radio. In these specific engineering communities, highly qualified engineers developed quite sophisticated analytical approaches to predict interference levels, taking into consideration the sources' parameters, the propagation media and the sensitivity of the pertinent telephone or radio receivers. The typical sources of EMI until the 1960s were natural atmospheric noise, motor commutators, fluorescent lights, automobile ignition systems and 50/60 Hz overhead lines (unintentional sources), plus jamming by other authorised (and sometimes unauthorised) radio transmitters.
The results of such interference ranged from a mere annoyance, such as poor telephone and radio audibility and disrupted TV reception, to a serious danger if the interfered transmission was critical, as in the case of an emergency service, an air navigation system, etc.
But let us say that, with proper handling of the frequency allocations and some rather simple constraints applied at national and international levels, the problem was acceptably under control. The "policing of the air waves" was a manageable task.

In the latter half of the century, an incredible proliferation of new RF sources has developed, including all computing and digitally operated devices (today it is very difficult to find a simple electrical home appliance that does not incorporate at least an 8-bit microprocessor), medical electronics, power switchers, machine tools, remote control systems, local area networks (LANS), etc. Most of these devices produce signal trains of discrete frequencies that can be a stable, continuous threat for radio communication.

At the same time, the legitimate users of the radio spectrum have grown in number and types of service. From a limited, identified number of radio, TV, navigation and

safety services before 1950, the number of RF spectrum users is still increasing, comprising satellite data transmission, telemetry and radio telescopes, CB and cellular telephones, vehicle positioning systems, etc.
As the number and variety of potential offenders increases, so does the number and variety of potential victims, and all growths are approximately exponential. Accordingly, the risk of interference increases astronomically, and it is little wonder that specifications and regulations have become more and more precise and stringent in an attempt to hold the problem to a manageable scale.
Equally serious is the problem of a system interfering with itself. This occurs due to ignorance or negligence regarding the EMI problem created by multiple sources and victims packed in close proximity.
Therefore, the subject of controlling interference, and more specifically radiated interference (the focus of this book), becomes both a matter of satisfactory performance of an equipment as well as its compliance with the non interfering norms of the military or civilian world. EMI control must be incorporated in the design stage of a product rather than postponed until the day of the prequalification or final acceptance testing. The latter invariably results in late and expensive fixes and retrofits.
This book will provide the necessary background and step-by-step procedures to design a product that neither radiates undesired signals in excess of the applicable specifications nor interferes with itself or other equipments in its vicinity. This desirable condition is known as electromagnetic compatibility (EMC).

1.2 BASIC UNDERSTANDING OF RADIATED EMI

Three players are needed for the interference act:
1. a source,or emitter
2. a victim, or receiver
3. a coupling path between the two

The coupling medium can be conduction or radiation. This dichotomy is, of course, overly simplistic: no conduction exists without some radiation, and vice-versa. However, it is true that certain couplings occur predominantly by conduction (through power mains, ground wires, etc.), while others occur mostly by space propagation. We will concentrate, in this book on radiation coupling.
Depending on their internal impedance, circuits can create, in their proximity, fields that are principally electric (given in volts per meter) or magnetic (given in amps per meter). At a greater distance from the source (farther than $\lambda/2\pi$), field is termed electromagnetic, no matter the source impedance, low or high.
In terms of radio frequency interference, a field of 1 V/m is a rather strong field, knowing that the field sensitivity of typical radio or TV receivers could be as low as $1 \mu V/m$. Therefore, it is foreseeable that circuits operating at high frequency, even on low-level signals, can create interference at substantial distances.

A simple calculation can give an instant feel of the problem. Let us assume a small computer consisting mainly of a CPU and memory board. The computer circuit includes 60 chips, each one consuming about 250 mW.

Assume also that only one fourth of the circuits resident on these chips are toggling synchronously at the internal clock frequency of 50 MHz, for instance.
It can be said that the total power switched at a given instant during a transition is:

$$(1/4) \times (60 \times 0.250) = 3.75 \text{ W}$$

Now assume that a minuscule fraction of this power is not dissipated by Joule effect in the chips, the wiring, and various resistances or displays, but is radiated instead. For instance (and this is quite optimistic) assume that on the 50 MHz fundamental, only 10^{-6} of the total switched power is radiated, that is 3.75 μW.
A simple formula gives the field strength of any given radiator:

$$E(V/m) = \frac{1\sqrt{30 \text{Pr}}}{D} \qquad (1.1)$$

where D = distance from source, in meters
P_r = radiated power (including antenna gain) in watts

At 3 meters distance, our 3.75 μW from the PC board will result in:

$$E = \frac{1\sqrt{30 \times 3.75 \times 10^{-6}}}{3} = 3.5.10^{-3} V/m = 3.5 mV/m$$

Expressed in standard units of EMI specifications for field strength, this is:

$$\text{E dB}\mu\text{V/m} = 20 \log (3.5 \times 10^3 \ \mu\text{V/m}) = 71 \text{ dB}\mu\text{V/m}$$

The minimum field strength required by TV and FM listeners for decent reception quality in remote areas is in the range of 50 to 60 dBμV/m. Therefore in the case of frequency coincidence (co-channel EMI) the computer clock may seriously affect radio/TV reception in its vicinity since, at a 3 m distance (assuming the computer has a plastic case with no shielding), the EMI field will be 3 to 10 times stronger than the sound or picture carrier. In this case, annoying interference is likely to exist 30m away, or even farther if there is some field enhancement caused by metallic structures, poles, etc. in the short radio path between the computer and the victim receiver antenna.

Of course, the basis of any sound design is:

1. to understand by which mechanisms a circuit devised to store and process data, or convert power, ends up transmitting radio
2. to have a numerical estimate of those mechanisms, in order to reduce or counteract them
3. to test the results as early as possible on prototype model

The two first items are really the design issues. Since no radiation can exist without voltages or currents, a large part of the quantitative approach will be spent on circuit design, waveforms analysis and layout recommendations.

1.3 EMI TERMINOLOGY AND UNITS

Due to the wide dynamic range we face in the EMI/RFI/EMC disciplines, logarithmic scaling is used extensively. Therefore most ratios (dimensionless) and magnitudes are expressed in decibels (dB).
Expressing the ratio of two powers becomes:

$$dB = 10 \log \frac{P_2}{P_1} \quad (1.2)$$

Often, data are measured in units of amplitude (e.g., voltage, current, field strength) instead of power.
Substituting $P = V^2/R$ into Eq. (1.2) yields:

$$dB = 10 \log \frac{(V_2)^2/R_2}{(V_1)^2/R_1}$$

$$= 20 \log (V_2/V_1) + 10 \log (R_1/R_2)$$

If $R_1 = R_2$, this becomes:

$$dB = 20 \log (V_2/V_1) = 20 \log (I_2/I_1) \quad (1.3)$$

Equations (1.2) and (1.3) are computed in Table 1.1 for all usual ratios.Corresponding negative dB equivalents are found by reciprocating any of the ratios.

The expression in dB for voltage or current is obtained by substituting 1Vor 1A into Eq. 1.3:

$$dBV = 20 \log V, \text{or}$$

$$dBA = 20 \log I$$

Retrieving voltage, current or field strength from its dB value is obtained by taking the antilog (log-1):

$$V_{volt} = \log^{-1}(\text{dBV}/20) = 10^{\frac{\text{dBV}}{20}}$$

If voltage has to be derived from a power in dBm (dB above 1mW), this is obtained by:

$$V_{dB\mu V} = 90 + 10\log(Z) + P_{dBm}$$

$$= 107 + \text{dBm, for } Z = 50\ \Omega \qquad (1.4)$$

For narrowband (NB) EMI, where only one single spectral line (i.e., a sine wave) is present in receiver (or victim's) bandwidth, the EMI signal can be expressed in:

I. Voltage: V, dB above 1 V (dBV) or dB above 1 μV (dBμV)
2. Current: A, dB above 1 A (dBA) or dB above 1 μA (dBμA)
3. Power: W, mW or dB above 1 mW (dBm)
4. E field: V/m, μV/m or dBμV/m
5. H field or magnetic induction: A/m, μA/m or dBμA/m
Tesla or Gauss (1 G = 80 A/m, 1 T = 10^4 G)
6. Radiated Power density: W/m^2, mW/cm^2 or dBm/cm^2

For broadband (BB) EMI, where many spectral lines combine in the receiver's bandwidth, the received EMI is normalized to a unity bandwidth:

1. Voltages: μV/kHz, μV/MHz or dBμV/MHz
2. Currents: μA/kHz, μA/MHz or dBμA/MHz
3. E fields: μV/m/kHz, μV/m/MHz or dBμV/m/MHz
4. H fields: μA/m/kHz, μA/m/MHz or dBμA/m/MHz

There are several ways to recognize NB or BB interference conditions. A simple one is this:
Given the passband or 3 dB bandwidth (BW) of the receiver (or victim's input amplifier) and Fo, the EMI source fundamental frequency, the interference will be:

$$\text{BB if BW} > F_o$$
$$\text{NB if BW} < F_o$$

TABLE 1.1 *: Ratios to dB Conversion*

Voltage, Current, or Field Amplitude Ratio	*Power Ratio*	*Decibels*
x 1.12	x 1.25	+1
x 1.25	x 1.6	+2
x 1.4	x 2	+3
x 2	x 4	+6
x 3.16	x 10	+10
x 5	x 25	+14
x 10	x 100	+20
x 1,000	x 10^6	+60

Examples:
1 μV = 0 dBμV
1 mV/m = 60 dBμV/m
50 Ω = 34 dBΩ
1 mW = 0 dBm
1 mW, in 50 Ω = 0.22 V = 107 dBμV

1.4 U.S. AND WORLDWIDE REGULATORY APPROACH AGAINST RADIATED EMI

Long ago, maximum emission levels were set by civilian commissions to protect broadcasting, as well as military organisations to insure optimum reception of vital radio communications, navigation and guidance system signals, etc. Of specific nature is the TEMPEST problem, which covers the possible eavesdropping on confidential data by unauthorised receivers. (TEMPEST, by the way, is a codeword for a classified program but not an acronym. Therefore, its letters do not "stand for" anything.) This is both a military/government concern for national security and a business/industry issue for the protection of sensitive and confidential data.
Civilian limits to radiated emissions are fairly severe but, generally, one could remark that military specifications are significantly more stringent due to the close proximity (sometimes less than 1 m) of sources and victims within aircraft, submarines, etc. Since this book is a design tool, and not an encyclopedia of specifications, we will review only briefly the basis of civilian and military standards that pertain to radiated EMI.

1.4.1 Worldwide Civilian Standards

Table 1 .2 gives a summary of the principal emission standards and national laws for the USA and worldwide. As indicated, many national laws are based on the CISPR documents.

In general, maximum emission levels have been set by an international commission on a device-by-device basis after establishing a need to protect radio communications from interference by a particular piece of equipment. In many

TABLE 1.2 Major Civilian Emission Standards & Regulations, Worldwide. Industrialized countries, including the ones not listed, generally follow CISPR recommendations for RF emissions.

	Computers & digital processing devices	ISM[4]	Electric appliances	Radio,TV receivers	Fluoresc. tubes & neon	Solid state variable speed drive	Engine ignition	Automob. electronics on-board	Cellular telephone systems (mob,base)	Unlicensed RF transmtrs remote contr. pagers, toys
CISPR[1] or Internatl	22	11	14	13	15	IEC 1800-3	12	25	ETS 300-342	
European[2] Norm,EN	55022	55011	55014	55013	55015		55012	EC Dir. 95/54		
USA[3]	FCC15-B	FCC-18		FCC15-B			SAE J551.C		FCC22	FCC15-C
Canada	ICES.003	ICES.001					ICES.002			
Japan	VCCI									
Australia, New Zeal.	AS3548	AS2064	AS1044	AS1053	AS4051		AS2557			
China & Taiwan	CNS 13438	CNS 13306		CNS 13439						

Notes: 1) CISPR = Comité Internationl Special des Perturbations Radio Électriques. 1 rue de Varembe,1211 Geneve, Switzerland
2) Each European country has translated, as required by Directive the Europ.Norms into a National Regulation, to have a legal base for enforcement. The National Regulation often has a different#, but is cross-referenced to EN or CISPR.
3)Although not exactly the same in certain frequency bands, FCC limits are close to CISPR22 or 11.
4) Industrial, Scientific or Medical equipment. For medical devices, the EMC requirements of IEC601/2 are quoting CISPR11

instances, a country such as the USA, Germany or the Netherlands has been the instigator and major player in the development of certain category of limits. But nevertheless, to avoid a myriad of different limits between countries, an international commission called the CISPR, part of the International Electrotechnical Commission IEC, is tasked to publish unanimously accepted limits and establish test methods for EMI emissions.

Once CISPR limits have been voted into effect by member nations of the IEC, they sooner or later become translated into national standards within the various Countries. Depending on the type of government in each country, these limits may remain as industry standards, more or less voluntarily applied, or turned into compulsory laws. The latter has been the case for many years in Germany, since 1980 in the USA, and more recently in all countries of the European Economic Community, where CISPR recommendations have been published as European Norms (ENs).

European Community has adopted a rather strict attitude regarding EMC: the 89/336 Directive states that no equipment, regardless if it is manufactured in Europe or imported, can be commercialized if it does not comply with low emission <u>and</u> immunity requirements. There has been a six year grace period for the industry to adjust and absorb the slack of existing products in stock. After January 1st 96, compliance has been mandatory, the conformity being attested by the CE Mark affixed on the equipment.

Failure to fulfill applicable test and certification requirements (even if no interference is actually created) is delictuous, and the manufacturer (not the user) can be prosecuted and penalized.

Small regulatory variations occur in some countries. However if a product is designed to meet CISPR or EN emission levels, it is likely that it will comply with interference laws for its specific class in most industrialised countries. Each country has a specific policy to ensure that manufacturers or vendors deliver products that meet emission limits. Some require test and certification by a national laboratory (for example, the VDE in Germany) and some accept manufacturers' self-certification.

In general, the standards consider two categories of potential interference signatures:

1. The RF signals emitted by intentional high-frequency sources (mainly covered by CISPR Publications 11, 13 and 22). These include HF industrial equipment, ovens, welders, oscillators, digital computing devices and, in general, any equipment which intentionally generates high-frequency signals but is not a licensed radio transmitter. In general, the frequency spectrum of these equipments contains a set of discrete, stable spectral lines (narrowband spectrum).
2. The RF signals emitted by non-intentional high-frequency sources (mainly covered by CISPR Publications 12, 14 and 15). These include motors, fluorescent lights, dimmers, car ignitions and so forth where production of RF energy is fortuitous. The frequency spectrum of these equipments is generally a dense series of random or correlated frequencies (broadband spectrum).

A lot of specific international EMC standards keep being issued as technology evolves. They are changing quite fast and listing them all in a book like this one would be tedious and often useless since they are perishable data. However, besides the well-known computer category, few of them are sufficently remarkable of a large population of equipment:

CISPR 25 RFI protection of receivers on board vehicles.
A vehicle (car, truck, bus) is an EMI nightmare in itself. A maze of noisy components like pulse-driven motors, Electronic Control Modules with fast clocks etc ... are in close proximity and share the same cable harness with sensistive radio receivers: car radio, cellular telephones, CB or emergency receivers, navigation / positioning systems. These receivers antennas are typically within a meter range of potential disturbers. This call for stringent emission control and, indeed, the CISPR 25 radiated limits are among the most severe in existence, sometimes tougher than Mil-std levels (see Chap.12).

IEC 1800-3 : Variable Power Drive Systems
The technical progress in power semi-conductors has generated a whole family of secondary power sources where the output is made adjustable by electronic control: light dimmers, adjustable speed drives etc... In the latter, an electrical motor from fractional HP to thousands kW can have its speed adjusted continuously without any mechanical or electromechanical device. Operating in fast switched mode as voltage or frequency converters, they are generally powerful sources of NB interference, since the wiring to the motor can run over significant lengths in a commercial building, a public site etc... IEC 1800-3 establishes conducted and radiated emissions limits, similar to those of CISPR22 for computers.

ETSI Standards
Telecommunications is a complex field which is more and more intermixed with digital techniques. The tremendous growth of mobile telephony and wireless data communication has exacerbated potential EMI situations. A specific European organisation, the European Telecom. Standard Institute has been founded in 1988 to develop ad-hoc specifications among which a certain number, labelled ETSxxxx are covering EMC aspects. Every time a system incorporates a telecommunication equipment, whatever it connects to a wired or radio link, the system has to comply with the relevant ETS norm, which are progressively replacing the national rules in the EC member countries. This, to some extent, becomes a pan-European version of the North American FCC Part 68 rules.

1.4.2 FCC Standards for Emissions

In the USA, Congress has delegated authority to the Federal Communications Commission (FCC) to regulate civilian radio and wire communication. This includes the issuance and enforcement of EMI regulations. General classifications are:

1. *FCC Part 15: restricted RF devices.* This is the most widely applied FCC EMI standard. It covers digital computing devices plus garage door openers, radio-

controlled toys, cordless telephones and other intentional low power transmitters. For computers, the FCC part 15 Sub-part "B"considers two classes of potential offenders:

A. Class A digital equipment that is marketed to be used only in industrial and commercial areas. For these, self-certification by manufacturer is permitted.
B. Class B digital equipment that can be used in a residential environment. Because of the higher probability of proximity with a radio or TV, Class B limits are approximately 10 dB more severe than Class A, and a formal procedure is required.The manufacturer must have its equipment tested in an accredited (FCC or NVLAP) laboratory.

Then he can choose either:

- the "Declaration of Conformity". The test report must be available any time, on electronic format, upon FCC request.
- the FCC Certification. Sending the test report, manufacturer applies for a FCC certification label and agreement #. This testifies its product as compliant.

2. *FCC Part 18: Industrial Scientific and Medical Equipment (ISM).* This part regulates spurious emissions from equipment and appliances that purposely generate RF energy for something other than radio or telecommunications. These include heating, ionization, ultrasonic process, medical treatment and diagnosis equipment, etc. Such devices have precise frequency allocations where no emission limit exists, but emission restrictions apply across the rest of the spectrum.

1.4.3 The Mutual Recognition Agreement (MRA)

In order to eliminate trade barriers and to comply with Worldwide Trade Organisation (WTO) rules, FCC and the European CENELEC have signed in 1997 an agreement by which the EMC qualification tests conducted in the USA by an FCC-approved lab will be accepted by European countries, and vice-versa. Therefore, US manufacturers and EMC laboratories can test and approve in USA an equipment bound for the European market.

1.4.4 Other US Government Standards (Nonmilitary)

Many industries and professional bodies have issued emission standards in areas that are not covered by FCC regulations, or where they believe that more stringent limits are needed. A few examples are described below.

FDA standard for medical devices

Issued by the Food and Drug Administration under the file name MDS-20 1-0004, these limits were designed to control emissions and susceptibility in medical electronics. This is of particular importance because interference in hospital (and especially intensive care units) can be directly life threatening. The emission levels are in the ball park of the MIL-STD-461. Around 1995, the emissions limits of this US Standard, which were found very hard to meet and required expensive hardware additions have been progressively superseded by the international requirements of

IEC 601-1-2, whose emission limits are those of CISPR11 (Industrial, Scientific or Medical devices).

NACSIM 5100

Better known as TEMPEST, this is a classified standard developed to ensure that confidential information cannot be captured by unauthorized receivers during electronic processing, handling or transmitting by government agencies. Special measurement techniques and limits are required to guarantee such a low level of emission that the clear, unciphered data is undetectable. These techniques and limits are, of course, classified.

SAE Standards
The Society of Automotive Engineers, among many engineering activities, has issued several EMI emission standards such as:
SAE J 551 radiated EMI from vehicles and associated devices
SAE J 181 Levels and measurement methods

RTCA Standards
The Radio Technical Commission for Aeronautics has issued a broad set of stringent standards for the critical domain of civilian aircraft. Environmental effect are addressed in DO 160 (present revision is D). EMI emissions are covered in Section 21. The emission limits are parallel to, although generally less stringent than, MIL-Std-461/ RE 01, 02, 03. Between rev "C" and "D", this standard has followed the same philosophy as MIL-Std 461, replacing the NB/BB dichotomy by a single limit.

1.4.5 Military Emission Standards

The military approach to EMI control is a remarkable example of a well structured, rather unforgiving, test program. The most prevalent EMC military standard is the MIL-STD-461 (up to 1993, its revision level was "C") and its parent document MIL-STD-462, which describes test methods. Its general organization tree is shown in Fig. 1.1.
Being a tri-service document (army, navy and air force) it is extremely versatile and can be tailored to any equipment, subassembly or part of a system. It is not applicable to an entire complex system, such as a fire-control system, a radar warning system, a vehicle, etc. Those need to be qualified on their site or carrier, per MIL- Std 464.

MIL-Std-461, due to its broad scope, has been recognized by many defense organizations outside the USA, as well as some nonmilitary agencies. These entities have more or less transposed (admittedly or not) the MIL 461/462 organization and values. This is the case for UK (Def-Stan), Germany (VGxxx), France (GAM EG 13), and many other countries. The NATO EMC Standards (STANAG) also apply the MIL 461/462 approach.

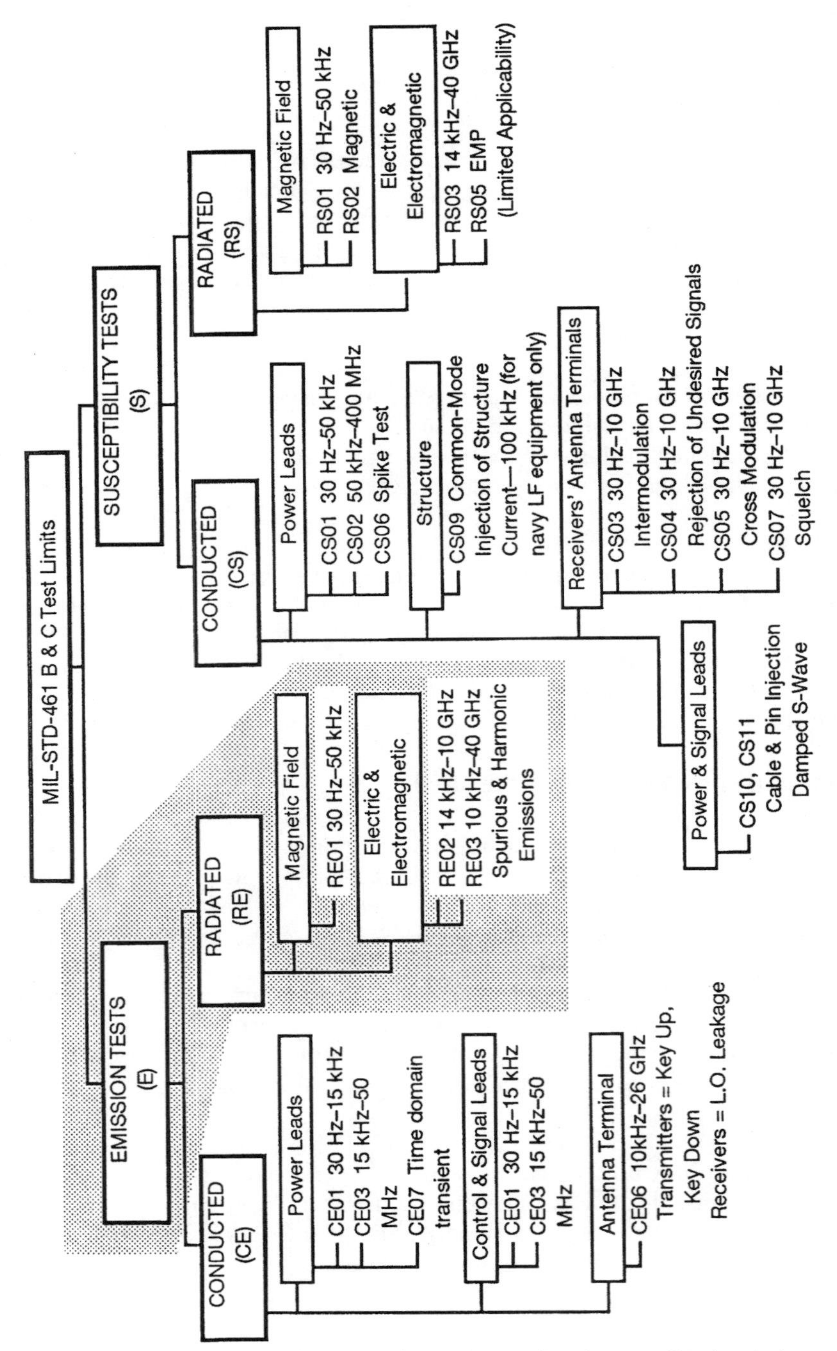

FIGURE 1.1 MIL-Std 461 Organization tree. Shaded portion shows radiated emissions

The emission aspect is covered by the left-hand branch of the Fig. 1.1 tree, which subdivides in two smaller branches:
1. the series of conducted emission tests (CE)
2. the series of radiated emission tests (RE)

STD-461 is to be regarded as a complete inventory of all the EMC tests that can be performed. This does not mean that every military deliverable item has to pass all 27 tests!
For each equipment, a test plan has to be written as part of the EMI control plan. Under supervision of the procuring agency, the test plan outlines which of the MIL-STD-461 tests will be performed. Depending on the nature of the equipment and its installation field, some tests are not applicable. Of the 27 tests, is not uncommon to see, only 10 or 12 being required. Therefore, it is important for the potential user of an available equipment which is claimed "compliant with MIL 461" to determine exactly what tests have been actually carried out.
In 1993, MIL-STD 461 and 462 incurred a major revision, becoming 461-D and 462-D. The changes are significant: for the emission tests, the NB vs BB discrimination has been suppressed, replaced by a single limit in a prescribed bandwidth. The conducted tests are performed only up to 10MHz, instead of 50MHz as before, and the conducted limit (CE102) is given as a voltage measured at a 50μH LISN port, instead of a current as in the "C" version. Although in many respects the rev."D" is an improvement in coherence, clarity and test practices, some of the changes were resented as a stepback in the degree of EMC protection, especially for analog radio receivers. As such, certain procurement specifications and test plans still quote the 461C (dated 1986).

1.4.6 Intrasystem vs. Intersystem EMI

A dichotomy appears when we attempt to define the borderline between a self jamming system (i.e., equipment that disturbs its own operation in a sterile electromagnetic environment) and one that is a nuisance to the external environment (see Fig. 1.2). These two types of EMI are referred to as intrasystem and intersystem EMI, respectively.
As far as radiated emissions are concerned, intrasystem compatibility requires two conditions:

1. None of the inside sources may radiate more than the field susceptibility of its neighbor components within the spectrum boundaries.
2. Condition 1 being met, the combined emissions of all sources together must remain below the field susceptibility threshold of any component within the system.

These are functional conditions. If they are not met, the system simply does not work properly, even in a noncritical environment. Therefore, satisfying intrasystem EMC is the prime, "selfish" goal of any designer who is concerned only with moving an operational product through the assembly line.

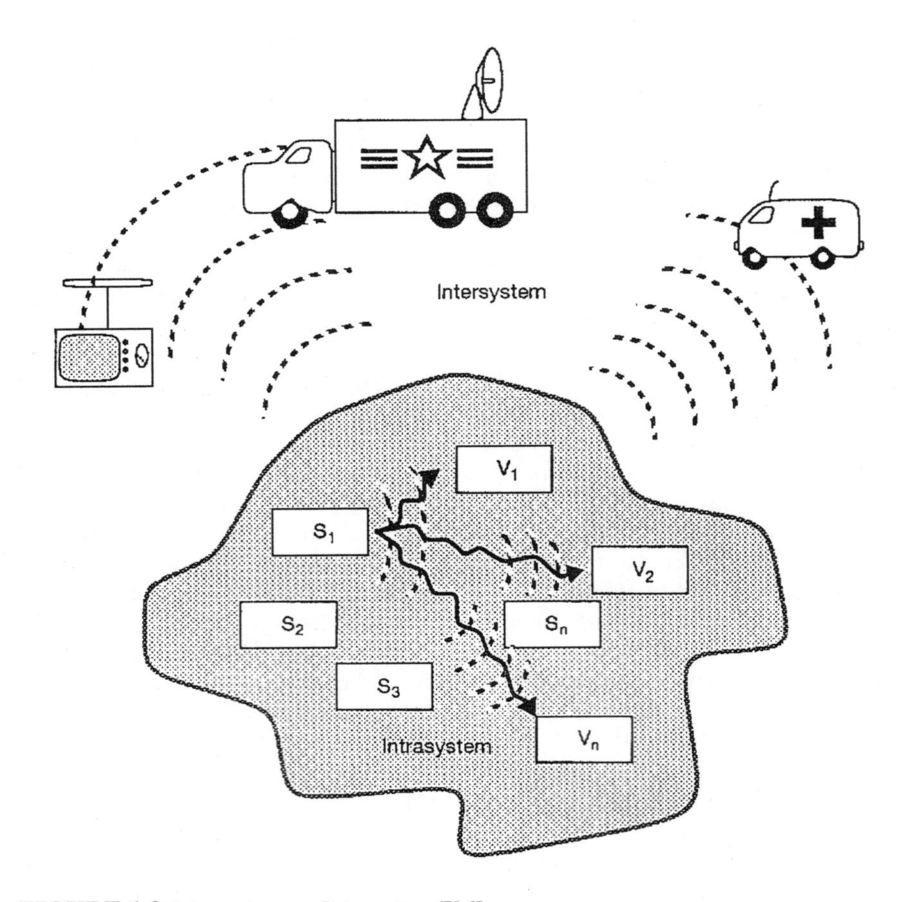

FIGURE 1.2 Intrasystem vs. Intersystem EMI

Quite differently, Intersystem EMC requires that when the whole system operates, no other system in its vicinity may be disturbed, especially by spurious radiated signals. Intrasystem EMC concerns only the performance of a single system, while Intersystem EMC deals with electromagnetic pollution of the environment.
With Intrasystem EMC, the culprit and victims are known and can be controlled. It is not absolutely required (although a safe practice) that interference be suppressed at the source. After all, if all victims were very well hardened, they could tolerate strong noise sources next door.
But with Intersystem EMC, the victims are unknown and out of our control. They are replaced by the emission limits which substitute for real victims. Thus, radiated emissions at source level should be suppressed for both intra- and intersystem EMC. Exceptions accounted, modeling and experience show that very often, if all sources in a system meet the radiated emission limits dictated by RF protection for the environment, these limits are stringent enough to guarantee that self jamming will be avoided within the system as well.

Chapter 2

Electric and Magnetic Fields from Simple Circuit Shapes

If one wants to avoid empirical recipes and the "wait and see if it passes" strategy, the calculation of radiated fields from electrical circuits and their associated transmission cables is of paramount importance to proper EMI control. Unfortunately, precisely calculating the fields radiated by a modern electronic equipment is a hopeless challenge. In contrast to a CW transmitter, where the radiation source characteristics (e.g., transmitter output, antenna gain and pattern, spurious harmonics, feeder and coupler losses, etc.) are well identified, a digital electronic assembly, with its millions of input/output circuits, printed traces, flat cables and so forth, is impossible to mathematically model with accuracy -at least within a reasonable computing time- by today's state of the art. The exact calculation of the E and H fields radiated by a simple parallel pair excited by a pulse train is already a complex mathematical process.

However, if we accept some drastic simplification, it is possible to establish an order of magnitude of the field by using fairly simple formulas. Such simplification include:

1. retaining only the value of the field in the optimum direction
2. having the receiving antenna aligned with the maximum polarization
3. assuming a uniform current distribution over the wire length, which can be acceptable by using an average equivalent current instead of the maximum value
4. ignoring dielectric and resistive losses in the wires or traces

The formulas described hereafter are derived from more complex equations found in the many books on antenna theory. They allow us to resolve most of the practical cases, which can be assimilated to one of the two basic configurations:

1. the closed loop (i.e., magnetic excitation)
2. the straight open wire (i.e., electric excitation)

2.1 FIELDS RADIATED BY A LOOP

An electromagnetic field can be created by a circular loop carrying a current, I, (Fig 2.1) assuming that:

• I is uniform in the loop.
• There is no impedance in the loop other than its own reactance.
• The loop size is $<< \lambda$
• The loop size is $<$ D, the observation distance.
• The loop is in free space, not close to a metallic surface.

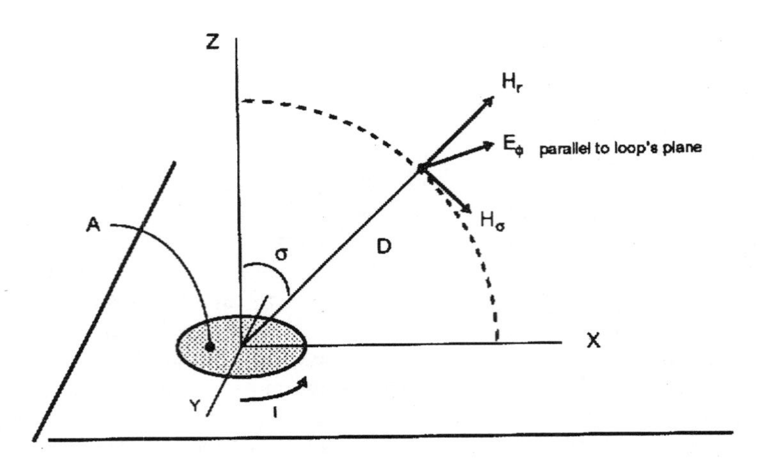

FIGURE 2.1 Radiation from a Small Magnetic Loop

E and H can be found by using the simple solutions that Schelkunoff derived from Maxwell's equations.
Replacing some terms by more practical expressions:

$$H_r \ \text{A/m} = \frac{IA}{\lambda}\left[\frac{j}{D^2} + \frac{\lambda}{2\pi D^3}\right]\cos\sigma \tag{2.1}$$

$$H_\sigma \ \text{A/m} = \frac{\pi IA}{\lambda^2 D}\sqrt{1 - \left(\frac{\lambda}{2\pi D}\right)^2 + \left(\frac{\lambda}{2\pi D}\right)^4}\ \sin\sigma \tag{2.2}$$

$$E_\phi \ V/m = \frac{Z_o \pi I A}{\lambda^2 D} \sqrt{1 + (\frac{\lambda}{2\pi D})^2} \ \sin\sigma \tag{2.3}$$

where,

I = loop current, in amperes
A = loop area in m^2
λ= wavelength in meters = $300/F_{MHz}$
D = distance to observation point, in meters
Z_0 = free space impedance = 120 Ω or 377 Ω

Comparing this with Fig 2.1, we see that for $\sigma = 0$, $E_ø$ and $H_ø$ are null ($\sin\sigma = 0$), while H_r is maximum ($\cos\sigma = 1$). Except near the center of a solenoid or a transmitting loop antenna, this H_r term in the Z-axis direction is of little interest because it vanishes rapidly, by its $^1/_D{}^2$ and $^1/_D{}^3$ multipliers. Notice also that there is no E_r term.

To the contrary, in the equatorial plane, for $\sigma = \pi/2$, H_r is null, and $E_ø$, $H_ø$ get their maximum value. So, from now on, we will consider systematically this worst-case azimuth.

Looking at Eqs. (2.2) and (2.3), and concentrating on boundary conditions, we see two domains, near-field and far-field, plus a transition region.

Near Field: for $\lambda/2\pi D > 1$ (i.e $D < \lambda/2\pi$ or $D < 48/F_{MHz}$)

Under the square root in Eqs. (2.2) and (2.3), the larger terms are the ones with the higher exponent Thus, neglecting the other second- or third-order terms, we have:

$$H_{A/m} = \frac{IA}{4\pi D^3} \tag{2.4}$$

$$E_{V/m} = \frac{Z_o IA}{2\lambda D^2} \tag{2.5}$$

We remark that H is independent of λ, i.e. independent of frequency: the formula remains valid down to DC. H falls off as $^1/_D{}^3$. E increases with F and falls off as $^1/_D{}^2$

In this region, called near-field or induction zone, the fields are strongly dependent on distance. Any move toward or away from the source will cause a drastic change in the received field. Getting 10 times closer, for instance, will increase the H field strength 1,000 times.

The E/H ratio, called the wave impedance (because dividing Volt /m by Amp /m produces ohms) is:

$$Z_w \text{ (near loop)} = Z_o \frac{2\pi D}{\lambda} \tag{2.6}$$

When D is small and λ is large, the wave impedance is low. We may say that in the near field, Z_w relates to the impedance of the closed loop circuit which created the field, i .e., almost a short. As D or F increases, Z_w increases.

Far Field: for $\lambda/2\pi D < 1$ (i.e., $D > \lambda/2\pi$, or $D > 48/F_{MHz}$)

The expressions under the square roots in Eqs. (2.2) and (2.3) are dominated by the terms with the smallest exponent. Neglecting the second- and third-order terms, only the value of 1 remains, so:

$$H_{A/m} = \frac{\pi I A}{\lambda^2 D} \tag{2.7}$$

$$E_{V/m} = \frac{Z_o \pi I A}{\lambda^2 D} \tag{2.8}$$

In this region, often called the *far-field, radiated-field or plane wave region* *, both E and H fields decrease as 1 /D (see Fig. 2.2). Their ratio is constant, so the wave impedance is:

$E/H = 120\pi$ or $377\ \Omega$

This term can be regarded as a real impedance since E and H vectors are in the same plane and can be multiplied to produce a radiated power density, in W/m^2.
E and H increase with F^2, a very important aspect that we will discuss further in our applications.

*However, "plane wave" does not have exactly the same meaning. There is another condition governing the near- or far-field situation which is related to the physical lenght of the antenna. If « l », the largest dimension of the radiating element is not small compared to distance D, another near-field condition exists due to the curvature of the wavefront. To have less than 1dB (11%) error in the fields calculated by Eqs. (2.7) and (2.8), another condition is needed, which is: $D > l^2/2\lambda$. However, if we keep the aerial dimension inferior to λ/2, we see that both length and distance far-field conditions are met for D (far-field) > λ/2π.

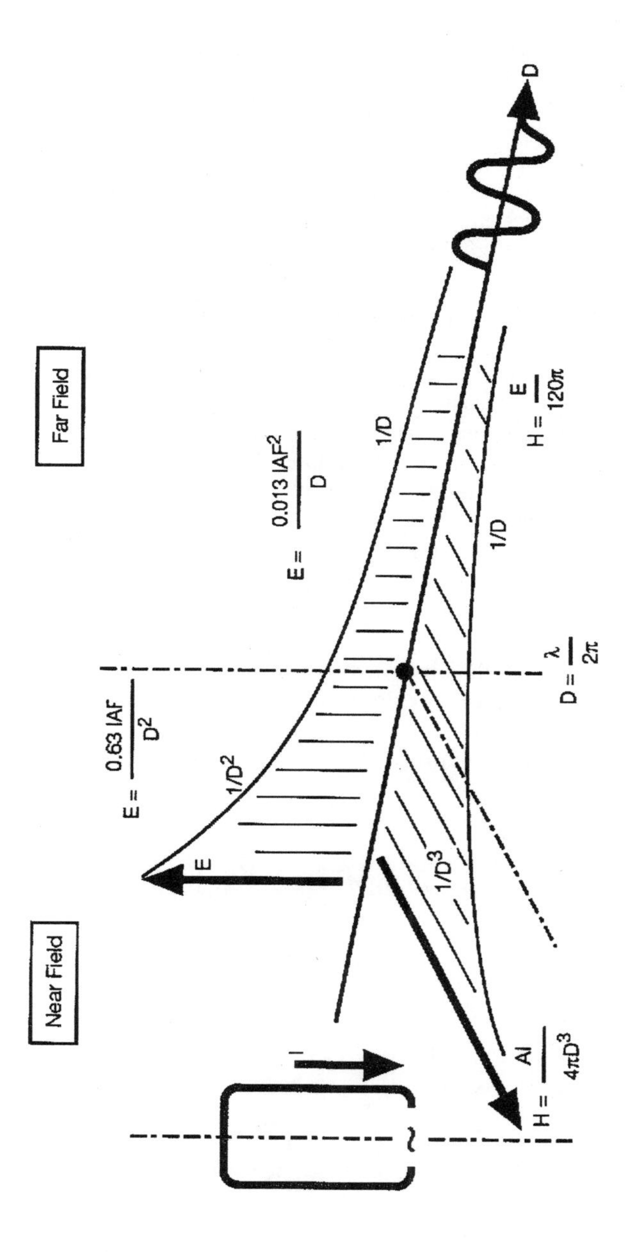

FIGURE 2.2 E and H Fields from a Perfect Loop

Transition Region: for $\lambda/2\pi D = 1$, or $D = 48 / F_{MHz}$
In this region, all the real and imaginary terms in field equations are equal, so all terms in 1/D, $1/D^2$ and $1/D^3$ are summed with their sign. This zone is rather critical because:
1. With MIL-461 testing (RE02 test), the test distance being 1m, the near-far field conditions will exist around 48 MHz, which complicates the prediction.
2. Speculations concerning the wave impedance are hazardous due to very abrupt changes caused by the combination of real and imaginary terms for E and H.

2.2 FIELDS RADIATED BY A STRAIGHT WIRE

It does not take a closed loop to create an electromagnetic field. A straight wire carrying a current, I, creates an electromagnetic field (most radio communication antennas are wire antennas). The only practical difficulty is that, in contrast to the closed loop, it is impossible to realize an isolated dipole with a DC current : only AC current can circulate in an open wire self-capacitance. Fields generated from a small, straight wire are shown in Fig. 2.3.
E and H can be derived from Maxwell's equations with the same assumptions as the elementary loop, i.e.:
- I is uniform.
- The wire length << λ
- The wire length < D, the observation distance.
- The wire is in free space, not close from a ground plane.

Using Schelkunoff's solutions for a small electric dipole, expressed in more practical units:

$$E_r = 60 I \ell \left(\frac{1}{D^2} - \frac{j\lambda}{2\pi D^3} \right) \cos\sigma \tag{2.9}$$

$$E_\sigma = \frac{Z_o I \ell}{2\lambda D} \sqrt{1 - \left(\frac{\lambda}{2\pi D}\right)^2 + \left(\frac{\lambda}{2\pi D}\right)^4} \sin\sigma \tag{2.10}$$

$$H_\phi = \frac{I \ell}{2\lambda D} \sqrt{1 + \left(\frac{\lambda}{2\pi D}\right)^2} \sin\sigma \tag{2.11}$$

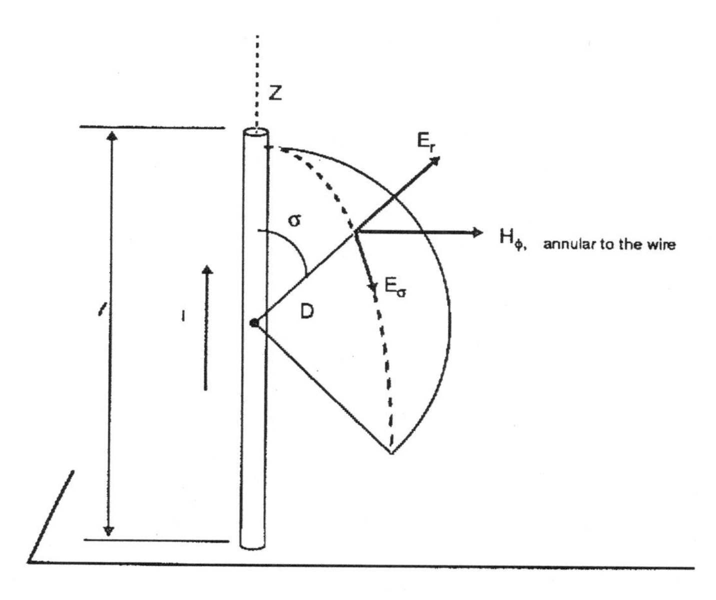

FIGURE 2.3 E and H Fields from a Small, Straight Wire

where,

> I = wire current, in Amperes
> l = dipole length in meters
> λ= wavelength in meter = $300/F_{MHz}$
> D = distance to observation point in meters
> Z_0 = free-space impedance = 120π or 377 Ω

As for the loop, we remark that for $\sigma = 0$, E and $H_ø$ are null ($\sin \sigma = 0$) while E_r is maximum ($\cos \sigma = 1$). E_r, in the axis of the wire, is of little interest because it drops off rapidly, as $1/D^2$, $1/D^3$

In the equatorial plane for $\sigma = 90°$, E and H have their maximum values. From now on we will consider this worst-case azimuth. In fact, for $\sigma = 90° \pm 25°$, the error would be less than 10%. As for the loop, we can see two domains plus a transition region.

Near Field: for $\lambda/2\pi D > 1$ (i.e., $D < \lambda/2\pi$)
As for the loop, the terms with the higher exponent prevail, under the square root.

Neglecting the other second or third terms:

$$H_{A/m} = \frac{I\ell}{4\pi D^2} \qquad (2.12)$$

$$E_{V/m} = \frac{Z_o I\ell\lambda}{8\pi^2 D^3} \qquad (2.13)$$

Here again,we notice that H, annular around the dipole is independent of F.
This formula holds down to DC, where it equals the well known result of the Biot and Savart law for a small element.

This time, it is to H to fall off as $1/D^2$ while E falls as $1/D^3$ Both are strongly dependent on distance. Since, for a current kept constant, E decreases when F increases, the wave impedance decreases when D or F increases:

$$Z_w = \frac{E}{H} = Z_o \frac{\lambda}{2\pi D} \qquad (2.14)$$

As for the loop, Z_w near the source relates to the source impedance itself which, this time, becomes infinite when F gets down to DC.

Far field: for $\lambda/\pi D < 1$ (i.e., $D > \lambda/2\pi$ or $D > 48/F_{MHz}$)
The terms with higher exponents can be neglected under the square root so:

$$H_{A/m} = \frac{I\ell}{2\lambda D} \qquad (2.15)$$

$$E\,(V/m) = Z0 \times I \times l\,/\,(2\lambda D) \qquad (2.16)$$

Both E and H decrease as 1/D. This ratio, as for the loop in far field, remains constant:

$$E/H = 120\pi \text{ or } 377\ \Omega$$

For a single wire, E and H increase as F (instead of F^2 for the loop).

Transition region: for $\lambda/2\pi D \approx 1$ or D $48/F_{MHz}$
The same remarks apply as for the loop.
Fig. 2.4 summarizes the evolution of Z_w for wires and loops as D/λ increases.

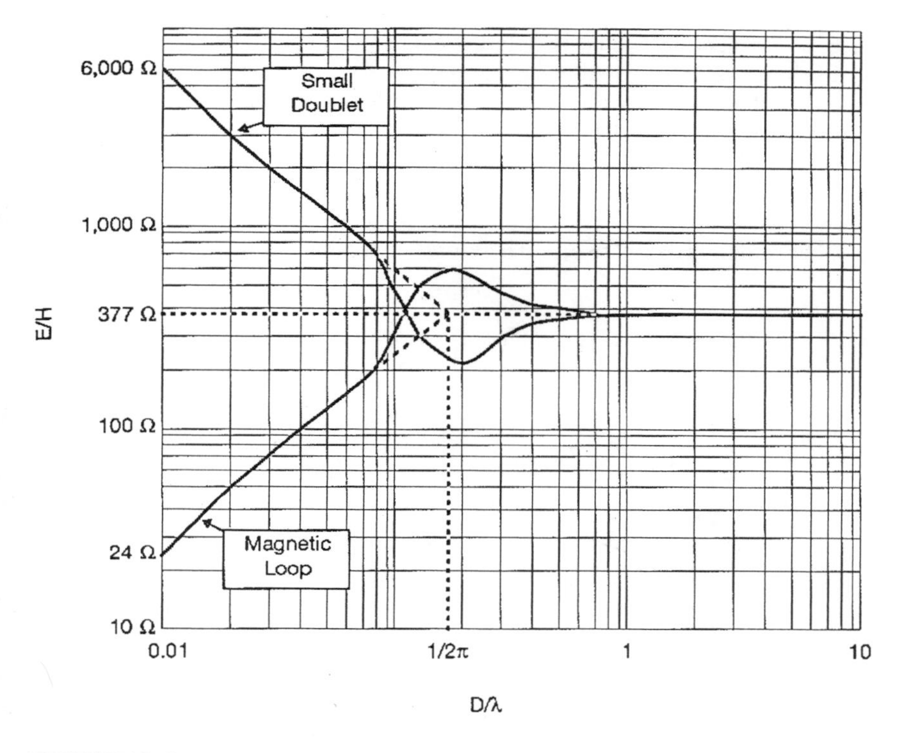

FIGURE 2.4 Wave Impedance vs Distance/Wavelength

2.3 EXTENSION TO PRACTICAL, REAL-LIFE CIRCUITS

Although theorically correct, ideal loop or doublet models have a limited practical applicability due to the restrictions associated with their formulas:

1. Distance D should be large compared to the circuit dimensions.
2. The circuit length should be less than $\lambda/2$, and preferably less than $\lambda/10$, for the assumption of uniform current to be acceptable.
3. The single wire model corresponds ideally to a piece of wire floating in the air, in which a current is forced, a situation seldom seen in practice.
4. The single wire model assumes that the circuit impedance is infinite in near field, or at least larger than the wire reactance alone; this condition is rarely met except in dipole or whip antennas.

5. Restrictions 3 and 4 seem alleviated if one switches to the loop model. The loop, indeed, is a more workable model for practical, non-radio applications because it does not carry the premise of a wire coming from nowhere and going nowhere. But

it bears a serious constraint, too: the loop must be a short circuit, such as the wave impedance, and hence the E field is only dictated by the coefficients in Maxwell's equational solutions. If this condition is not met (and it is seldom met except in the case of a coil with only one or few turns and no other impedance), the H field found by Eqs. (2.4) and (2.7) will be correct, but the actual associated E field will be greater than the calculated value.

In reality, we deal with neither purely open wires nor perfect loops, but with circuit configurations which are in between. Therefore, predictions in the near field would produce:

- a higher E field than reality, if based on open-wire model (pessimistic error)
- a lower E field than reality, if based on ideal loop model (optimistic error)

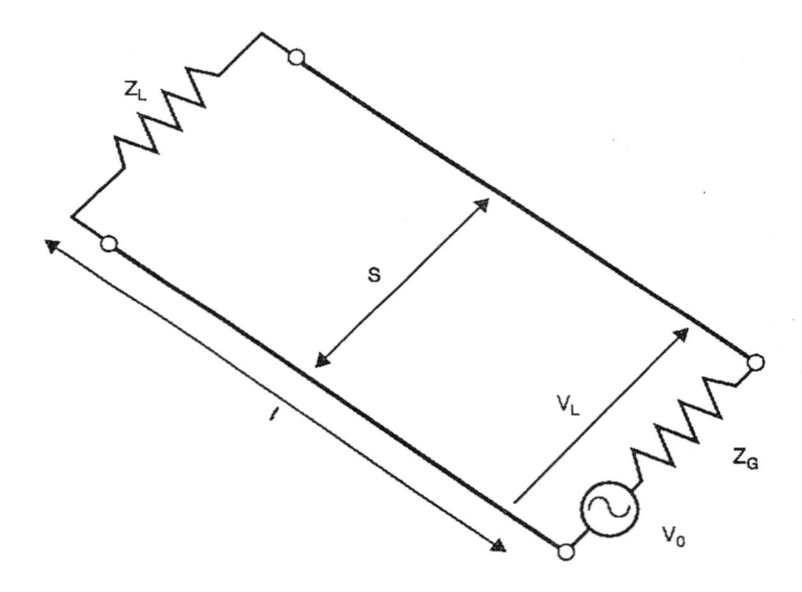

FIGURE 2.5 The Modified Single-Wire Model

Measurements have proven that the latter can cause underestimates as large as 60 dB or more. Therefore, certain adjustments need to be made. Assuming these adjustments, the modified equations and models can be usable by the practician for most of the actual circuits and cable configurations encountered.
The core of this modeling is the "modified single wire model" of Fig. 2.5 where, instead of a straight wire or circular loop, we have a more practical vehicle, where an area l x s can be treated by the loop equation or regarded as two single wires with a radiation phase shift equal to $\sin(2\pi s/\lambda)$.
Depending on the circuit impedance, we will use one or the other, as explained next. The basis for the simplification is that, in near field, the wave impedance Z_w = E/H is "driven" by the circuit impedance Z_c every time this circuit impedance is in-between the ideal dipole Z_w (high) and the ideal loop Z_w (low).

In the near field, given the total circuit impedance Z_c (wiring plus load):

$$Z_c = Z_g + Z_L$$

A. If $Z_c > 7.9 \times D_m \times F_{MHz}$, we will use the modified wire model:

$$E_{V/m} = \frac{VA}{4\pi D^3} \tag{2.17}$$

where,

V = driving voltage in V (the actual line voltage, not the open circuit voltage)
A = circuit area l x s, in m^2
D = observation distance, in m

Very often, more practical units are welcome:

$$E_{(\mu V/m)} = \frac{7.96\ VA}{D^3} \tag{2.17a}$$

for V in volts, A in square centimeters and D in meters.

B. If $Z_c < 7.9 \times D_m \times F_{MHz}$, we will use the ideal loop formulas, since the circuit impedance is low enough for this model to hold.

$$E_{V/m} = \frac{0.63\ IAF_{MHz}}{D^2} \tag{2.18}$$

for current (I) in amperes, A in square meters and D in meters.

Using more convenient units:

$$E_{(\mu V/m)} = \frac{63\ IAF_{MHz}}{D^2} \tag{2.18a}$$

for current (I) in amperes, area in square centimeters and D in meters.

Because, with such low-impedance radiating circuits, this is the H field which is of concern, we can employ a straightforward application of Eq. (2.4):

$$H\ A/m = IA/(4\pi D^3)$$

for A in square meters and D in meters. Notice that this near-field expression is exactly the mirror image of the high-impedance loop E field in Eq. (2.17).
The product I x A is often referred to as the "magnetic moment". Using more convenient units:

$$H_{(\mu A/m)} = \frac{7.96 \times I \times A}{D^3} \qquad (2.19)$$

for current (I) in amperes, A in square centimeters and D in meters.

In the far field, regardless the type of excitation (i.e., circuit impedance), E and H are given by Eqs. (2.7) and (2.8), which we will express in terms of frequency rather than wavelength:

$$E_{V/m} = \frac{0.013\ VAF^2_{MHz}}{D \times Z_c} \qquad (2.20)$$

with V in volts, I in amperes, A in square meters and D in meters.

$$H_{A/m} = \frac{E}{120\pi} = \frac{35.10^{-6} \times I \times A \times F^2_{MHz}}{D} \qquad (2.21)$$

Using more standard units of EMI measurement,

$$\boxed{E_{(\mu V/m)} = \frac{1.3}{D} \times \frac{V}{Z_c} \times A \times F^2_{MHz}} \qquad (2.22)$$

for V in volts, A in square centimeters and D in meters.

At this point, a few remarks are in order:

1. We now have an expression for E fields that can be calculated by entering the drive voltage, which often is more readily known to the circuit designer than the current.
2. Except for very low impedance loops (less than 7.9 Ω @ 1MHz, less than 7.9 mΩ @ 1kHz), i.e. low-voltage circuits carrying large sinusoidal or pulsed currents, it is generally the wire pair model (Eqs. 2.17 and 2.19) that applies.
3. In the near field, for all circuits except low-impedance loops, E is independent of frequency and remains constant with V. At the extreme, if Z_c becomes extremely large, I becomes extremely small but Zw increases proportionally, keeping E constant when F decreases down to DC.
4. In the far field, radiation calculated for a two-wire circuit (the single dipole formula times the weighing factor sin $2\pi s/\lambda$ due to the other wire carrying an opposite current) would reach exactly the same formula as the one for a radiating circular loop. Therefore, as long as its dimensions are $\ll \lambda$, the actual circuit shape has virtually no effect on the radiated field in the optimum direction. Only its area counts.
5. For $l \geqslant \lambda/4$, the circuit begins to operate like a transmission line or a folded dipole. Current is no longer uniform and, in the equation, the length "l" must be clamped to $\lambda/4$, i.e. l_m is replaced by 75/FMHz. In other words, the active part of this fortuitous antenna will "shrink" as F increases. Furthermore, if the circuit does not terminate in its matched impedance, there will be standing waves, and the effective circuit impedance will vary according to transmission line theory. Incidentally, the radiation pattern will exhibit directional lobes.
When s is not $\ll l$, i.e. the loop is not a long rectangle but is closer to a square, the upper bound is reached (Keenan, Ref.1) when $(l + s) = \lambda/4$, i.e.,
Fmax = 7,500 /$(l + s)$, for F in megahertz and l, s in centimeters. Furthermore, with conductors partly in a dielectric, the velocity is reduced by a factor of:

$$\frac{1}{\sqrt{1 + (\varepsilon_r/2)}}$$

For PVC or Mylar cables, this gives: $F_{max} \approx 5{,}300 / l_{cm}$. For printed traces, Fmax $\approx$ 4,400 $/l_{cm}$. So an average value $F_{max} \approx 5{,}000 / l_{cm}$ could be retained, above which the physical length should be replaced by $5{,}000/F_{MHz}$.

6. In the far field, if $Z_c > 377\ \Omega$, the value of 377 Ω must be entered in Eq. (2.20). This acknowledges the fact that an open-ended circuit will still radiate due to the displacement current.
7. In the far field, E increases as F^2 for a loop or a pair. This is a very important effect that we will address in the application part of this book. Equations (2.17a) to (2.22) are plotted in Figs. 2.6a, 2.6b and 2.7 for a "unity" electrical pair of 1V - cm^2 and a unity magnetic moment of 1A - cm^2. They show E and H at typical test distances of 1m and 3m.

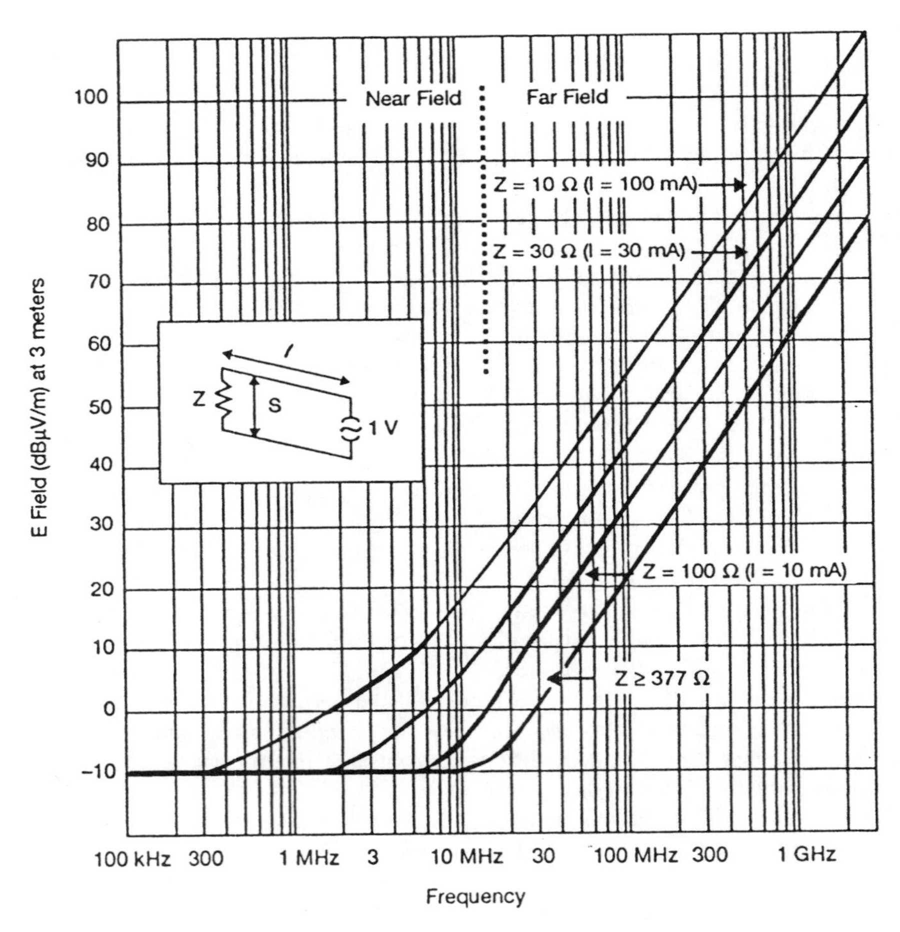

FIGURE 2.6a E field from a 1cm2 loop, driven by 1V, at 3m. For other voltages and l x s values, apply correction: 20 logV+20 log l x s.

USE This Chart

adjust for Voltage
" " CM2

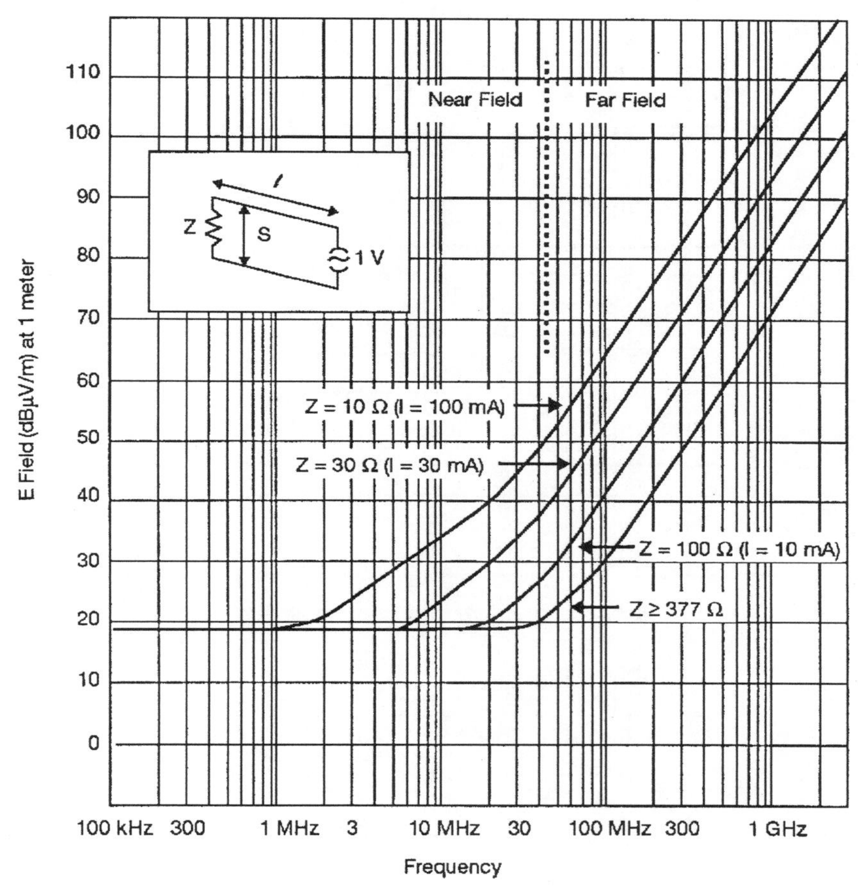

FIGURE 2.6b E field from a 1cm2 loop, driven by 1V, at 1m. For other voltages and l x s values apply correction: 20 logV+20 log l x s.

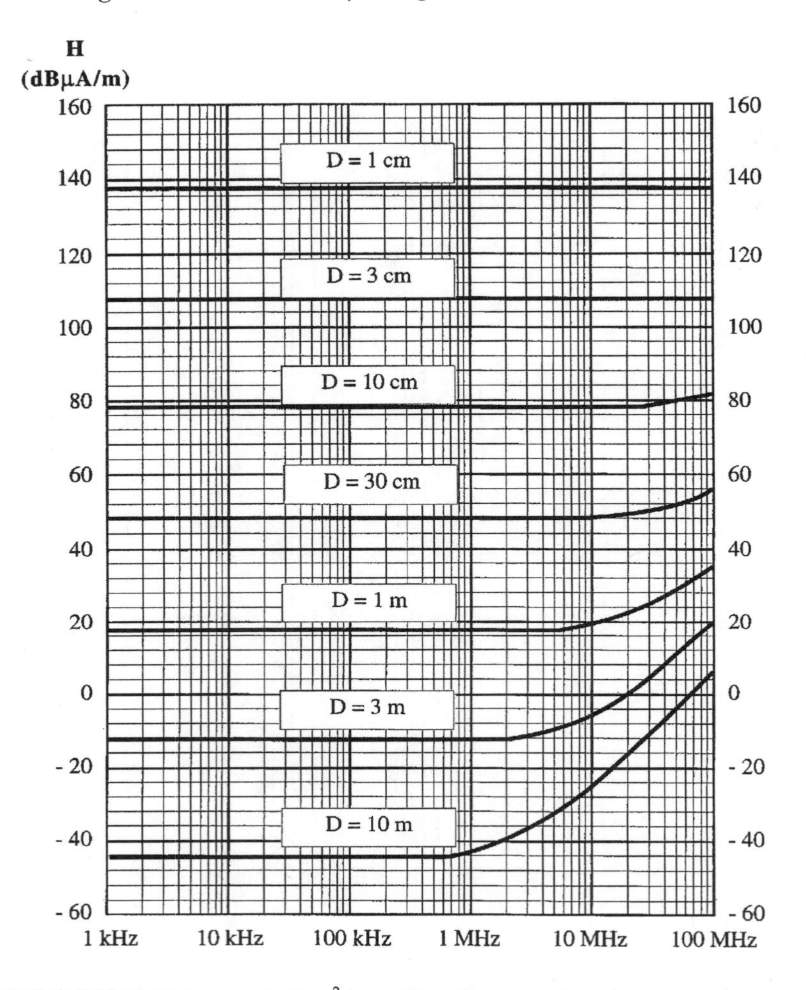

FIGURE 2.7 H Field from a 1A, $1cm^2$ loop. For other currents and areas, apply correction: $20 \log(I \times A_{cm}^2)$

2.4 DIFFERENTIAL-MODE RADIATION FROM SIMPLE CIRCUITS

The simplest radiating configuration we will encounter in practice is the small differential-mode radiator, whose largest dimension, *l* is smaller than both the observation distance, D, and the height above ground. Such circuits (see Fig. 2.8) are found with:

- printed circuit traces
- wire wrapping or any hard-wired card or backplane
- ribbon cables
- discrete wire pairs (for $l \ll D$)

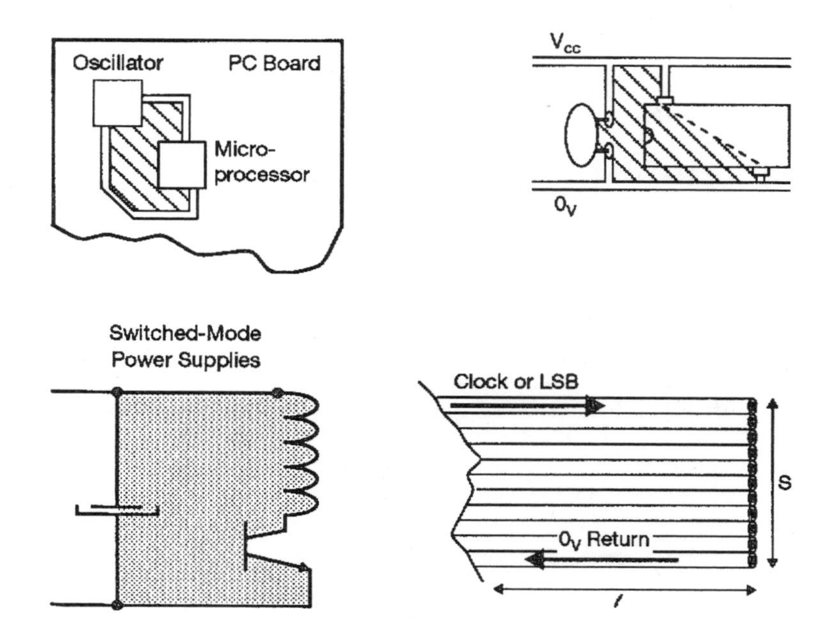

FIGURE 2.8 A Few Typical Differential-Mode Radiators

The culprit source exciting such circuits can be a digital or analog signal, a switching transistor, a relay, a motor creating transient spikes, etc. There is also a possibility that the differential pair is simply a carrier of an EMI signal that has been generated in the vicinity and coupled to it through power supply conduction or nearby crosstalk.

The procedure is then:

1. Determine V_{diff} or I_{diff} at the frequency (or frequencies) of interest, and the circuit impedance.
2. Check for $D_m \geqslant 48 / F_{MHz}$ (far-field conditions).
3. If far-field, use the curves in Fig. 2.6 or Eq. (2.22).
4. If near-field, determine if the circuit belongs to the low-Z, loop model (for $Z_c < 7.9 \times F \times D$) or to the wire model($Z_c > 7.9 \times F \times D$).
5. Check if $l_{cm} > \lambda/4$, or 5,000/F. If l is larger, replace l by 5,000/F for entering area correction.
6. Repeat step 5 for wire or trace separation, s.
7. Calculate $A_{cm}{}^2 = l \times s$, and determine area correction, 20 log A, using adjustments 5 and 6, if needed.
8. Find the E field by: $E_{dB\mu V/m} = E_o$ (from curves) + 20 log A + 20 log V
9. If H field calculation is desired instead, use the H-field curves and add corrections.

Example 2.1

A video signal crossing a PC board is to be switched to different displays. The carrier is 100 MHz with a line voltage of 10 Vrms. The PCB has the following characteristics:

- single layer (no ground plane)
- average video trace length, l = 6 cm
- average distance to ground trace, s = 2.5 cm

Calculate the E field at 1 m versus the RE02 limit of MIL-STD-461 when:

1. The circuit is loaded with 75 Ω.
2. The circuit is "on" but stand-by, open-ended

We will assume, as starting condition, that there is no box shielding.

- V_{diff}= 20dBV (normal) or 26dBV (when open-ended, the voltage will double).
- At 100 MHz, the near-field/far-field transition distance is:

$$D_{N\text{-}F} = 48/100 = 0.48 \text{ m}$$

Therefore, at 1m we are in far-field conditions. We can use Fig. 2.6 or Eq. (2.22).

- Surface correction: 20 log(6 x 2.5) = 24 dB. The 6 cm length is << $\lambda/4$ at 100 MHz.
- For the 75Ω load, we can interpolate between the 30 and 100 Ω curves. For the open circuit ($Z=\infty$) we will use the curve for $Z \geqslant 377$ Ω.

The calculations steps are detailed below:

Frequency	100MHz (with Z=75Ω)	100MHz (open circuit)
E_0 (1V-1cm^2)	44dBμV/m	30dBμV/m
Amplitude correct.	20dBV	26dBV
Area correct (cm^2)	24dB	24dB
E_{final}	88dBμV/m	80dBμV/m
E specification limit	29dBμV/m	29dBμV/m
Δ dB	59	51

The specification limit is exceeded by 50-60 dB. Such an attenuation can only be obtained, in practical terms, by using a multilayer board or a single-layer board with a ground plane (this reduces the radiating loop width by a 40-times factor i.e., 32 dB) and a correctly designed metal housing to provide about 30dB of shielding at 100 MHz. Both solutions will be discussed further in this book.

Example 2.2

A 5 V/20 A switching power supply operates at the basic frequency of 50 kHz. In the secondary loop (formed by the transformer output, the rectifier and the electrolytic capacitor), the full-wave rectified current spikes have a repetition

frequency of 100 kHz and an amplitude of 60 A (peak) on the fundamental. The circuit loop dimensions are 3 x 10 cm. The loop impedance at this frequency is 0.2 Ω
Calculate E and H at 100 kHz for a 1 m distance.

1. The 60 A amplitude corresponds to 36 dBA.
2. At 100kHz, the near-far transition distance is D_{NF}=48/0.1 = 480m.
At 1m, we are in very near field.
3. With Z_c = 0.2 Ω we meet the criteria for Z_c < 7.9 x F x 1 m.
4. The area correction is 20 log 30 cm^2 = 30 dB.

We will use the ideal loop model, i.e. Fig 2.7 for H field, or Eqs. (2.18a) and (2.19). The field is computed as follows:

F:	0.1MHz	
H_0 (1A.1cm^2)	17dBμA/m	
E_0 (1A.1cm^2)		15dBμV/m
Amplitude correct. (Amp)	36dB	36dB
Area correct (cm2)	30dB	30dB
H_{final}	83dBμA/m	
E_{final}		81dBμV/m

2.5 COMMON-MODE RADIATION FROM EXTERNAL CABLES

External cables exiting an equipment are practically always longer than the size of the equipment box, so it is predictable that they will be the major contributors to radiated emissions (just as they would be for radiated susceptibility). Cables radiate by the differential-mode signals that they carry, as discussed in the previous section, but also by the currents circulating in the undesired path; that is, the ground loop.

Ground loop CM currents are due to the unbalanced nature of ordinary transmitting and receiving circuits, the imperfect symmetry of the differential links and, more generally, the quasi-impossibility of avoiding some CM return path, whether the loop is visible (circuit references grounded at both ends to chassis and/or earth) or invisible (floated equipments or plastic boxes). This phenomenon of common-mode excitation of external cables causing radiated emissions is one of the most overlooked one in computers and high frequency devices interference (Ref. 3).
The very simple example of Fig. 2.9 shows the unavoidable generation of a common-mode current. Assume that over a cable length l, the wire pair (untwisted) separation is s = 3mm, and the cable height above ground is h = 1m. When a signal is sent from equipment #1 to equipment #2, although the designer believe in good faith that current is coming back via the return wire, we see no reason why some of the current (i_3) could not return by the unintended path, i.e. the ground loop.

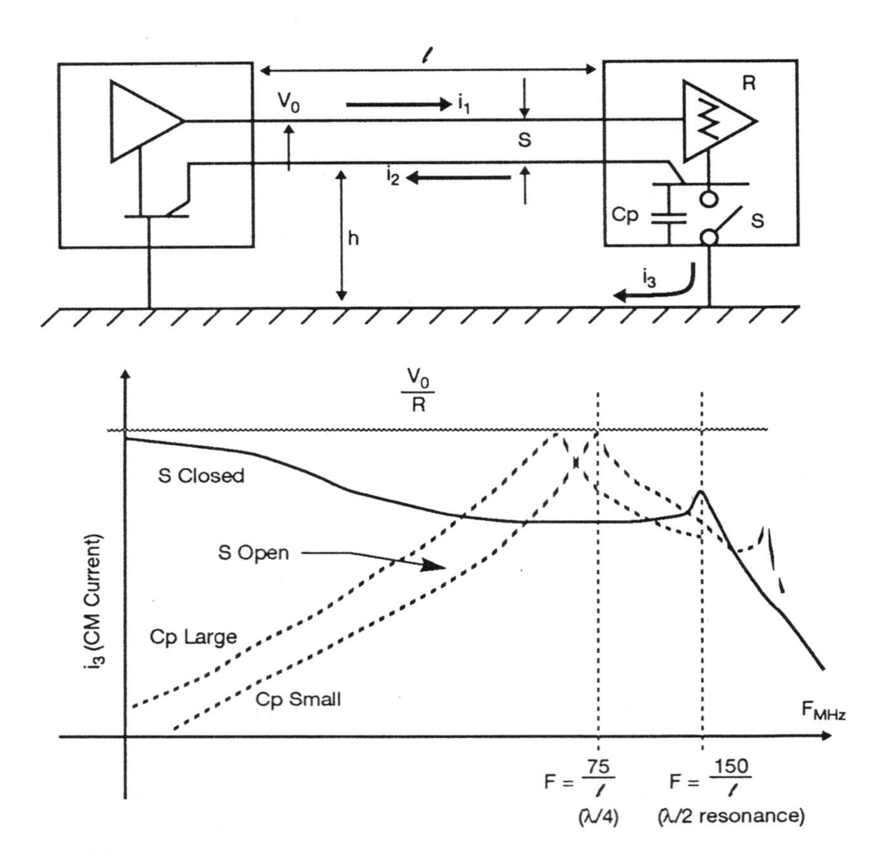

FIGURE 2.9 Conceptual view of CM Current Generation by a Differential Signal

Currents i_2 and i_3 will split to the prorata of the respective impedances. If the wires in the pair are in close proximity, the return inductance by the pair is significantly less than the return impedance via the ground. But less does not mean none. Let us assume that only 10 percent of the current is returning by the ground loop.
The differential-mode (DM) radiation is related to 0.9i x l x s. The CM radiation is related to 0.1i x l x h. The ratio of the two magnetic moments is:

$$CM / DM = (0.1i \times l \times 1\ m) / (0.9i \times l \times 0.003m) = 37$$

$$20 \log 37 = 31dB$$

The CM loop, although the corresponding current is regarded as a side effect, radiates 31dB above the DM loop (notwithstanding that this latter can be further reduced by twisting).

Opening the switch S, i.e. floating the PCB 0V reference, would reduce I_{cm} at low frequencies (say, below a few megahertz for a 10 m cable). In this range, radiated EMI is generally not a problem; but the problem would aggravate at first cable resonance because we now have an oscillatory inductance-resistance-capacitance (LRC) circuit with a high Q (low R). The hump in CM current (Fig. 2.9) depends on the value of C_p, the PCB stray capacitance to chassis. At this occasion, we see that the traditional recipe of grounding the PCB only at one box (star grounding) is useless in the frequency domain of most radiation problems and can even be slightly worse at some specific frequencies.

The undesired CM currents that are found on external cables can be some percentage of the signal currents that would normally be expected on this interface but, more often, cables are found to carry high-frequency harmonics *that are not at all part of the intentional signal* (see Fig. 2.10). Rather, they have been picked up inside the equipment by crosstalk, ground pollution or power supply DC bus pollution. Since the designer did not expect these harmonics, it is only during FCC, CISPR, MIL-STD-461 or other compliance testing that they are discovered.

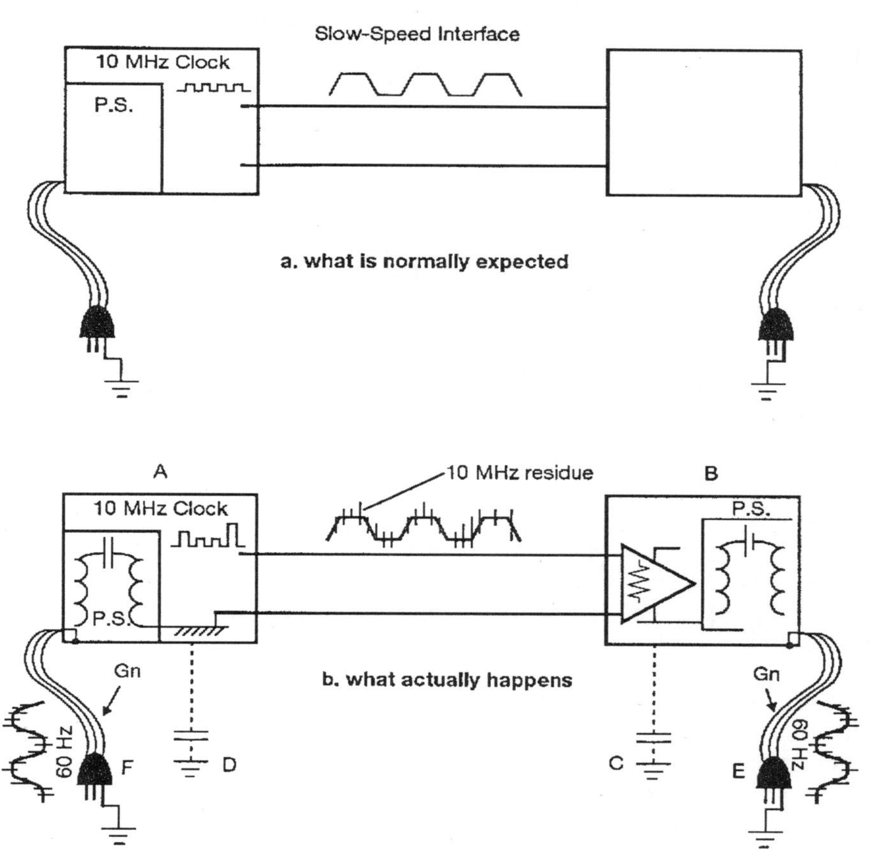

FIGURE 2.10 Contamination of External Cables by Internal HF circuits

In Fig 2. 10, we see a contrast between (a) what is normally expected-power line carries only 50/60 or 400 Hz currents, I/O cable carries a slow serial bus, and the 10 MHz clock is used only internally, and (b) what real life provides: I/O pairs or ribbon cable carry 10 MHz residues from the clock, picked up internally; their spectrum extends easily to 200 or 300 MHz. Because of the primary-to-secondary capacitance in the power supply transformer, power wires (phase, neutral and ground) are also polluted by 10MHz harmonics. The radiating loops can be ABCD, ABEF or combinations of all.

Predicting such radiated emissions from external cables will consist in:

- measure, or estimate the CM currents driving the external cables
- estimate the geometry for the CM - driven antenna
- apply simple, appropriate antenna formulas (loop or open wire)

2.5.1 How to estimate CM currents on cables

As said before, the CM currents found on external cables can have essentially two origins. The mechanisms which are causing such currents are reviewed next.

a) Intentional signals
By this, we mean the portion (generally undesired) of signal current which is closing by the external ground loop. Let us review the four principal cases.
- ***single wire transmission***
The useful signal is carried on a single wire, the chassis or other structural system ground forming the return path. This type of link is, of course, highly detrimental for EMC and almost never used any more. Few exceptions are found in automobile, aircraft or helicopter applications where certain signals are still carried between a hot wire and the vehicle body, to save on copper weight. In this case, the full signal spectrum is driving the single wire antenna.
- ***unbalanced, two-wire transmission*** (RS 232, RS 423 etc ..).
This time, the CM current is the % of the signal current which "elect" to return by the external path instead of using the return wire. Above few kHz, the mutual inductance between the two wires is strong enough to attract most of the current into the intended path i_2 (Fig. 2.9), leaving only 20 - 30% of the current closing by the ground loop (i_3). Unless one knows the exact value of the mutual inductance, which itself varies with the cable height and wires separation, a conservative value is to assume $i_3 = i_1$ -10 dB. This is of course assuming an unshielded cable (see Chap. 11 for the additional suppression by a shield).
- ***unbalanced coaxial transmission*** (RF, Video, Ethernet, etc..)
This type of link is inherently low radiating, due to the high % of current - typically more than 99% - returning by the shield. This, too, is covered in Chap. 11.
- ***differential transmission*** (RS422, RS485, Mil.1553, Diffl.SCSI, CAN, LVDS etc ..). With such links, the transmitter and receiver circuits have been designed to force balanced currents into the wire pair. In addition to excellent immunity and low crosstalk, this provides also a low CM current generation, hence less EMI radiation.

Along with the good symetry of the driver & receiver, a good symetry of the wire pair is also required, such as the overall balance of the transmission is within the 1 - 10% range.
Unless it is known exactly, we can conservatively assume 5%, i.e: $i_3 = i_1$ - 26 dB.

b) Non-intentional signals
As shown on Fig.2.10, these are residues of clock frequencies, switch mode regulators etc..., coupled onto PCB I/O traces by:

- ground traces /plane pollution
- Vcc pollution (this in turn is partially transferred to I/O lines by some "transparency" of the driver or receiver Ics)
- crosstalk

They can also be coupled by near-field radiation from the PCB hot spots onto the exiting cables, because of box shield leakages around the connector area. These leaks are causing the two (or n) wires to be CM-driven like a single conductor.

The first difficulty is to evaluate the amplitude of these undesired components. A pragmatic approach is to measure their voltage or current spectrum directly on the cable itself. This is easy to do at a diagnose-and fix level, but it requires at least a representative prototype at the design stage. How can one do this when the machine does not yet exist?
A deterministic approach would consist in calculating every possible internal coupling between the inner circuitry and the leads corresponding to I/O ports. This is feasible but takes a considerable length of time. A crude but effective solution is to make the following assumption by default:

"Unless one knows better, it is logical to assume that the noise picked up by internal couplings is just below the immunity level of the circuits interfacing the external link in question."

The rationale for this is that designers will at least make their product functionally sound; if worst-case values were exceeded (e.g., 0.4V for a TTL noise margin), the system simply would not work.

Example 2.3
A serial link between a computer and its peripheral operates at a 20 kb/s rate. Internally, the microprocessor uses a 20 MHz clock with Schottky logic, whose characteristics are:

- amplitude = 3.5 V
- rise time (STTL) = 3 ns
- noise margin (worst case) = 0.3 V

Lacking of any other data, we can at least make the following worst-case assumptions (see Fig. 2.11):

- The amplitude of any parasitic coupling from a clock-triggered pulse to a nearby trace or wire will not exceed 0.3 V.

• Because crosstalk and ground-impedance sharing are all derivative mechanisms, the pulsewidth of the coupled spike on the victim trace will be in the range of the STTL transition time of 3 ns. This is a worst-case guess since pulse stretching and ringing will occur due to distributed parameters of the victim line.
• If crosstalk is expected, this is a differential excitation (trace-to-Gnd), and a DM-to-CM reduction factor must be applied to the driven cable,, like -10dB for an ordinary unbalanced link (see previous paragraph. a)

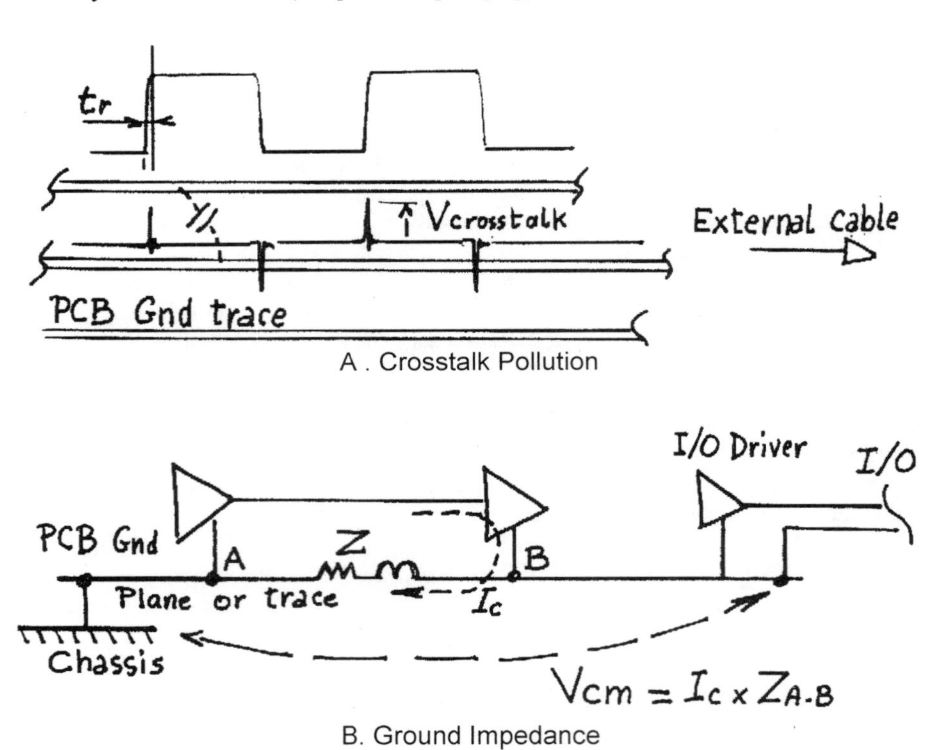

FIGURE 2.11 Coupling of Internal Clock Transitions by Crosstalk to I/O Traces

2.5.2 Approximating the proper radiating geometry

Having determined the CM current driving mechanism, we now have to figure-out what is the driven antenna: closed loop or open wire?
Although they consist in gathering many configurations in two rather crude configurations, resorting to these two simple models give remarkably good results when compared to actual validation test data.

a) Closed Loop

By this term, we do not imply necessarily that the loop is physically closed; the signal grounds (0V ref.) may or may not be connected to chassis or earth reference both ends. If the I/O cables are connecting to metallic equipment cases (and therefore, most likely, are grounded), the CM radiating loop is geometrically identified by its size *l* x h (Fig. 2.12). If the ground references are floated inside the box, this will increase the low-frequency impedance of the loop, yet its size remains. Therefore, depending on the loop impedance and the near-field or far-field conditions, we will apply either the loop equations or the two-wire equations of Section 2.3.

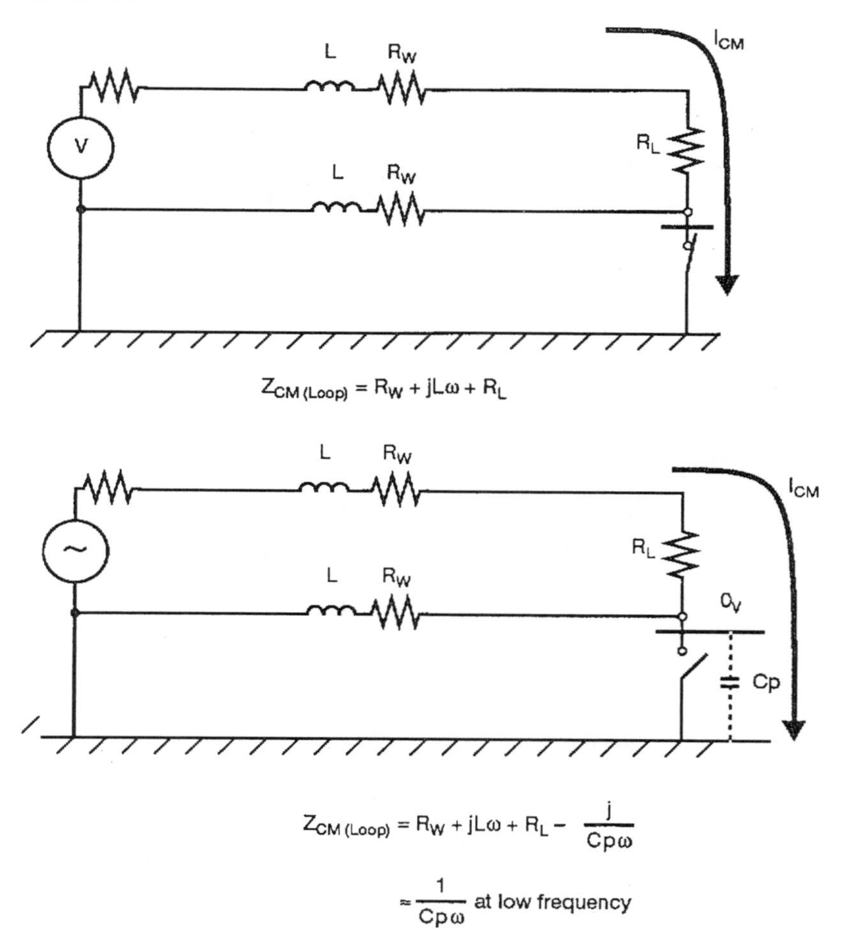

FIGURE 2.12 Loop Impedance of Radiated External Cables

For all - grounded ends (PCB to chassis and chassis to earth), the loop impedance seen by the current I_{cm} is:

$$Z_{CM} = R_{wire} + j\omega L + R_{load} \qquad (2.23)$$

where L is the wire self-inductance above ground plane. A typical value for L, with heights in the 0.1 to 1m range is 1 - 1.5µH/m, and R_{wire} (typ. 0.1Ω/m) can be neglected, so Eq.2.23 is written in a more convenient way:

$$Z_{CM} \approx R_{load} + j\ 7.5\ \Omega \times \ell_m \times F_{MHz} \qquad (2.23a)$$

For a floated end, the loop impedance seen by the I_{cm} current is:

$$Zcm = Rload + (\ Rwire + j\omega L + Rload\ - j\ /Cp\omega\ ) \qquad (2.24)$$

where C_p is the PCB-to-chassis stray capacitance (use 30 pF for small boxes, 100 to 200 pF for large cabinets). The bracket simplifies to 1/Cpω for low frequencies.
When cable length exceeds λ/2 (for both ends grounded) or λ/4 (for one end floated), Z_{cm} can be approximated by the characteristic impedance of the cable above ground:

$$\text{in the air, } Z0 = 60\ Ln\ (4h\ /d) \qquad (2.25)$$

where,
Ln = natural logarithm
d = cable diameter (average contour of whole wire bundle)

For practical h/d range of 3 to 100, this give a span of Z_0 from 150 to 360Ω. The low value would correspond to a typical MIL-Std461 test setup, the high value being the extreme for a table top equipment in an FCC or CISPR test. A typical real-life value would be 300 Ω.

Example 2.4

A 5 MHz clock is used on a short-haul parallel bus. For a 5V pulse, the ninth harmonic, at 45 MHz, has an amplitude of 0.3 V. The characteristics of the I/O cable between the two metallic equipments are:

1. cable length, l =1.20 m
2. height, h = 0.30 m
3. inductance of cable above ground L = 1.2 µH/m
4. terminating resistor = 120 Ω

Calculate the 45 MHz E field at 3 m, against the FCC class B limit for (a) both PCBs grounded to chassis and (b) one PCB ground (0V) floated, with a total stray capacitance of 30 pF.

Solution:

- $D = 3$ m, $> 48 / 45$ MHz., so we are in far-field conditions.
- l_m and $h_m < 75 / 45$ MHz, so we are below cable antenna resonance.

We can use directly Eq. (2.20) or the curves from Fig 2.6 for 3 m distance:

Area = 120 cm x 30 cm = 3,600 cm^2 = 72 dBcm^2

For a grounded condition (the cable is mostly an inductance), impedance is calculated from Eq. (2.23a):

$$Z_{cm} = 120\Omega + j\,(7.5 \times 45\ \text{MHz} \times 1.2\ \text{m}) = 420\ \Omega$$

For a floated condition, the cable inductance resonates with the floating PCB capacitance at:

$$\begin{aligned} \text{Fres} &= 1 / 2\pi \sqrt{(LC)} \\ &= 1 / 2\pi \sqrt{(1.2\text{m} \times 1.2.10^{-6}\text{H} \times 30.10^{-12}\ \text{F})} = 24\text{MHz} \end{aligned}$$

Therefore, due to resonance downshifting caused by the stray capacitance, the cable is now beyond resonance condition. We will use a typical characteristic impedance of 300Ω, governing the average value of the current along the cable.
Calculation steps:

Frequency	45 MHz	
1. E_0 in Fig.2.6(for 1V-1cm^2)		
. for Z = 420Ω	8 dBμV/m	
. for Z = 300Ω		10 dBμV/m
2. Area correction	+72	
3. Amplitude corr. (0.3V)	-10	
4. E = 1 +2 +3	70 dBμV/m	
		(72 for floated)
FCC limit, class B	40 dBμV/m	
Off specification	30 dB	(32 dB for floated)

For the all-grounded case, a quick analysis of the circuit in Fig. 2.12 at top, with the variables of the example, give the following current split:

. i_1 (upper wire) = 1.4mA
. i_2 1.1mA
. i_3 0.3mA, which is 12dB below i_1

This is due to the hair-pin inductance of the wire pair being approximately a third of the large loop inductance. Notice that we have passed the point where a floated PCB could be of any use. The exact calculation of $I_{average}$ could be made using transmission line theory and would give slightly different results for each resonant condition. To reduce this excessive emission will require one of the several solutions (e.g., CM ferrites, cable shield, balanced link) that we will examine later.

b) Open Wire, Monopole or Dipole

We now examine the case where no geometric loop can be identified (Fig. 2.13). The external cable terminates on a small, isolated device (sensor, keypad, etc.) or into a plastic, ungrounded equipment. It may even terminate nowhere, waiting for a possible extension to be installed. No finite distance can be measured to a ground plane. In this case, we use the single-wire radiation model described in Section 2.2. In a sense, we can say that the floating, open wire is the maximum radiating aerial that can be achieved when the height of a ground loop increases to infinity. To calculate the radiated field using Eq. (2.13) or (2.16) requires that the CM current in the wire be measured or calculated. Measurement with a high-frequency current probe is easy, but only if a prototype is available. Otherwise, one can simply use the cable self-capacitance of ≈ 10 pF/m for low-frequency modeling, and the cable characteristic impedance (Eq .2.25) with high values of h, above the first resonance.

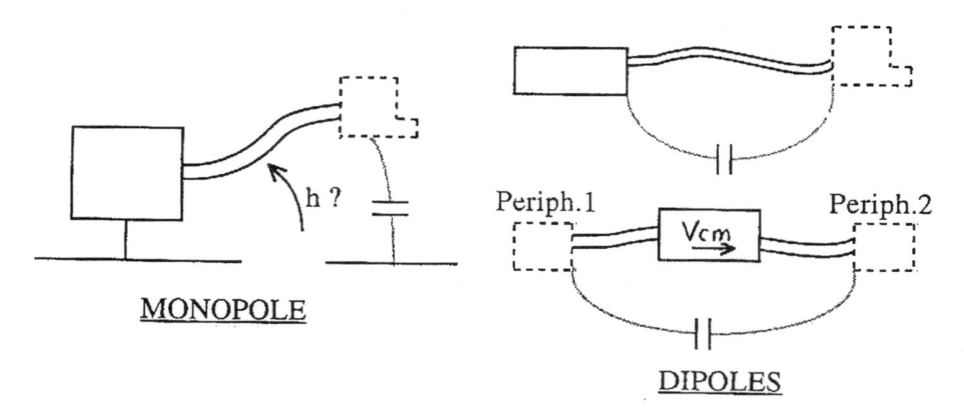

FIGURE 2.13 Common-driven Monopole or Dipole

Using more realistic units, single wire radiation is expressed as:

$$\text{In the near field, } E_{(\mu V/m)} = \frac{1,430 \times I_{\mu A} \times \ell_m}{D^3 F_{MHz}} \tag{2.26}$$

$$\text{In the far field, } E_{(\mu V/m)} = \frac{0.63 I_{\mu A} \times \ell_m \times F_{MHz}}{D_m} \tag{2.27}$$

If the cable interconnects two units that are completely floated and not close to any ground, the length, l is regarded as a radiating dipole length. If one of the two units is grounded or is in a metallic case close to ground, the cable has to be regarded as a radiating monopole whose length, l radiates like a dipole twice as long. In this case, 2 x l should be entered in the formula.

Of course, wire lengths may exceed $\lambda/2$ ($\lambda/4$ for a monopole). In such case, the current can no longer be regarded as uniform over the wire length. But the "active" segment of the radiator cannot exceed a length of $\lambda/2$ - the other $\lambda/2$ segments create fields that mutually cancel due to phase reversal (except for the field propagation delays, which are unequal). Everything behaves as if the antenna were electrically "shrinking" as F increases. In this case, l is replaced by $\lambda/2$ in the formula.

Applying a correction factor averaging I_{max} over the length, we have, for free space:

$$E_{(\mu V/m)} = \frac{60\ I_{\mu A}}{D} \tag{2.28}$$

Interestingly, we observe that E becomes independent of F and l. This formula is extremely useful, and we will employ it frequently.

Example 2.5

For the same 5 Mb/s signal as in Example 2.4, assume that the cable now terminates into a plastic equipment on one end. The cable is far from any ground plane. What is the maximum CM current tolerable on this cable to meet the FCC (B) 3 m limit of 100μV/m at 45 MHz (harm #9) and 85 MHz (harm #17)?

• At 45 MHz, $l < 75/F$. From Eq. (2.27), remembering that we have a monopole, a length (2 x l) is entered as radiating length:

$$E = 0.63I \times (1.20 \times 2) \times 45\ \text{MHz}/3\text{m}$$

Solving for I,

$$I < E/22.5 \sim \text{so}\ (I < 4.4\mu A)$$

• At 85 MHz, $l > \lambda/4$. We will use Eq. (2.28):

$$E = 60\ I/D$$

$$I \leq E.D/60\ ,\ \text{so}\ (I < 5\mu A)$$

Therefore, before running an exhaustive radiated EMI test, a simple measurement on the cable with a high-frequency current probe will indicate whether the equipment has a good chance of meeting the specification.

c) Simple Voltage-driven Nomogram for Open Wire

If the CM current cannot be measured on a prototype, instead of computing the open wire impedance at each frequency, the nomogram of Fig. 2. 14 for isolated wires can be used for a quicker estimate. All that is needed is the value of the voltage(s) driving the monopole, or dipole. The curves are based on cable capacitance, for electrically short lines, and a conservative 150Ω CM impedance otherwise. Notice that, at each exact resonance, a 6dB hump has been accounted for the 75Ω impedance of a tuned dipole. Results are given in dBμV/m for one μV (0dBμV) of CM excitation.

Example 2.6

Taking the same equipment as example 2.5, assume that the 5MHz CM current is not known, but the harmonic #9 CM driving voltage is 100mV. We will extrapolate from the curve for a 1m monopole at 45MHz.

Test distance 3m, Frequency:	45MHz
1. E_0 for 2m dipole & 0dBμV	-26 dBμV/m
2. Length corr. * 20 log $(1.20/1)^2$	+3
3. Amplitude corr. (10^5μV)	+100
E = 1 +2 +3	77 dBμV/m (37dB above classB)

* for a voltage-driven wire, and below resonance, radiation efficiency increases like the square root of length (relates to equivalent radiating area).

We see that with approximately the same CM current, the field is approximately twice larger than with a closed loop, 0.30m above ground. This can be considered as the upper bound for a CM-driven cable infinitely far from ground. Notice how little voltage it takes to excite an open wire above the radiated limits.

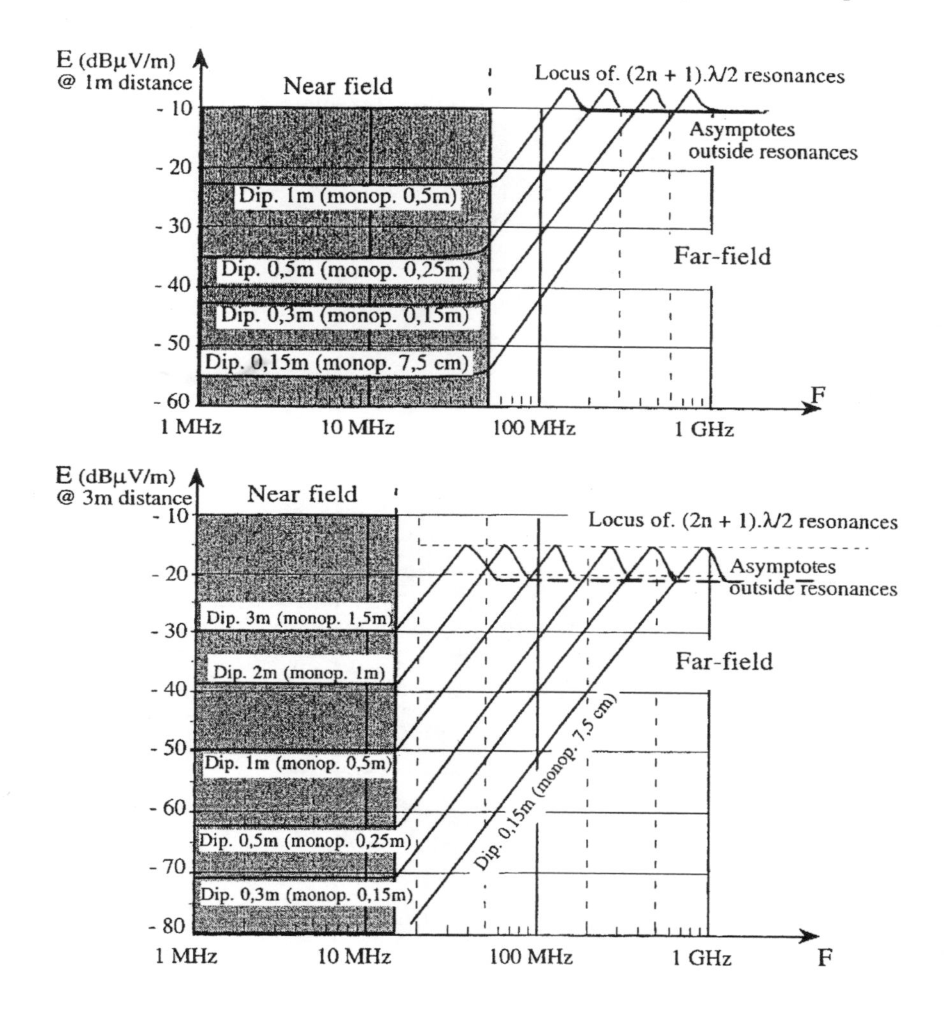

FIGURE 2.14 E-field radiation from voltage-driven, open wires. Field is given for 0dBµV of CM drive voltage. Horizontal locus correspond to the odd multiples of $\lambda/2$.

d) Influence of a Nearby Ground Plane

If there is a conductive plane near the cable, this proximity causes a reflected wave with a phase shift (see Fig. 2.15). If the plane is sufficiently close, this shift is always at phase reversal with the directly radiated wave, and the total field equals E_o- E_r. It is not necessary that the source or load be referenced to this plane, but the plane must be quasi-infinite; i.e., in practice, it must extend far enough around the cable projection. The radiation reduction, for $h < 0.1\lambda$ (i.e, $h_m < 30/ F_{MHz}$), is:

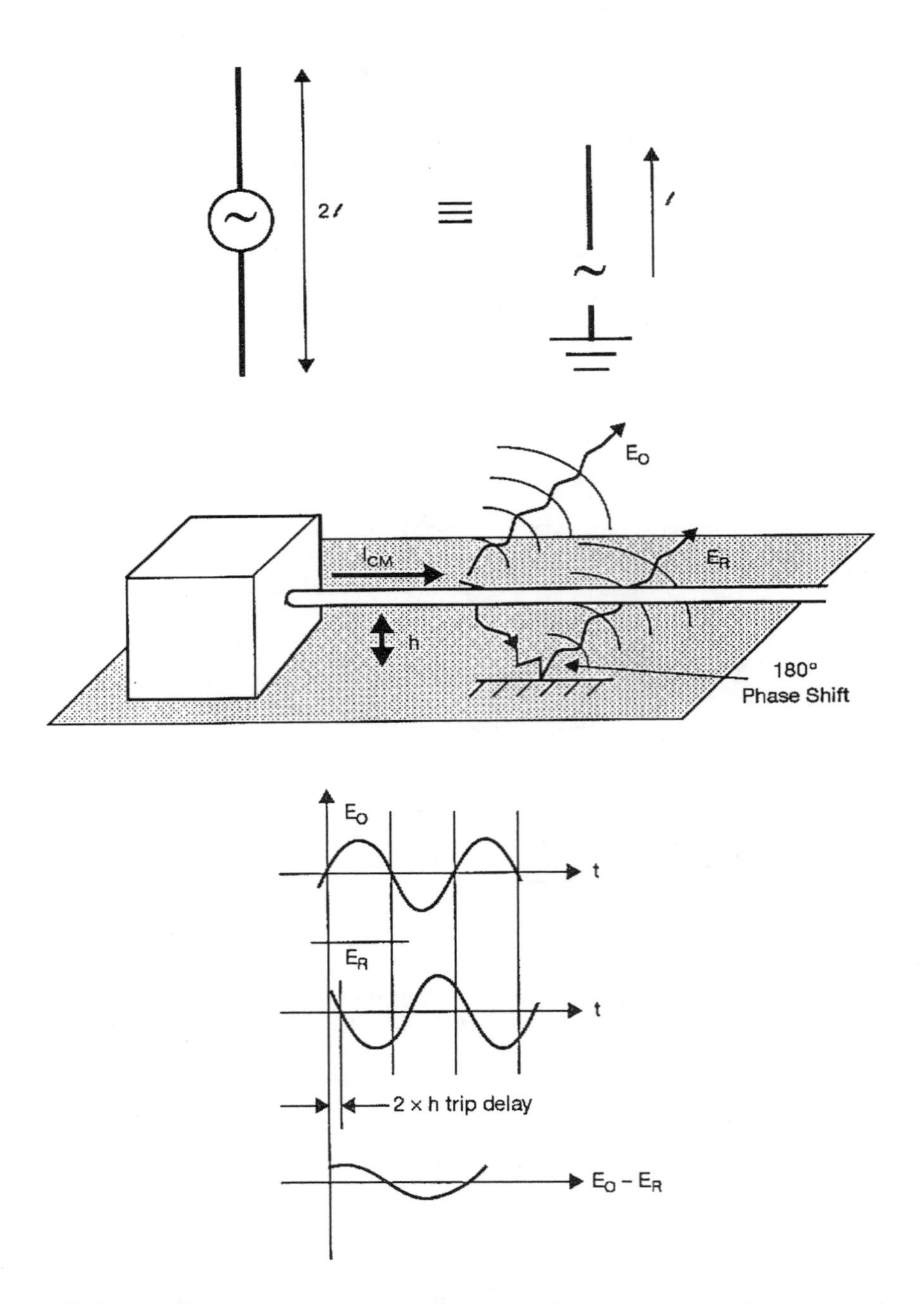

FIGURE 2.15 Equivalent Antenna for Wire Floated at Both Ends (dipole) or Grounded at One End (Monopole). Effect of a Nearby Ground Plane for $h<\lambda/10$.

$$\frac{E_{total}}{E_o} = \frac{h}{0.1\lambda} \text{ or } \frac{10h}{\lambda} \tag{2.29}$$

Entering this factor into Eq. (2.27), for far field:

$$E_{\mu(V/m)} = \frac{0.021 \times I_{\mu A} \times \ell_m \times h_m \times F^2_{MHz}}{D_m} \tag{2.30}$$

If h > 30/F, the reflected field is alternatively additive or subtractive, and the field is not reduced but doubled at certain frequencies. Incidentally, Eq. (2.30) shows a similarity with the radiation from a loop, since E now also depends on the area l x h, and F^2.

Example 2.7

For the equipment of Example 2.5, recalculate the maximum value for I_{cm} with the cable now located 5 cm from a ground plane. The criterion to meet is MIL-Std 461 RE02, at 1 m distance. At 45MHz, the limit is 25dBμV/m. At 85MHz, E_{max} is 29 dBμV/m. We can accept the far-field assumption for both frequencies.

Solution:

• At 45 MHz, λ = 6.6m, so the 1.20m cable length is < λ/4. Since h is 0.05 m, it is < 30/F. Per Eq. (2.30), and remembering that we have a monopole:

$$E = 0.021 \text{ x I (1.20m x2) x } 0.05 \text{ x } 452 = 5 \text{ I} = \text{I dB}\mu\text{A} + 14 \text{ dB}$$

Therefore.

$I \leqslant 25$ dBμV/m - 14

$I \leqslant 11$ dBμA

• At 85 MHz, l > λ/4, and h is still < 30/F. From Eq. (2.28), corrected by h/0.1λ:

$$\begin{aligned} E &= (60I/D) \text{ x } 5.10^{-2} \text{x } 85/30 \\ &= 8.5 \text{ I} \equiv \text{I} + 18 \text{ dB} \end{aligned}$$

Therefore,

$I \leqslant 29$ dBμV/m - 18 dB

$I \leqslant 11$ dBμA

e) Ground Reflection with FCC / CISPR Tests

The reflected field addition, practically doubling the value of E, is typical of FCC-15B or CISPR 22 emission testing. The equipment under test (EUT) and its cables are located above a ground plane while the receiving antenna is cranked up and down to find the maximum level. A typical test configuration is the following:

- EUT + cables : 0.5 - 1m above ground
- test distance : 3m
- antenna set for Vert. and Hor. polarization, with height scan 1 to 4m

The corresponding correction for ($E_{dir.}$ + E_{refl})max versus free space field is given below (Ref. 4)

	Horiz. Polar.		Vert. Polar.	
EUT height	Corr. Δ dB		Corr. Δ dB	
above ground :	0.5m	1m	0.5m	1m
F_{MHz}				
30	+5	+4	-8	-4
50	+5	+4	-5	-1
70	+5	+3	-2	+2
100	+4	+2	0	+4
150	+3.5	0	+3	+5
200	+3		+4	+5

Below 100MHz, the worst reflection is detected with horizontally polarized sources (seldom the case with CM-excited cables), typical of DM sources inside the EUT. Above 100MHz, the additive reflections are caused by vertically polarized sources, adding a maximum of 5dB to the theoretical free-space value. We therefore can retain, as a reasonable worst case for all FCC or CISPR-like measurements a ground reflection adder of +5dB in our emission calculations.

As a concluding remark, Fig. 2.16 is showing the respective field contributions of PCB vs external cables in two situations. The first plot, at top, correspond to a PCB with 30MHz clock circuit, attached to a 1.5 m external cable carrying typical spurious contents. As explained in this section, the cable CM radiation dominates the PCB DM contribution by 25-30dB, and is the sole violator of FCC/CISPR limit. In the bottom plot, the same geometry with a 150MHz PCB shows a different split. Above 50 -100 MHz, the spurious spectrum carried by the cable start decreasing because:

- the maximum possible crosstalk coefficient in the PCB has been reached.
- cable starts exhibiting HF losses.
- around these frequencies, the cable antenna efficiency for 1.2 -1.5m lengths has reached its maximum.

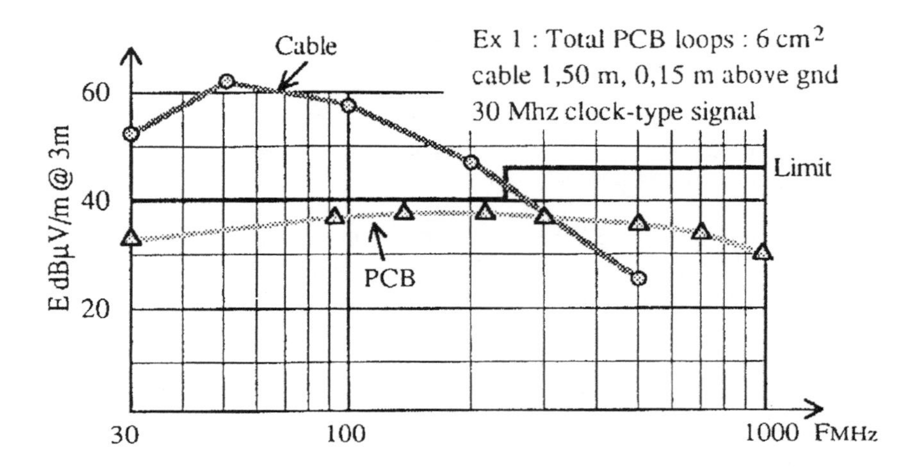

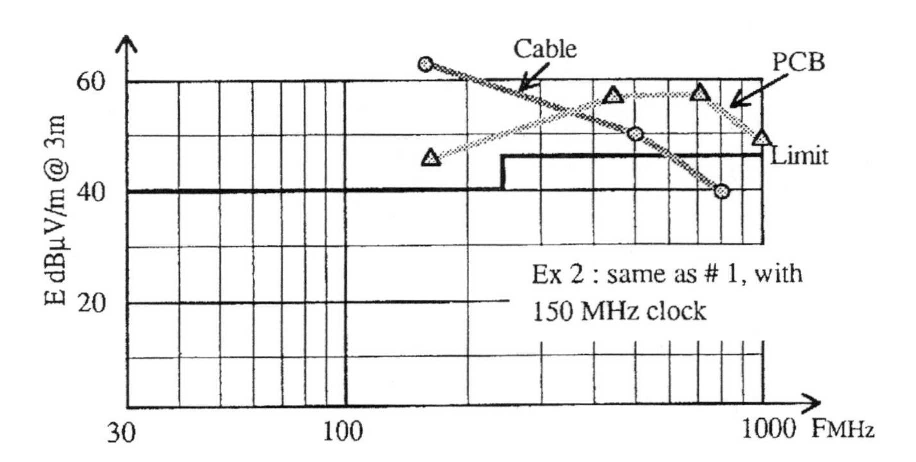

FIGURE 2.16 Comparison of PCB vs cable contribution to radiated spectrum.

2.5.3 Radiation from a Long Wire

The distance restriction imposed for using the loop model or the wire models (i.e., $l < D$) rapidly becomes an obstacle in many configurations where cables lengths exceed a few meters. In this case, the physical length of the wire is such that it cannot be considered as a small element with respect to the observation distance. We can use a practical expression, taking into account the wide viewing angle from the observation point to the cable (infinite wire model):

$$I_{A/m} = \frac{I_A}{2\pi D} \tag{2.31}$$

Only the H field can be correctly determined by this, Ampere's law. An "equivalent" E field could be derived using a $120\pi\ \Omega$ wave impedance, but it would be inaccurate in this proximal zone from the antenna.

Eq. (2.31) assumes the wire is far from a ground plane, with respect to observation distance (i.e., in practice, $h >> D$). If the wire is close to a ground plane, and the observation point is at the same height, $h << D$:

$$H \approx \frac{I}{2\pi D}\left(1 - \cos\frac{2h}{D}\right) \tag{2.32}$$

The criterion to decide when a wire has to be considered to be "long" in comparison to D is simple: the maximum field is reached when $l \geqslant \pi D$. If one takes $l > D$ as criteria, the error would be only 16 percent (1.3 dB). This is because a length increase from $l = D$ to $l = \pi D$ corresponds only to an or viewing angle variation from $\cos\alpha = 0.84$ to $\cos\alpha = 1$.
If the long wire itself becomes electrically long (i.e., its physical length exceeds $\lambda/2$—see Eq. [2.28]), we are back to the situation where, as the antenna electrically "shrinks", the only effective radiation comes from the $\lambda/2$ segment facing the observation point. At this juncture, there is no longer a difference between the "infinite" wire and a smaller wire. They are both limited by the $\lambda/2$ clamp, and Eq. (2.28) would apply in either case.

Case of the Long Wire Pair (DM Radiation)

The infinite wire equation, when transposed to a long wire pair carrying equal and opposite currents, becomes:

$$H\ \text{A/m} = 0.16\ I\ [\ s\ /\ (D^2 - s^2/4)] \tag{2.33}$$

with,
D = pair to receiver distance, measured from pair axis
s = wires separation

Chapter 3

Fields Radiated by Nonsinusoidal Sources

The simplified equations briefly described in Chapter 2 can be extrapolated to the majority of practical cases where the excitation signal is not a pure sinewave but a repetitive signal with a known period and waveform. This chapter will describe the two most common cases of nonsinusoidal sources: periodic signals with narrowband (NB) spectrum, and periodic signals with broadband (BB) spectrum.
In both cases, the prediction process consists of first performing a Fourier analysis of the source signal, then treating each harmonic (or group of harmonics in a given bandwidth) as a sinewave for calculating radiation.

3.1 FREQUENCY SPECTRUM, AND RADIATION FROM PERIODIC PULSES

Fourier theory states that a periodic signal can be expressed as a series of sine and cosine signals, at frequencies that are multiple integers of the pulse period. On the other hand, the standard emission limits typically range from 10 kHz to 10 GHz for military applications and 30 MHz to 1 or 10 GHz for most civilian ones. Thus, if one were to take a periodic signal and perform a rigorous Fourier computation, thousands or more discrete terms would have to be sorted out.

Instead, we will use the Fourier envelope method for voltages or currents. Figures 3.1a and 3.1b show the Fourier envelope shapes and equations, corresponding to few typical waveforms, as they would appear on an oscilloscope with a sufficient bandwidth. (To make sure that the oscilloscope does not distort the rise times, use an oscilloscope bandwidth $BW_{MHz} > 350/t_r(ns)$, with $t_r(ns)$ representing the fastest rise time of the observed pulse, measured in nanoseconds).

The Fourier envelope is the locus of the maximum harmonics, without considering their phase. To draw this envelope, we must know:

• the peak amplitude A (volts, amperes, etc.)
• the pulse width, τ
• the period, T, where the signal reproduces itself
• the rise time, t_r, at 10 percent to 90 percent crossings. If fall and rise times are different, select the shortest of the two for t_r.

For a periodic signal, the frequency spectrum is composed of a series of discrete sinewave harmonics, consisting of the fundamental ($F_o = 1/T$) and integer multiples of F_o.

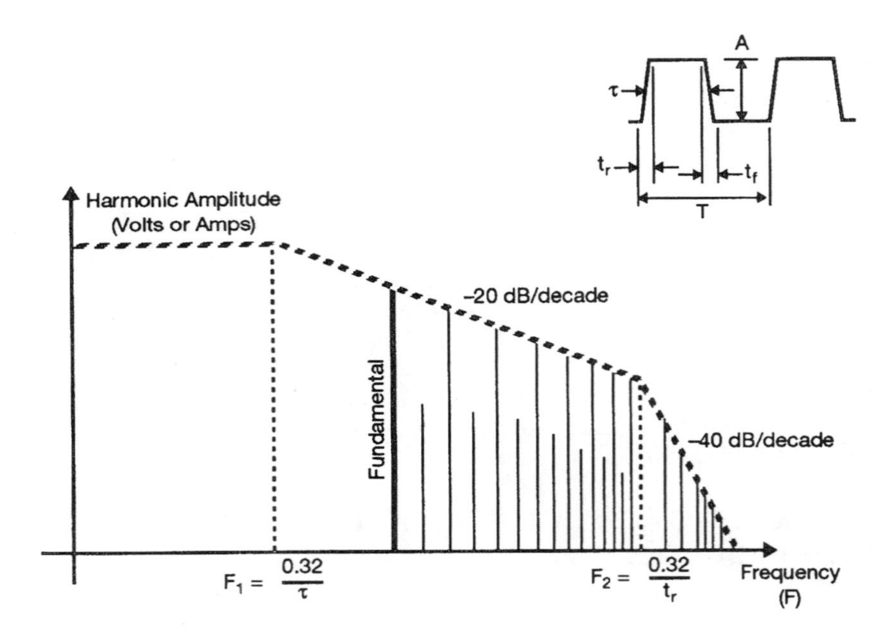

$$\text{for } F_x < F_1,\ V \text{ (or I)} = \frac{2A(\tau + t_r)}{T}$$

$$\text{for } F_1 < F_x < F_2,\ V \text{ (or I)} = \frac{0.64\ A}{T \times F}$$

$$\text{for } F_x > F_2,\ V \text{ (or I)} = \frac{0.2\ A}{T \times T_r \times F_x^2}$$

(If t_r [risetime] $\neq t_f$, use the shorter of the two.)

FIGURE 3.1a Fourier Envelopes for Narrowband Spectral Amplitudes, Trapezoidal Pulse Train (any duty cycle). Continued next page.

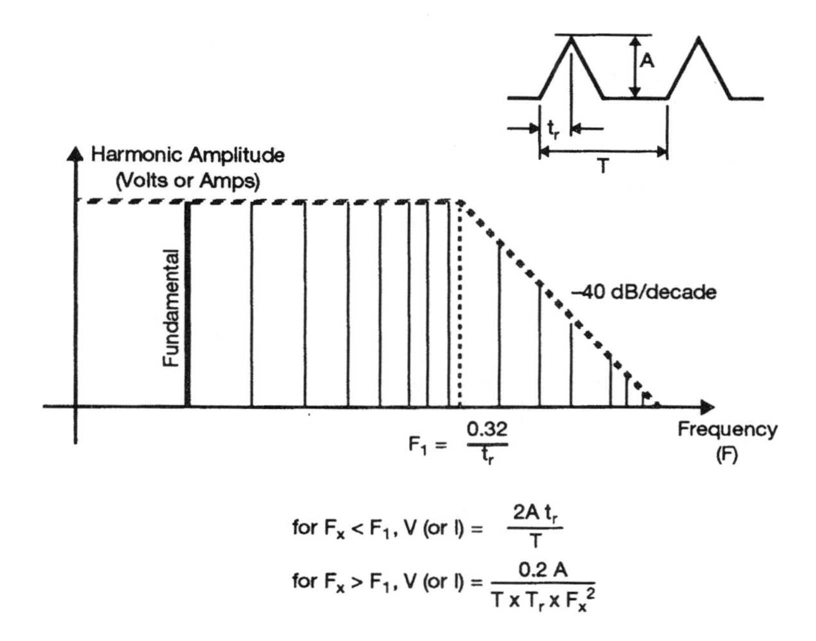

FIGURE 3.1b Fourier Envelopes for Narrowband Spectral Amplitudes, Symmetrical (Isosceles) Triangular Pulses

Besides the fundamental, F_o, the two key frequencies in constructing the envelope are F_1 and F_2:

$$F_1 = 1/\pi\tau$$

Above F_1, the locus of the maximum amplitudes rolls off with a 1/F slope (-20 dB/decade).

$$F_2 = 1/\ \pi t_r$$

Above F_2, the spectrum decreases abruptly, with a $(1/F^2)$ slope (-40 dB/decade).
Notice how critical this frequency is: the shorter the rise time, the higher the spectral occupancy.
A nomogram has been constructed (Fig 3.2) to provide easy and quick approximation of the spectral envelope. Once the envelope is drawn, the worst-case amplitude of any harmonic can be found.

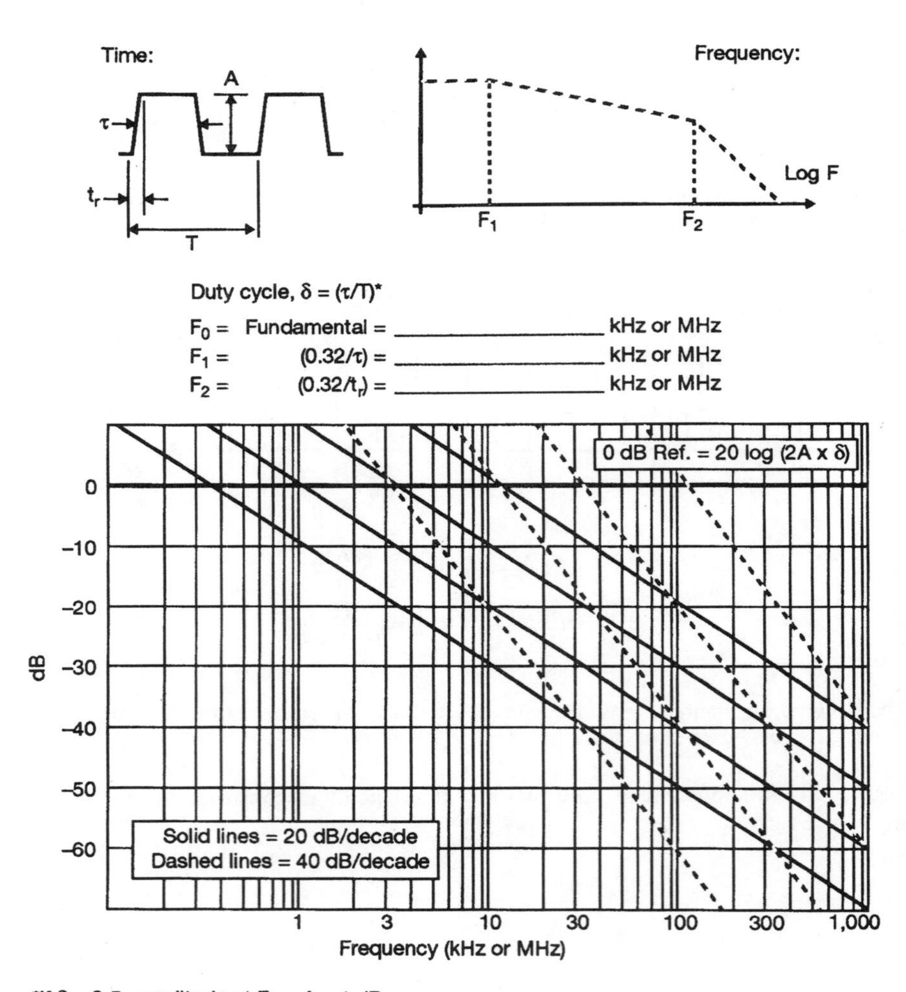

*If δ = 0.5, amplitude at F_0 = A – 4 dB

Harmonic #	Frequency	Ref. Amplitude (dB V or dB A)	Slope Decrease dB	Amplitude of Harmonic #

FIGURE 3.2 Template for Quick Frequency Spectrum Plot

Example 3.1
Find the Fourier envelope of the signal in Fig. 3.3.
The first step consist in identifying the three acting frequencies:

$$F_0 = 1 / 0.066\mu s = 15MHz$$

$$F_1 = 1/\pi\tau = 9.5MHz$$

$$F_2 = 1/\pi tr = 45MHz$$

Notice that, at frequency F_1, such a pulse train, with a 50 percent duty cycle, has no frequency component, yet. However, we need F_1 to construct the envelope.
The starting amplitude for the reference line is:

$$20 \log (2A\tau/T) = 20\text{Log} (2 \times 5V \times 0.5) = 14 \text{ dBV}$$

The frequency F_1 (9.5 MHz) is marked on top of the 0dB reference line. Then a 20 dB/decade slope is drawn, using the parallel grids, until F_2 (45 MHz) is encountered. From this point, a 40 dB/decade slope is drawn. The amplitude of any harmonic, in dBV or dBA, can be found by simply subtracting the slope decrease in dB, from the reference amplitude.
Notice the small ringing of the waveform at the edges due to a slight mismatch reflection (assuming it was not caused by the scope probe). This ringing of about 2 percent of amplitude is almost unnoticeable on an oscilloscope trace but causes a 6 dB hump around 150 MHz on the frequency spectrum. The +6 dB corresponds to the normal harmonic amplitude of -20 dBV, adding to the ringing amplitude of 0.1 V peak. This is shown below in Table 3.1 for few typical harmonics.

TABLE 3.1 Amplitude Calculation for Example 3.1
*parasitic edge resonance

Harmonic #	Fundamental	#3	#5	#10	#20
Frequency (MHz)	15	45	75	150	300
Reference Amplitude (dBV)	14	14	14	14	14
Slope Decrease	–4	–14	–22	–34 + 6*	–46
Amplitude (dBV)	10	0	–8	–14	–32

*parasitic edge resonance

To calculate the radiation from a circuit carrying such spectrum, we simply need to apply the equations or graphs for loop surface or wire length.

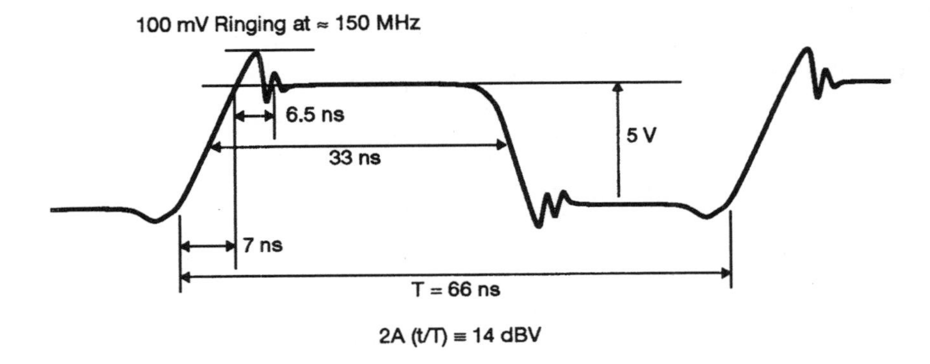

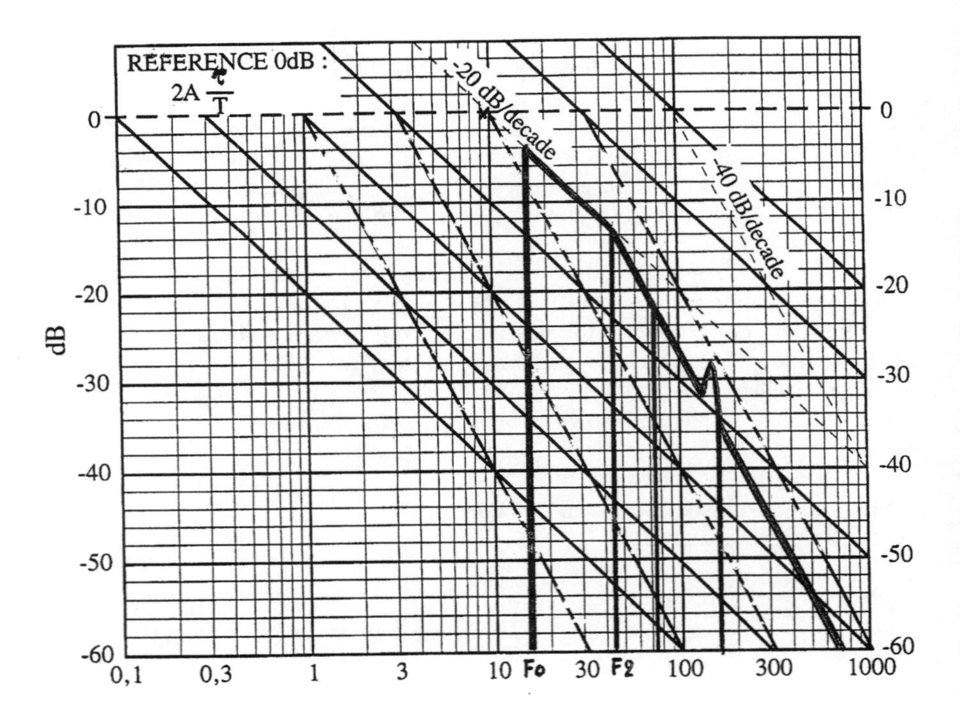

Figure 3.3 Signal from Example 3.1 in Time and Frequency Domains, Showing Envelope and a Few Spot Frequencies

Example 3.2

The clock pulse train of Example 3.1 is distributed to several daughter cards via the mother board of Fig. 3.4. The critical radiating circuit has the following characteristics:

- five clock runs with discrete ground return traces (no ground plane)
- clock trace to return trace separation, s = 2.5 cm
- trace length, l = 40 cm
- terminal resistance at line end = 300 Ω
- approximate line characteristic impedance, Z_0= 200 Ω (to use above λ/4)

Calculate the field strength at 3 m against FCC class B limits. The measuring bandwidth for this test is 120 kHz, so each harmonic is measured one at a time (narrowband situation).

Using the curves of Fig. 2.6, we will apply the following corrections:

Voltage correction: from Table 3.1
Area correction: 20 log (40 cm x 2.5 cm) = +40 dB
Number of synchronous loops: 20 log N = 14 dB
(Viewed from 3 m, the five loops radiate approximately as a single one with five times more current or five times more area.)
The quarter wavelength limitation (see Section 2.4) for current distribution on PCB traces is reached when l = 5,000 cm/F_{MHz} (i.e.here, 125 MHz for l =40cm). Above this frequency, l is replaced by 5,000/F in the area correction. Beyond this point, each harmonic will drive a gradually diminishing effective area: the efficiency of the "antenna" decreases, as does the voltage spectrum. As a result, the total radiation profile collapses. At the same time, the load resistance of 300Ω will be replaced by the 200 Ω line characteristic impedance. The calculation steps are shown in Table 3.2.

TABLE 3.2 Calculation Steps for Example 3.2

F_{MHz}	15	45	75	150	300	500
Amplitude (dBV)	10	0	–8	–14	–32	–40
$E_o(1\ V,\ 1\ cm_2)$						
(for Z_{load} = 300 Ω)	–10	+10	+18	+30		
(for Z_{load} = 200 Ω)					45	53
Area Correction	+40	+40	+40	+38	32	28
Number of Synch. Loops	14	14	14	14	14	14
Total Field (dBμV/m)	54	64	64	68	59	54
FCC Limit	NA	40	40	43	46	46
Off Specification		+14	+24	+25	+13	+8

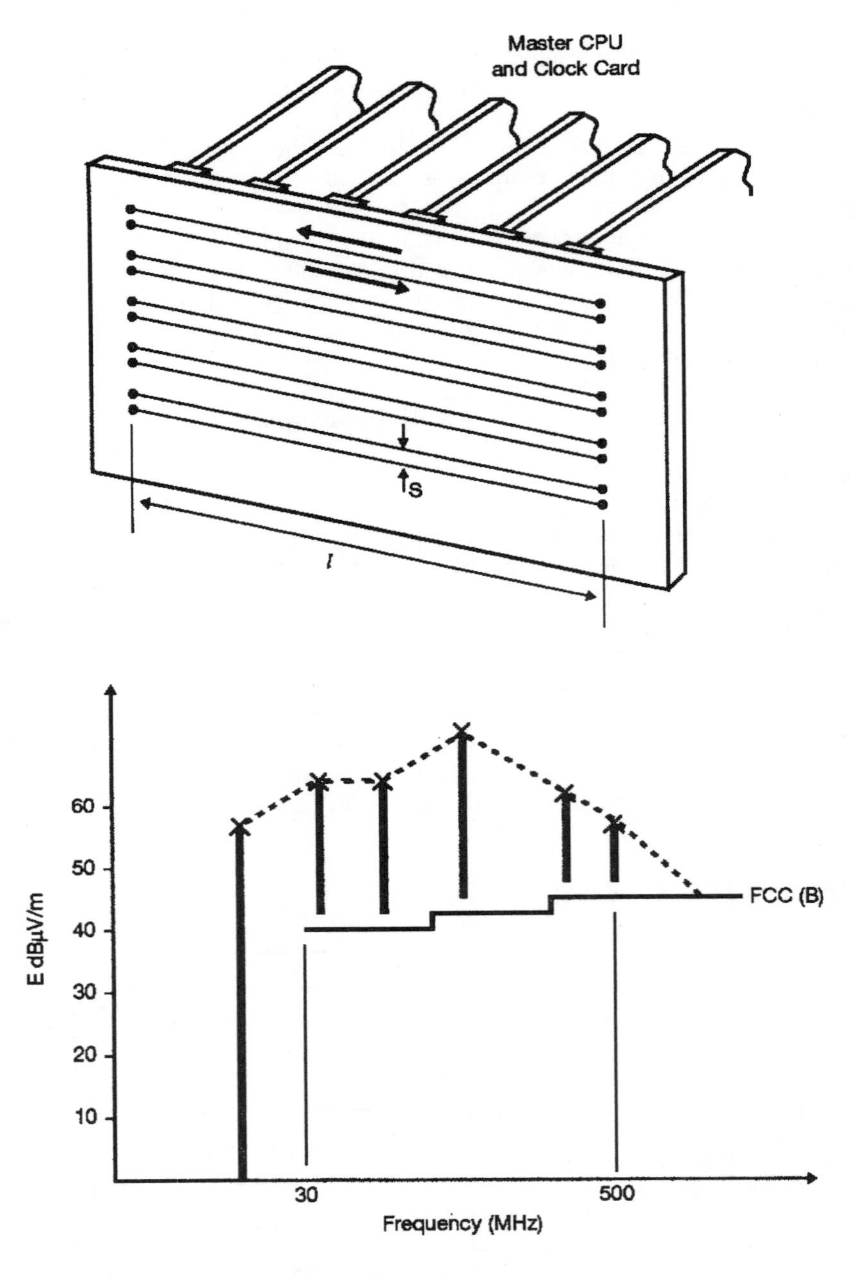

Figure 3.4 Motherboard of Application Example 3.2

The FCC limit, accounting for no shielding at all from the box and covers, is violated by 25dB up to 150 MHz. Some action has to be taken, e.g., reducing the loop size (bringing the clock traces closer or using a ground plane) or shielding the box.

For the sake of precision, it should be remarked that:

1) The Fourier series gives peak values for each harmonic, while EMI receivers are scaled in rms terms, so expect a +3 dB prediction error (pessimistic).

2) The FCC procedure calls for scanning the equipment height with the antenna to search for maximum readings, which typically causes up to 5 dB increase at frequencies where the ground - reflected wave come in phase addition (Ref.. sect. 2.5.2e).

3) The model assumes lossless propagation in cables and circuit traces. In reality, line losses start becoming significant above 200 to 300 MHz in meter-long circuits. This causes an increase in rise times, making the signal spectrum roll off at 40 dB/decade sooner than calculated. On the other hand, the model takes into account that $\lambda/4$ resonance is reached about 0.7 times earlier.

The opposite effects of items 1) and 2) can be handled as follows: the - 3dB correction for peak-to-rms adjustment is almost offset by the + 3-5dB ground reflection, such as in all our subsequent calculation examples, *we will simply disregard these two corrections*. If one wants to perform more exact predictions (knowing that actual measurement uncertainty will cause a greater error...), he can still use the corrections.

3.2 SPECTRUM AND RADIATION FROM BROADBAND SOURCES

The previous NB analysis, which examines one harmonic at a time, is inapplicable when the radiating circuit is carrying signals with a BB spectrum (i.e., signals whose frequency harmonics are closely staggered in the Fourier spectrum). This is especially true when the repetition frequency is lower than the receiving instrument (or victim) bandwidth. The same periodic waveform as in Fig. 3.1 can be displayed by its spectral density (see Fig. 3.5). The starting voltage amplitude is:

$$V_{(\mathrm{dBV/MHz})} = 20\ \log 2A\tau \tag{3.1}$$

All the same, the current spectral density is derived from the circuit impedance Zc(f) at the frequency of interest:

$$I_{(\mathrm{dBA/MHz})} = 20\ \log \mathrm{dBV/MHz} - 20\ \log Z_c(f) \tag{3.2}$$

For a trapezoidal pulse, at any frequency, F_x, across the spectrum, the BB voltage in a unity bandwidth of 1MHz will be equal to:

For $Fx < 1/\pi\tau$:

$$V(V/MHz) = 2\ A \times \tau \quad \text{or,} \quad V\ (dBV/MHz) = 6 + 20\log A\tau$$

For $1/\pi\tau < Fx < 1/\pi tr$:

$$V\ (dBV/MHz) = 20\ \log A - 4 - 20\log F_X\ (MHz)$$

For $Fx > 1/\pi tr$:

$$V\ (dBV/MHz) = -\ 14 + 20\log A - 20\log tr - 40\log\ F_X\ (MHz)$$

for,

V = voltage amplitude in time domain
τ = 50 percent pulse width in microseconds
t_r= rise / fall time in microseconds

If the results are desired in a bandwidth B_i different from 1MHz, use 20 log B_i(MHz) for correction. Then the calculation steps are similar to a narrowband excitation, except that the resulting field will be expressed in dBμV/m/MHz.

Example 3.3

Consider the train of timing pulses in Fig. 3.6. This pulse train is carried over two parallel wires on a flat ribbon cable, with:

l = 1 m
s = 0.5 cm

Circuit load impedance is matched to 50 Ω. The cable is installed at 5 cm over a ground plane and terminates onto a small plastic keypad, isolated from ground, with a parasitic capacitance of 10 pF.
Calculate the radiated BB field by the DM and CM radiation against the RE02 Broad band limit (Air Force and Navy) at one meter and few specific frequencies.

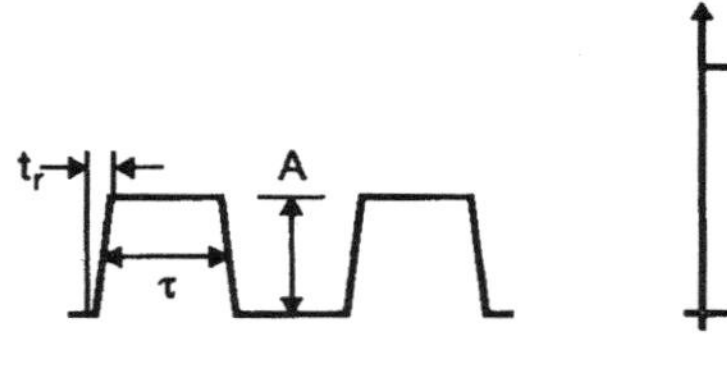

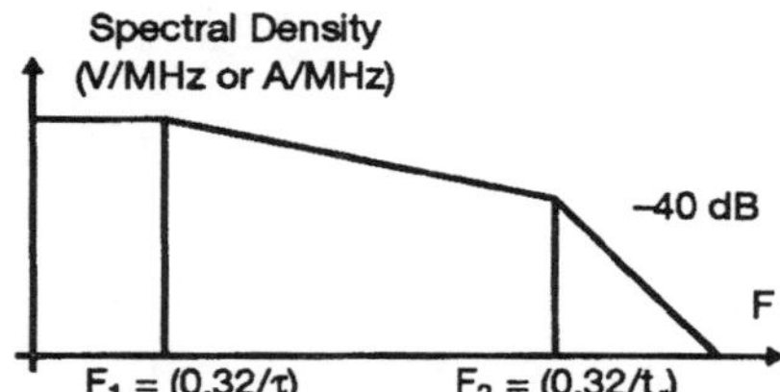

for $F_x < F_1$: $V\ (V/MHz) = 2A\tau$ $dBV/MHz = 6 + 20 \log (A\tau)$

$F_1 < F_x < F_2$: $V\ (V/MHz) = (0.64A/F_x)$ $dBV/MHz = 20 \log (A) - 4 - 20 \log (F_x)$

$F_x > F_2$: $V\ (V/MHz) = 0.2A/(t_r \times F_x^2)$ $dBV/MHz = 20 \log (A/t_r) - 14 - 40 \log F$

(t, t_r in μs)

a. trapezoidal pulses (any period)

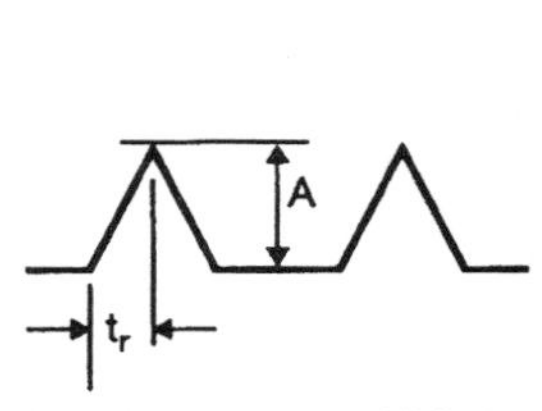

–40 dB

F

$F_1 = \frac{0.32}{t_r}$

for $F_x < F_1$: $V\ (V/MHz) = 2At_r$ $dBV/MHz = 6 + 20 \log (At_r)$

$F_x > F_1$: $V\ (V/MHz) = 0.2A/(t_r \times F_x^2)$ $dBV/MHz = 20 \log (A/t_r) - 14 - 40 \log F$

(t, t_r in μs)

b. isosceles triangular pulses (any period)

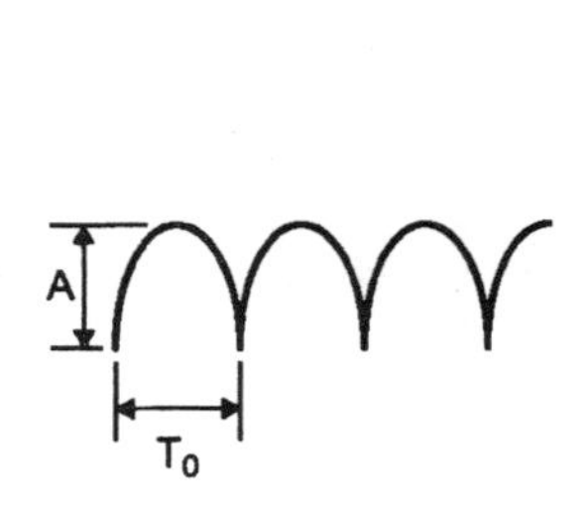

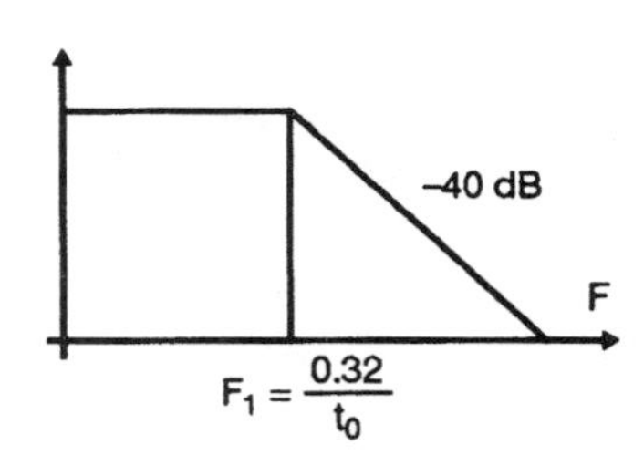

for $F_x < F_1$: $V\ (V/MHz) = 1.2AT_0$ $= 20 \log AT_0 + 2\ dB$

$F_x > F_1$: $V\ (V/MHz) = 0.32A/(T_0 \times F_x^2)$ $= 20 \log (A/T_0) - 10\ dB - 40 \log F_x$

c. full-wave rectified sine wave (for half-wave, subtract 6 dB from amplitudes)

FIGURE 3.5 Fourier Envelopes for Broadband Spectral Amplitudes. Notice that spectral density is independent of pulse repetition rate and remains true for a single pulse (Fourier integral).

Calculation of key parameters:

$F_0 = 100$ kHz
$F_1 = 1/\pi\tau = 320$ kHz
$F_2 = 1/\pi tr = 32$ MHz
$2A \times \tau = 2 \times 10\ V \times 1\ \mu s = 20$ V/MHz = 26 dBV/MHz

Area corrections:

DM loop = 100 cm x 0.5 cm = 50 cm^2 = 34 $dBcm^2$
CM loop = 100 cm x 5 cm = 500 cm^2 = 54 $dBcm^2$

For CM radiation, the line consists of wires above ground; i.e., most of the propagation media is air, not a dielectric. The propagation velocity can be taken as that of free space. So, the $\lambda/4$ length clamp occurs for a monopole when:

$l_m = (300/F)/4 = 75/F$, or F = 75 MHz

At some frequencies, the CM loop impedance (open wire) will be:

For 10 MHz, $Z_{cm} = 1/C\omega = 1/(10.10^{-12} \times 2\pi \times 10^{-7}) = 1.5\ k\Omega$
For 30 MHz, $Z_{cm} = 500\ \Omega$

Above 75 MHz, Z_{cm} will be replaced by the characteristic impedance of this cable above ground, i.e. $\approx 200\Omega$.
Table 3.3 shows the calculation steps for DM radiation. The BB limit is satisfied with at least a 12 dB margin.

Calculations for CM Radiation
For CM radiation, we will use the loop equation, with the capacitive impedance. The calculation results are shown in Table 3.4. Profiles of the DM and CM radiation are shown in Fig. 3.7. The BB limit is exceeded up to 10 MHz, and the margin is small up to 300 MHz.

3.3 RANDOM VERSUS PERIODIC SPECTRA

Whenever there are stable frequency devices in an equipment, test results invariably show that the NB and coherent BB components (those with a stable repetition frequency) are the dominant signatures in the measured spectrum. Although other noise components are found in a radiated spectrum, the BB field from diode noise, motor brushes, fluorescent tubes, ignition and so forth is generally inferior, when measured in a bandwidth-limited receiver.

TABLE 3 3 Calculation Steps for DM Radiation

F_{MHz}	1	10	30	150	300
Reference Amplitude (dBV/MHz)	26	26	26	26	26
Slope Decrease	–10	–30	–40	–68	–80
Amplitude (dBV/MHz)	16	–4	–14	–42	–54
E_0, $1\ V \times 1\ cm^2$, $50\ \Omega$ @ 1 m, dBμV/m	18	18	26	54	66
Area Correction (dB)	34	34	34	28	22
E_{tot} *(dBμV/m/MHz)*	68	48	46	40	34
Spec RE02 (BB)	80	68	64	55	58

TABLE 3.4 Calculation Steps for CM Radiation (The BB Limit is Exceeded up to 10 MHz)

F_{MHz}	1	10	30	150	300
VdBV/MHz	16	–4	–14	–42	–54
Z Loop	15 k	1.5 k	500	200	200
E_o $1V \times 1\ cm^2$ *for (Z)*	18	18	18	42	54
Area Correction (dB)	54	54	54	48	42
E_{tot} *(dBμV/m/MHz)*	86	68	58	48	42
Spec RE02 (BB)	80	68	64	55	58

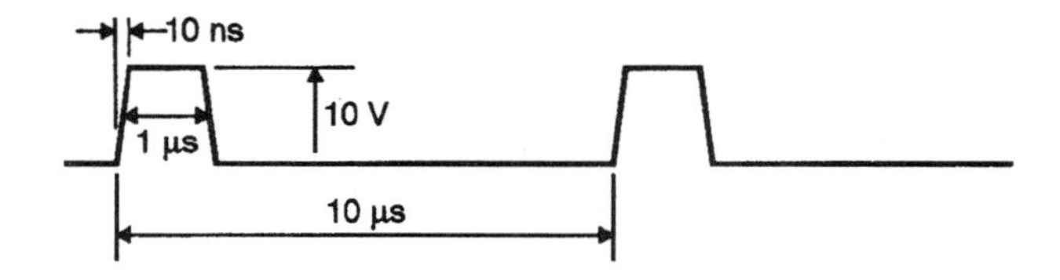

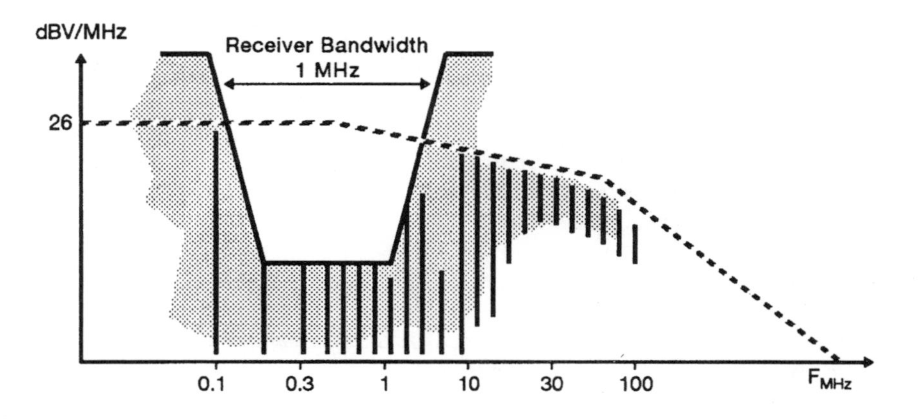

FIGURE 3.6 Example of Pulse Train with Low Duty Cycle, from Example 3.3

As a result, when a radiated limit is exceeded by a device containing digital cicuits (hard to find one which does not..), the major violators are the clock harmonics.
For instance in the actual test results of a digital circuit (see Fig.3.8), the clock signal being a frequency-locked, perfectly stable pulse train, generates a coherent spectrum. In contrast, the digital signals have durations which constantly vary depending on the transmitted message, producing only a noncoherent spectrum with no stable harmonics. Therefore, it is the clock harmonics, switching power supply harmonics and the like that will dominate the radiated profile. Because of this, clock circuits on PCBs, motherboards and cables should be treated very carefully, as will be described further on.

However, with the growing EMC awareness in PCB design teams, clock circuits and their PCB traces are treated more and more carefully, while at the same time, high speed parallel buses with frequencies of 50 or 100 MHz are now commonly used. The result is that fast digital equipments may show limits violations not only by their clock harmonics, but also by the BB contents (generally random), of their address/data buses, as described hereafter in 3.4.1.

3.4 PECULIAR ASPECTS OF SOME FREQUENCY SPECTRA

Certain pulse trains produce Fourier spectra which are not as simple as the immaculate 50% duty cycle clock of our former examples. However, provided a few adjustments, they can be treated by the same straight forward envelope method.

3.4.1 Random signals with Narrow Band contents

NRZ digital signals, although they are synchronized with a fixed clock rate, exhibit constantly varying duty cycle depending on each byte content. In theory, they can only create a random type of BB spectrum, whose average spectral density is proportional to the square root of receiver bandwidth (10 Log BW), instead of 20 Log BW as in coherent BB (see Par.3.2). However, even random pulse trains often contain repetitive patterns which are causing some NB contents. Examples of possible NB contents buried in a BB random spectrum are (Ref 5):

-the least significant bit (LSB) lines on a parallel bus, whose state is almost constantly toggled at the clock rate or submultiples of it.

- the logic circuits using RZ (return to zero) Manchester or similar codings.
- certain signalling codes (ex. in ISDN) with an alternating mark-space pattern.
- clock recovery in data transmission.
- certain cyclic software sequences.
- « transparency » of ASIC and μP cells to clock activity inside the chip.

They appear as a random pulse modulation of discrete harmonics in the emitted spectrum. With typical μProcessor address or data bus, this can be estimated on a probabilistic basis, by some worst case assumption of the data stream.
Based on the probable spectral density,Fig. 3.9 shows the amplitude of a BB spectrum of random NRZ pulse train with 10 and 100Mb rates, when received in a

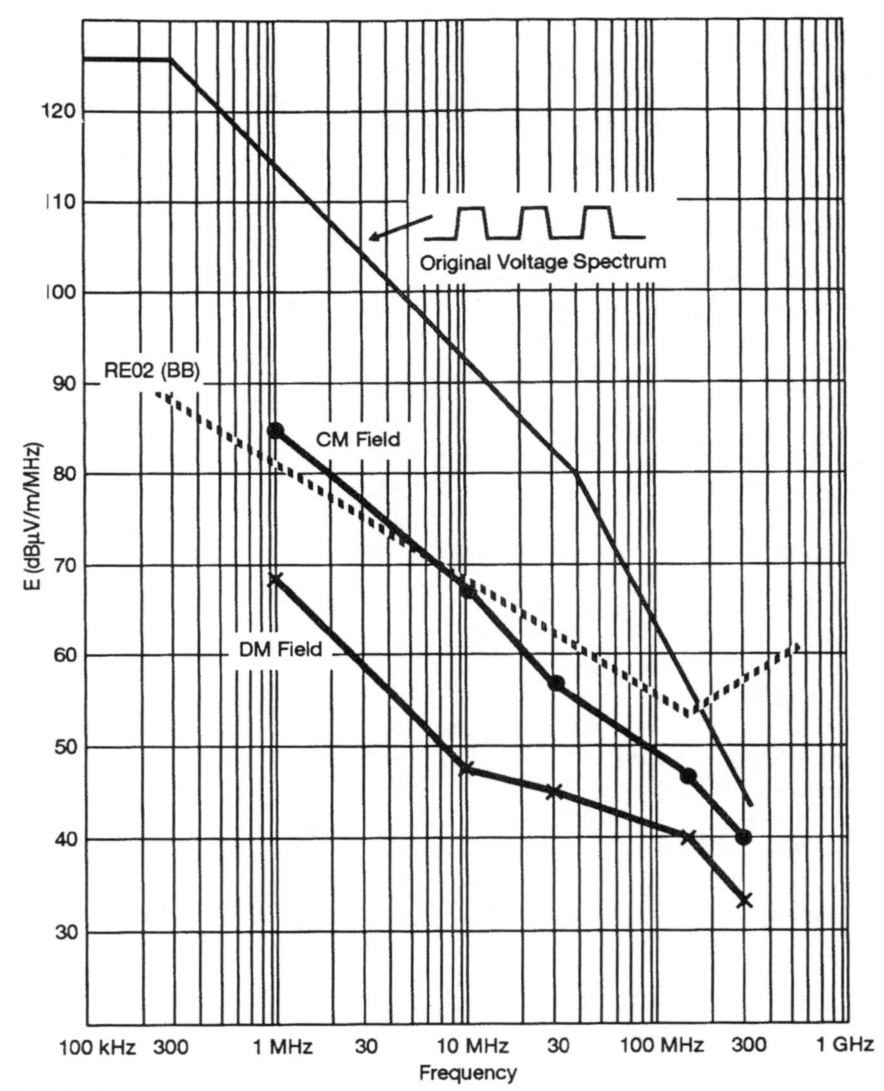

FIGURE 3.7 Profiles of DM and CM Radiation, from Example 3.3

120kHz bandwidth. Compared to the original spectrum of a clock with the same rate, we see that clock residues (NB) will emerge from an otherwise BB spectrum if the amplitude is Δor less below the original clock spectrum. This can be re-stated as folllows :

« if PCB traces, ribbon cables, etc... are polluted by clock residues, these can be examined as prime candidates to spec. violation if their amplitude is greater than : original clock spectrum - ΔdB »

The Δ term depends on the ratio of the clock pusewidth compared to the elementary data bit width, and the measuring receveir bandwidth. For instance, in Fig. 3.9, Δ is 30dB for the 10MHz / 10Mb case, and 50dB for the 100MHz/100Mb case.
If the clock has been extremely well segregated and filtered from the rest of the circuits, with a decoupling greater than Δ, the BB spectrum sources can be examined as the next potential candidates. The two candidates are not mutually exclusive.

Example 3.4
Calculate the BB spectrum and the peak amplitude, in a 100KHz bandwidth, for the following random NRZ pulse trains :

a) 10Mb, 1V ; one bit width, τ(mean) = 50ns, t_r, t_f = 5ns
b) 100Mb, 1V ; one bit width, τ(mean) = 5ns, t_r, t_f = 0.5ns

Solutions

Rep. Rate	10Mb	100Mb
Mean spectr . density		
2 Axτ(V/MHz)	0.1V	0.01V
dBμV/MHz	100	80
1st corner freq.	6MHz	60MHz
2nd corner freq.	60MHz	600MHz
Peak amplitude in 100kHz BW		
BW correction for random BB		
10 log (0.1MHz/1MHz)	-10	-10
Amplitude up to 6MHz	90dBμV	70dBμV
Amplitude @ 60MHz	70dBμV	70dBμV
Amplitude @ 600MHz	30dBμV	50dBμV

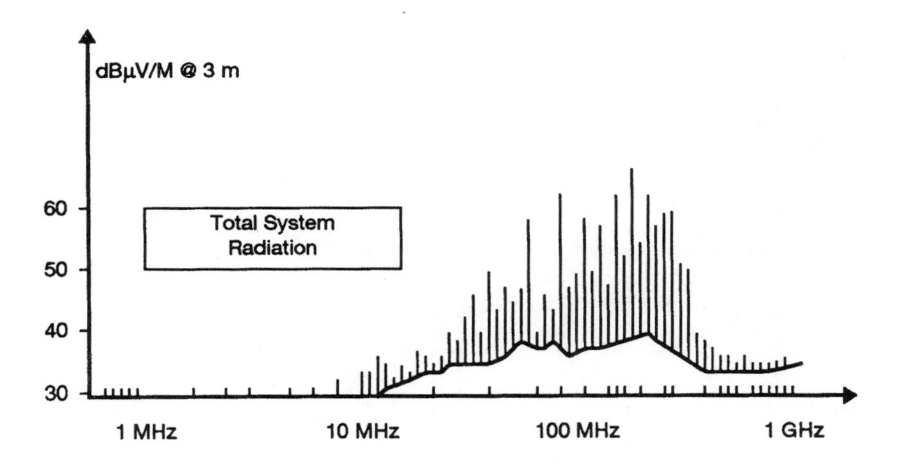

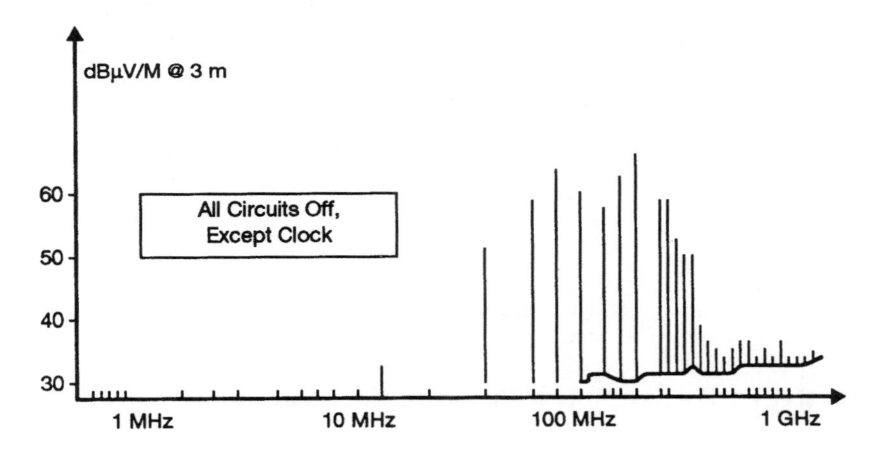

FIGURE 3.8 Periodic vs. Random Radiated Signatures

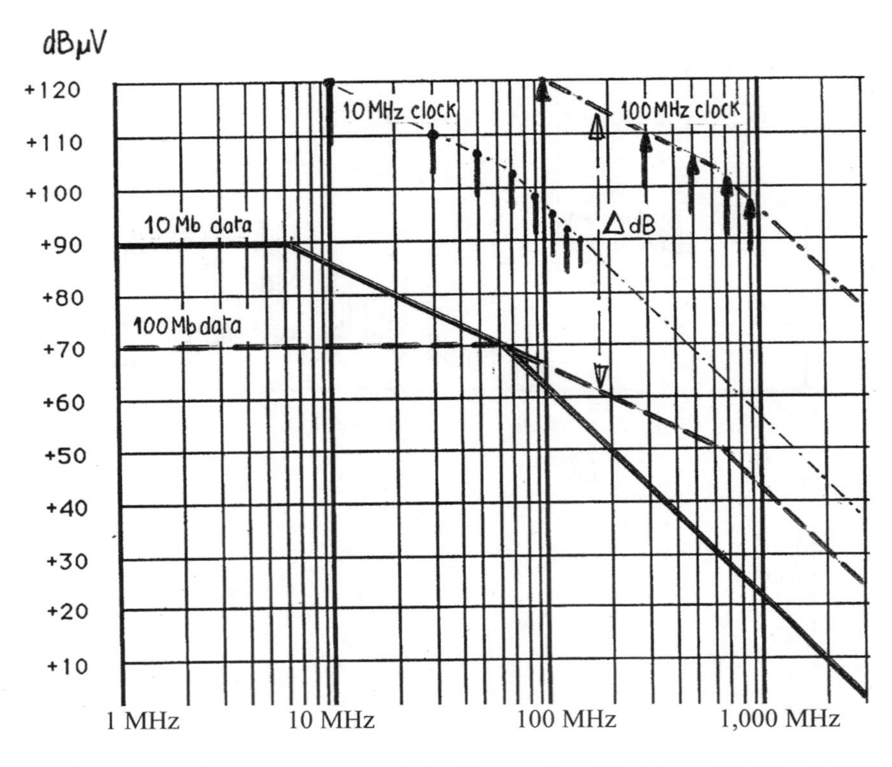

Fig. 3.9 Magnitude of the BB spectrum of 1 Volt random NRZ data with 10 and 100MB when measured in a 100 kHz BW. The discrete spectrum of V clocks with same frequencies is shown for comparision. Rise times are assumed 10% of elementary bit width.

3.4.2 Repetitive Symbols Enhancing EMI Radiations

High definition CRT displays use pixel clock oscillators at 100MHz or more, with video signal rise times < 2ns. This pixel frequency correspond to the smallest picture segment, i.e. the elementary pixel duration:

Pixel clock freq. = Number horiz. pixels x Number vert. pixels x refresh rate

In some EMI analysis or testing, it is sometimes useful to force the highest symbol rate, i.e. 0.5 Fclock, corresponding to Mark-Space-Mark-Space etc. In that respect, one of the busiest test pattern for CRTs is a full screen of "H" characters, generating a high level of discrete harmonics (Ref. 5A).

3.4.3 Spread Spectrum Clocks (SSC) for reduced EMI

An interesting concept was developed by Bush (Ref.6) whereas the digital processor clock, instead of a stable frequency, is using a frequency modulated clock, such as the actual spectrum occupancy for each one of the clock harmonics is spread over wider bandwidth. If the modulation is Δf on the fundamental, each consecutive nth harmonic is spread over a bandwidth +/-n x Δf around the theoretical narrowband frequency. If Δf is choosen to be greater than the typical receiver bandwidth, the resulting net amplitude will be less than with an ordinary NB spectrum. The decrease in measured amplitude is approximately equal to $\sqrt{n}/2$, i.e. 10 Log n /2, or 10 (Log n) - 3dB

For example, with a FM deviation of 125 kHz and a standard EMI receiver bandwidth of 100 kHz, the reduction brought by the spread spectrum for a 20 MHz clock, (odd harmonics) is:

harmonic #	Δ
3	- 2dB
5	- 4dB
9	- 6,5dB
etc...	

One key factor for the improvement is that the modulating signal is not a sinewave itself but a smoothed sawtooth whose time derivative at zero crossing is minimal. This technique is applied on some recent processors of the Pentium and Celeron families.

3.4.4. Even vs Odd harmonics when duty cycle is not exactly 0.5

The general expression for each spectral term A_n of a trapezoidal pulse train is:

$$A_n = A_o[\sin(n\pi\delta)/n\pi\delta] \times [\sin(n\pi t_r/T) / n\pi t_r/T] \qquad (3.3)$$

a) When $\delta = \tau/T$, the duty cycle, is exactly 0.5 as we often assume in the simplified spectrum envelope of a clock signal, a quick look at the "A_n" function shows that:

- for each even value of "n", the term sin ($n\pi\delta$) equals zero: only odd harmonics are present
- for small values of n (1, 3, 5, etc...), the first term equals ($1/0.5n\pi$) while the second term is approximately equal to 1, resulting in the 1/n decrease (-20dB /dec) we show beyond first corner frequency F_1
- for higher values of n, the [sinus] takes any value between 0 and 1 while $n\pi t_r/T$ can become very large. So, the product of the two [ ] terms falls-off like $1/n^2$.

b) If δ is not exactly equal to 0.5 (as often the case, due to pulse pedestal distorsions), even harmonics will take place. Related to the fundamental amplitude (A_o), the even harmonics value is:

$$A_{even}/A_0 = \sin n\pi p/n \qquad (3.4)$$

where $p = \delta - 0.5$ i.e the deviation from an ideal 50% duty cycle.
As long as $n\pi p < \pi/6$ (i.e. where $\sin x \approx x$) the even harmonic amplitude is given by:

$$A_{even}/A_0 \approx \pi p \qquad (3.5)$$

Thus, although their level is lower at the beginning than the odd ones, the result is that the even harmonics are keeping a constant amplitude when F increases, as long as condition of Eq (3.5) will hold, reaching eventually the maximum spectral envelope. Then, even harmonics will fall-off at - 40 dB/decade, like the odd terms. This does not change our maximum field prediction based on the envelope (in fact it justifies it) but the designer must be prepared to find, among the radiated emissions, even harmonics with an amplitude as large as the odd ones (Fig. 3.10).

Example 3.5 *:*
Let us take a 10 MHz digital pulse train whose the duty cycle is slightly asymetrical: $\delta = 0.5 - 2\%$, i.e τ = 49ns for T = 100ns

Assume t_r = 5ns, so F_2 = 64 MHz
The pulse amplitude will be normalized as 1 volt

A precise calculation of each term is shown below:

n	A_{odd}, volts	(dBV)	A_{even}, volts	(dBV)
1	0.64	(-4)		
2			0.02	(-34)
3	0.21	(-14)		
4			0.02	"
5	0.13	(-18)		
6			0.02	"
7	0.08	(-22)		
8			0.02	"
9	0.05	(-26)		
10			0.02	"
11	0.03	(-30)		
12			0.02	"
13	0.025	(-32)		
14			0.02	"
15	0.019	(-34)		
16			0.02	"

Beyond this point, even terms fall-off like $(^1/_F)^2$, staying at $\approx$ same amplitude as odd terms

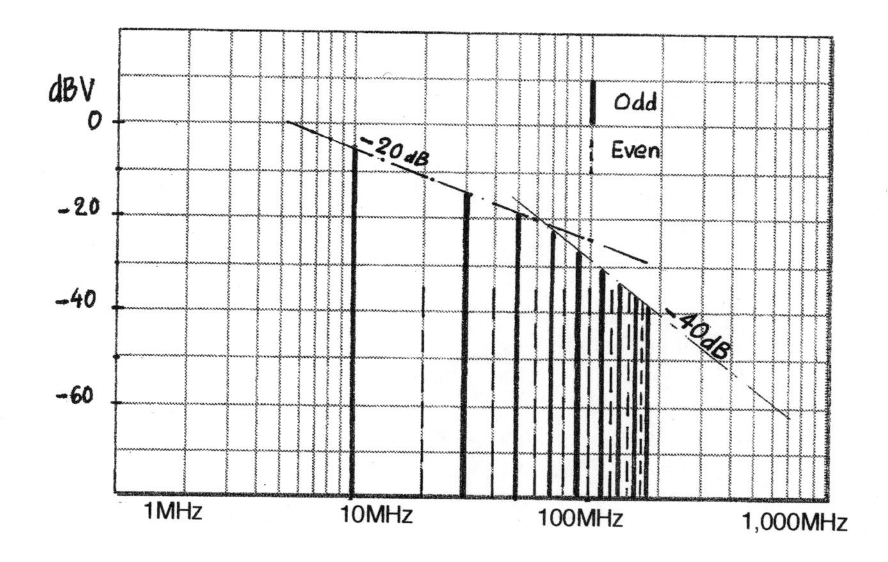

Fig. 3.10 Presence of even harmonics in a 10MHz pulse train with 49% duty cycle

Example 3.6
As a recapitulation exercise, we will run the following example of cable radiation with a combined NB + BB spectrum.
An external cable with unknown balance is carrying differential data at 25Mb rate. The equipment I/O port is also polluted by small 50MHz clock residues riding over the digital pulses. The characteristics for the wanted and unwanted signals are as follows:

1) 25Mb Diffl signal	2) 50MHz Clock spurious
	rep. period, T = 10ns (every clock transition)
Amplitude = 5V	Amplitude= 50mV
rise time = 3ns	pulse width and tr,tf = 2ns

Cable length ⩾ 1.50m, cable out.diam.= 1cm, height above grnd ⩾ 0.80m
Calculate the worstcase radiation from this long cable, high above ground, compared to FCC classB limit, measured in 120kHz bandwidth.
Solution :

1) Calculate first the CM characteristic impedance of this cable:

$$Zcm = 60 \text{ logn } (4\ h/d) \approx 300\Omega = 50\text{dB}\Omega$$

2) Voltage Spectrum envelope

	BB (25Mb)	NB
Mean bit width:	$\tau = 20$ns	$\tau = 2$ns
1st corner freq.	16 MHz	160 MHz
2nd corner freq.	100 MHz	160 MHz
Pedestal amplitude:	$2\,A\tau = 0.2$ V/MHz = - 14dBV	$2A\tau/T = 0.02$V = - 34dBV
Correction for random pulse train: 10 log BW(MHz):	= - 9dB	N.A.
Spectrum pedestal, in 120kHz BW :	- 23 dBV	- 34dBV

Since we are calculating the worst possible radiation,we will estimate the CM current which can drive such long wire configuration. By default, we assume a 10% cable unbalance.

F,MHz	30	60	100	150	300
BB diff. voltage, dBV in 120kHz:	-28	-34	-38	-45	-57
NB diff. voltage, dBV	-34	-34	-34	-34	-46
Larger of the two, or rms combin.	**-28**	**-31**	**-33**	**-34**	**-46**
CM /DM derating for 10% unbal.	-20	-20	-20	-20	-20
Corresponding CM voltage drive, dBV	-48	-51	-53	-54	-66
Conversion into dBμV: +120	+72	+69	+67	+66	+54
CM current, dBμA = dBμV - 50dBΩ:	**+22**	**+19**	**+17**	**+16**	**+4**
Maximum I_{cm} criteria (Ref. sect. 3.) for meeting class B	+10	+10	+10	+10	

The composite voltage spectrum envelope is shown on Fig. 3.11.

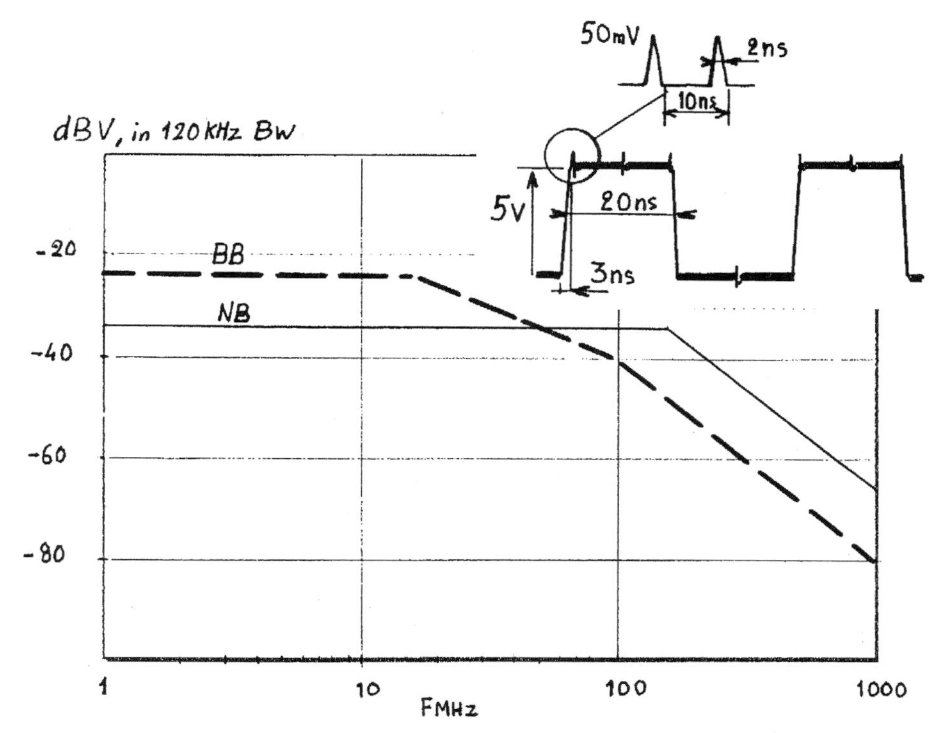

FIG. 3.11 Composite Voltage Spectrum of Example 3.6

We can see that the CM current criteria is exceeded from 30 to ≈ 200MHz. The larger CM current is due to the wanted BB signal up to 50MHz. In the mid-interval 50-100MHz, the BB spectrum is undercarrying the NB one which progressively takes over, and this is their rms addition which is seen. Above 100MHz, the NB spectrum is the dominant mode.

The problem requires a rather moderate >12dB improvement, which can be obtained by a better balancing (changing from 10% to 2%) or ferrite loading. Notice that filtering the clock residues would not be sufficient.

Once again, we see how little it takes to exceed the radiated limit. The 50mV, shown to scale on Fig. 3.11 would go almost un-noticed on a simple oscilloscope display of the differential 5V signal.

Chapter 4

General Strategy for Designing a Low-Radiation Product

Compliance with radiated emission limits does not occur by chance. With the fairly severe specs in force today, and with the number and types of fast switching or continuous wave (CW) devices to be found in any equipment, there is practically no example of a product which, having been designed with complete disregard to these aspects, clears through the certification test with an OK label. Therefore, for sake of self-compatibility (i.e., no self jamming) as well as compliance with applicable standards, it is necessary to make EMI control a design parameter and to integrate it within the normal process of design activities and reviews.
We will examine briefly, at a managerial level, the following topics:
• system design decisions that impact EMI emissions
• areas of design activity for keeping emissions under control
• EMI control milestones to be observed during the design process

4.1 BASIC SYSTEM DECISIONS THAT IMPACT EMI CONTROL

The designer must make a lot of fundamental decisions in the early stages of a product design, many of which have an impact (good or bad) on the product's EMI characteristics. Therefore, it is of prime importance that an EMC specialist (or someone in design engineering management with an EMC background) could present a clear picture of how all of the given technical options will affect EMI performance. The principal items on this "product design checklist" are as follows (an asterisk indicates that radiated EMI levels are affected):

1. <u>Interface choice</u> (e.g., IEE 488, RS-232, RS422, RS485, IBM, Ethernet, I^2C etc..)
• Balanced or unbalanced link. A balanced pair with a symmetrical driver/receiver reduces CM loop coupling, for both emission and susceptibility.*
• Connector pin assignment. Typical interface standards impose a fixed pin assignment. Some imposed assignments are rather poor with regard to crosstalk,

radiated emission and susceptibility, immunity to ESD and fast transients induced on cables.*
• Data rates and maximum permissible rise times. These considerations have a direct effect on the spectrum profile and the ability or inability to use signal filtering.*
• Handshake protocols, error detection, error recovery, etc.. These affect the device's susceptibility to transients and ESD.
• Connector type and mounting style. Being generally imposed by the standard interface, this has a direct impact on emissions and susceptibility. Depending on the style (e.g., sub-D, circular, Micro-D, Centronics, BNC), the locking method (threaded, bayonet-mount, screw-mount), and material and finish, bonding between cable shields and the equipment cabinet will range from good to poor.*

2. Type of I/O cable used (e.g., twisted pair, shielded twisted pair, shielded twisted pair plus overall shield, ribbon, shielded ribbon, coaxial). This choice affects:
• Radiated and conducted emissions*
• Radiated and conducted susceptibility

3. Cabinet /housing style (e.g., nonconductive plastic, conductive plastic, metallic with or without seams & aperture shielding). The choice of the envelope is usually is driven by aesthetics, weight and manufacturing considerations, but there are significant related EMC implications. They include
• Radiated and conducted emission*, and susceptibility. Although cables are the predominant radiators, box emissions are frequently found in a range beginning above 100 or 150 MHz. Even with minimal box emissions, a metallic surface may be needed to correctly terminate the I/O cable shield and shielded connectors, and to bond EMI filters.
•Internal EMC. A metal cabinet facilitates the mounting of internal shielded compartments to prevent self jamming.

4. Internal technologies and clock rates. Here again, the choices are driven by functional necessities but have a serious (maybe the most serious) impact on radiated EMI. Quite often, a designer chooses a technology with a super-fast clock frequency and corresponding rise times for the whole circuitry. Then, after EMI emissions have haunted the engineers, and the prototype fails evaluation testing, it is found that 90 percent of the internal lines must (and can) be decoupled to slow down the rise times.

5. Power supply type (e.g., linear, phase control, switcher with or without a front-end transformer, switcher with a multiple isolated outputs, resonant switching power supply). The EMC implications related to power supplies include:
• Conducted and radiated (magnetic) emissions*. Switcher frequency and topology directly affect conducted and radiated EMI levels. There is also the possibility of self-jamming.
• Maximum permissible ground leakage current. This is a safety issue that relates to the type of filter allowed by applicable regulations.
• Primary-to-secondary isolation*. The topology and type of transformer also affect clock and other harmonics that return and reradiate via the power cord.

• Isolated multiple-outputs*. These may be required when there is a mix of analog, RF and digital circuits in the same equipment. This approach will avoid ground pollution of one circuit by another. This affects self jamming as well as external emissions.

6. Applicable mandatory EMC standards (e.g., FCC Class A or B, CISPR/EEC, MIL-Std 461, DO160, TEMPEST). These standards are dictated by market regulations (for a civilian product) or the procurement contract (for a military or government supply). Some products that were liberally quoted as "FCC Class A" devices at the design phase have been determined to be relevant to Class B requirements shortly before the first customer shipment. This distinction translates into a different administrative route (self-verification versus certification) and tougher limits.

7. Frequency management and frequency plan.* This item applies only in cases where the equipment is part of a larger telecom or radiocommunication system or is installed near to one. In such a case, all functional and accessory oscillator frequencies inside the equipment should be checked for possible coincidental co-channel interference with the host installation.

4.2 DESIGN CONTROL AREAS FOR RADIATED EMISSION REDUCTION

Once the designer has made the appropriate choices enumerated in Section 4.1, in full knowledge of the EMI implications they bear, he should proceed to a comprehensive EMC analysis in the following areas (four of which are illustrated in Fig. 4.1):

1. Digital ICs and high-frequency generating components. Determine dV/dt, dI/ dt, instant power demand, package type (DIP, SO, SMC, PGA, plastic or metal can). Establish noise margin and decoupling capacitor needs. For the larges ICs with complex internal functions (MCMs, ASICs, Processors), try to obtain from the manufacturer some EMI characterization. These can be gathered from specific component emission testing like SAE J1752/3 (TEM cell method), VDE or IEC 61967 conducted signature methods.

2. PC board. Examine power distribution and return channels. Consider single vs. multilayer board design, the need for impedance matching, and the desirability of separate ground planes to reduce crosstalk. Examine radiated emissions at card level.

3. Motherboard. Considerations include wrapping, single or multilayer design, surface vs. buried signal traces, and ribbon cables (with or without a ground plane). Also examine DC voltages and 0V distribution, and pin assignment to daughter cards.

4. Internal packaging. Determine the need for compartmental shielding around noisy or especially sensitive components, and review I/O port placement and decoupling. Also relevant are power supply location and filtering (an excellent discussion of this subject can be found in Ref. 26). Possible grounding schemes include isolated ground (0Vref), grounded 0 V, or a mix of the two. The choices must be justified.

5. Housing design and shielding. If calculated predictions or early tests show that a conductive housing is necessary, it has to be consciously designed rather than thrown in at the last minute. Conductive plastics require a careful analysis of their

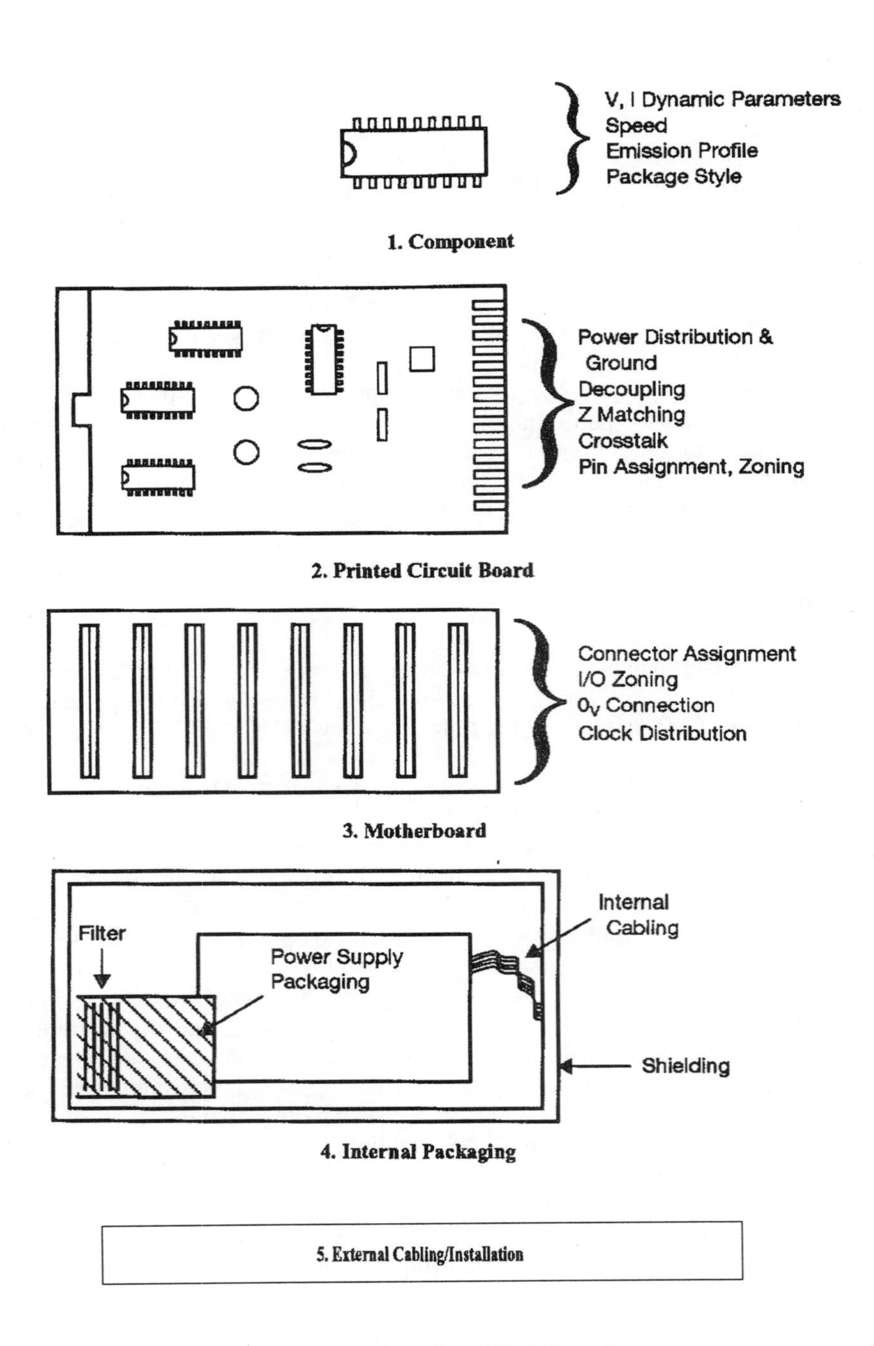

FIGURE 4.1 Areas of Design Strategy for Radiated EMI Control

internal shapes and ribs for coating adhesion and component grounding, as well as for the integrity of the mating edges. Metal housings need a proper tolerance analysis and knowledgeable choice of surface treatment to ensure effective, continuous cover contact.
6. Installation and external cabling. Depending on whether the system supplier must provide the external cables and installation guidelines, this may be another area where EMC must be considered. The designer must properly specify (1) external cable type, (2) vendor reference and (3) any acceptable substitutes. Otherwise, unacceptable "look-alike" cables may be installed.
7. Testing and certification.

The remainder of this book will review in detail the items enumerated above.

4.3 DESIGN MILESTONES FOR RADIATED EMI CONTROL

Radiated EMI is but one of several facets of the EMC discipline, which also includes radiated susceptibility, conducted emissions and conducted susceptibility.

Because these facets are interrelated, there is no reason for the designer to address them separately; that is, one should not first address radiated emissions, then conducted EMI, then electrostatic discharge, etc. There is a strong reciprocity between emissions and susceptibility, so they should be considered in concert. A calendar of EMI design milestones can incorporate all aspects of EMI reduction, so the reader should keep in mind that the following discussion applies to interference reduction in general, not just to radiated EMI.
The following milestones are recommended, for a methodical radiated EMI control:

1. Calculate an approximate radiation profile plot for PCBs and internal parts (DM radiation). This is done using the Fourier envelopes described in Chapter 3, and the DM loop models of Section 2.3 (especially Fig. 2.6), using the following procedure (see Fig. 4.2):

- Make a list of all the printed circuit cards and other building blocks generating or using pulsed periodic signals, including the signal waveforms and the circuit dimensions.
- For each flow of signal S_1, S_2, S_3, and so forth listed in the above step, count the number of traces or wires in a given PC board, power supply or other component. For a purely digital card, this can be approximated quickly :
number of traces for signal S = 0.5 (number of modules x average number of clocked or
synchronous pins per module).
- For each signal S_1, S_2, S_3, etc., evaluate the average $A = l \times s$ product of the radiating loop. A rigorous evaluation would require one to physically measure each wire length, but a good approach is to estimate quickly via the following equation:

$$l_{average} = (\Sigma \text{ all } S_1, S_2, S_3 \text{ etc run lengths}) / \text{N runs}$$

An even quicker approach is to take the coarse approximation: *l* average = 1/4 (PCB diagonal dimension). Dimension s is the average distance to the next parallel ground trace (for single-layer boards) or the dielectric thickness to the Gnd plane (for cards with a ground plane, or multi layers).

- Add the following to the previously calculated l x s areas:
  a. module package areas ($0.4cm^2$ for a 14-pin DIP, $0.1cm^2$ for a 14-pin SMT, etc.)
  b. area of decoupling capacitors to IC pins
  c. areas of I/O wire pairs for related signal
- For each signal S_1, S_2, S_3, etc., perform a quick plot of the NB or BB Fourier envelope. We know that for each circuit carrying S_1, S_2, . . . S_n, the radiated field is calculatable from the electromagnetic moment Volt x cm^2 or Amp x cm^2 per a function:

At frequency F_x: $E_1\ (\mu V/m) = V_1 \times A_1 \times K(F_x)$

$E_2\ (\mu V/m) = V_2 \times A_2 \times K(F_x)$

etc..

where K(Fx) is the loop-to-field transfer function for $1V\text{-}cm^2$ (See Eq. [2.22]), a constant for a given frequency , except for the circuit impedance variation if less than 377 Ω.

- For the desired test interval F_{min}, F_{max} (e.g., 30 MHz - 1,000 MHz, for FCC regulations), we will fill in Table 4. 1 .

If, while still at the drafting level, the design meets the previous test, there is nearly a 100 percent chance that the box itself (without I/O signal cables and power cables) will comply. This is due to the several small overestimations that were made intentionally.

2. If step 1 shows excessive radiation, the designer must consider the following options while the design is still flexible. The appropriate one will depend on which signal and which circuit has caused the specification violation.

a. Enlarge 0V ground areas.
b. Provide better IC decoupling.
c. Use a multilayer board.
d. If all else fails, add box shielding.

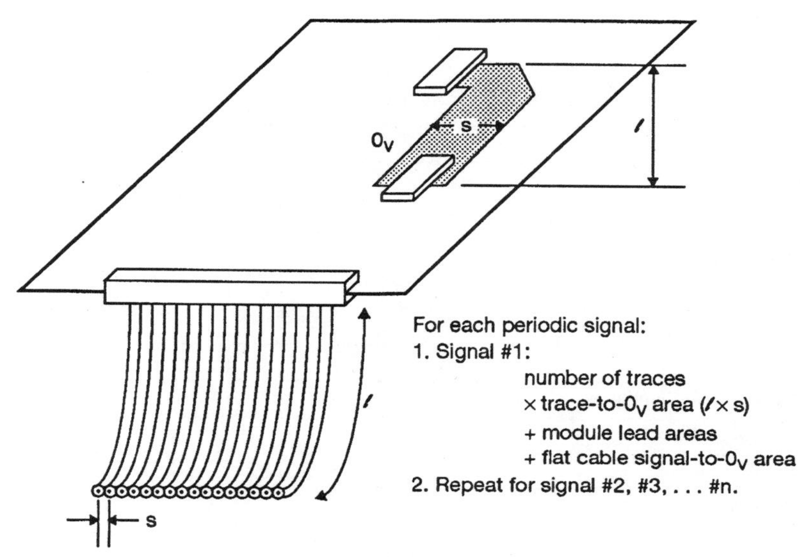

3. Plot each signal spectrum, starting from the lowest rep. frequency.

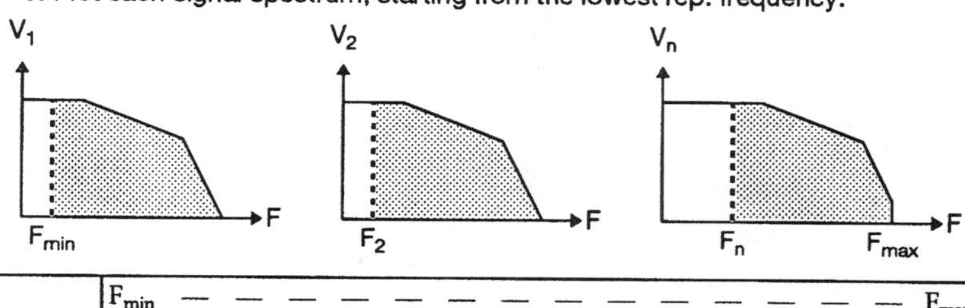

	F_{min}	—	—	—	—	—	—	F_{max}
Signal #1	V– cm^2	V– cm^2	V– cm^2	V– cm^2	V– cm^2			
Signal #2	X	X	V– cm^2	V– cm^2	V– cm^2	V– cm^2	V– cm^2	
. . .								
Signal #n	X	X	X	X	V– cm^2	V– cm^2	V– cm^2	V– cm^2
ΣV–cm^2								

4. Sum up all V-cm^2 across the spectrum.

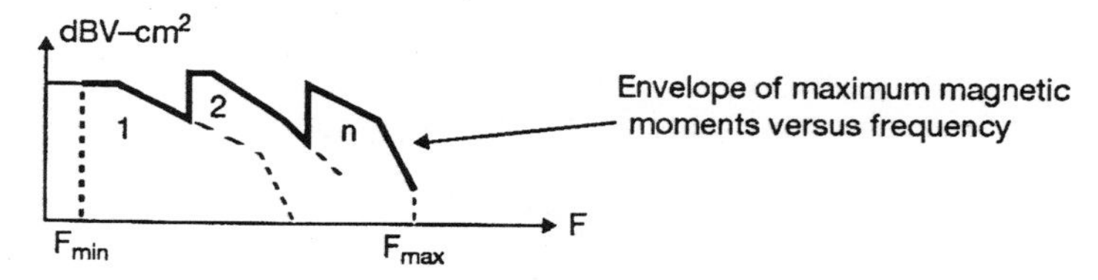

5. Use dBV – cm^2 to calculate composite field across frequency.

FIGURE 4.2 Method for Making a Coarse Estimate of Cumulative Fields from All DM Loops

Table 4. 1 Calculated Radiation Profile Data Sheet

F_{min}	F_1	F_2	F_3	$-F_3$	F_n	F_{max}
Signal #1						
V_1, harmonic amplitude (dBV)						
Area, A_1 (dBcm^2, including λ/4 limitation)						
*dB current adder = 20 log(377/Z_L) for Z_L<377 Ω						
M_1 (dBV–cm^2)						
Signal #2						
V_2, harmonic amplitude						
Area, A_2 (dBcm^2)						
dB current adder						
M_2 dBV–cm^2						
Signal #n						
V_n harmonic amplitude						
Area, A_n						
dB current adder						
Mn (dBV–cm^2)						
$K_{dB\mu V/m}$ for 1 V_x 1 cm^2 and 377 at frequency F_x						
**E_{total} = K + log [$10^{M1/20} + 10^{M2/20} + ... + 10^{Mn/20}$]						
Specification to be met (dBμV/m)						
Δ = E – specification						

* The current multiplier is to account for the E field increase, especially in far field, for circuits with low load impedance, i.e. higher currents than V/377 Ω (or 2.6 mA per volt)
**A quicker way is to retain, in each frequency column, only the larger value of M, provided it dominates the others by >20 dB. This is not always possible.

3. As soon as prototypes of the major subassemblies are available, run a coarse emission test in an anechoic chamber, at 1m distance, to see if a sufficient margin exists at individual block levels. For some components (e.g., microprocessor cards, master clocks, oscillator cards, I/O drivers), it may be convenient to power the prototype by a battery pack strapped under the card instead of using a bench-type power supply.
If the limit to be met is specified at 3 or 10 m (FCC, for instance), apply 1/D conversion factor to the results, above ≈ 30 MHz. If a same card is to be present N times in the future equipment, add 20 log N to the measured field, for coherent

addition, 10 Log N for random addition (above 2 -3 identical cards, the quasi-peak detector used for FCC/CISPR makes that the addition is seldom coherent).
4. Up to this point, we have focused mostly on controlling DM radiation from PCBs and internal circuits. As soon as a breadboard prototype of the complete product exists, check it in the lab (which need not be a shielded room if ambient RF noise, especially from FM and television stations, is low enough) for CM current leakage on I/O cables. This can be done relatively easily with a current probe and a spectrum analyzer. Apply the following criteria for each spectral line:

a. For FCC or CISPR/EEC class(B), $I_{cm} < 10$ dBμA at $F > 30$ MHz
b. For FCC or CISPR/EEC class(A), $I_{cm} < 20$ dBμA)
c. For MIL-Std461 RE02 (NB) (the most severe limit curve),
$I_{cm} < 10$ dBμA at $F > 48$ MHz

If the criteria are not met, add filters to I/O ports, use ferrite beads or cable shielding, and reiterate until Icm decreases sufficiently. More details on this evaluation method are given in Chapter 13.

4.4 INFLUENCE OF THE RESIDENT SOFTWARE ON RADIATED EMI

Although radiated emissions above 30MHz are generally dominated by the clock frequency spectrum (see Chap.3, Fig. 3.8), amplitudes variations up to 10dB have been recorded for a same product with two versions of the software. The version which was radiating the most had simply few more instructions and a particular shift in a sequence of the main program (ex. changing from address 7FFF to 8000, where 16 bits are switching simultaneously). Because they kind of "underlay" the clock spectral lines, data and address bus are non-negligible contributors to total radiation.

The software dependency of their contribution depends on how they are multiplexed. In the main operating program, instructions which are constantly cycling in a repetitive mode will contribute to measured EMI, since typically the radiated test is made with the spectrum analyzer / receiver in a peak/hold mode. Several software precautions can reduce this contribution (Ref. 7):

• To access frequently-used variables, use as much as possible the RAM internal to the processor.
• Do not send hi-rate bus signals on peripheral devices if these signals are only needed few times /sec. For instance, on a display, use a software latch such as multiplexed data are sent to the display flat cable only when needed.
• A keyboard does not need 4MHz multiplexing. A multiplexing made at < 1kHz still allow keystroke detection, while generating less average radiation.

Chapter 5

Controlling Radiated Emissions at the Chip and Integrated Circuit Level

The term "controlling" necessarily implies "understanding". To start with, let us say that it takes two things to constitute a radio transmitter, intentional or otherwise: a signal source and an antenna. We have discussed the fortuitous antenna in Chapter 2. Here, we will address the most elementary building block in the system with regard to its ability to act as an RF source. Choosing ICs designed with low RF signature makes the whole product EMC compliance an easier target to hit. Interestingly, most of the parameters and behaviors that we will discuss are also major players in the susceptibility and self-compatibility of our equipment.

5.1 LOGIC FAMILIES

Table 5.1 lists the typical characteristics of the most popular digital families, including the parameters of interest for radiated emission. They correspond to one simple active device (an elementary gate, or a driver) in a module that may contain four to thousands of these devices. This elementary gate will be the switching device we will always consider in our discussion of noise generation.

The bandwidth is shown as $1/\pi t_r$, with t_r,t_f being the rise and fall times. For bipolar technologies, rise and fall times are generally different, but the equivalent bandwidth has been calculated from the shortest of the two. Rise and fall times also depend on loading, with the faster transition (e.g., the highest bandwidth) corresponding to the lowest loading. The table values for t_r or t_f correspond to a moderate loading (20 to 40 pF) to represent reasonably fast conditions.

The column captionned "output resistance" is indicative of the worst-case internal resistance of a single device output, in Low or High state. Logic gates being non-linear, these dynamic resistances have been derived from a piecewise approximation of the V_{OL}-I_{OL} and V_{OH}-I_{OH} curves. For immunity analysis, the designer is concerned with the highest value of $R_g = \Delta V / \Delta I$, since this corresponds to a high impedance condition for crosstalk and E field coupling. For emission control, our concern here,

Table 5.1 Essential Characteristics of Logic Families

Logic Name	*Volt.Swing* ΔV_{L-H} (V)	t_r/t_f (ns)	*Equivalt Bwidth* (MHz)	*Cap.(pF)* (pF)	*Input /Output Resist.* L/H(Ω)	*DC Noise Margin* (V)
CMOS	5	50/50	6.5	5	500/350	1.2
LV-MOS	3.3	3/3	110	"	50 /100	0.6
HC00,AHC	5	3.5/3.5	95	4.2	50 /80	1.3
HCT	5	3.5/3.5	95	4.2	50 /80	0.7
ACT	5	0.7/0.7	500	3	10/15	0.4
TTL7400*	3	10/8	40	5	30/100	0.4
LS	3	10/5	65	5	20/40	0.4
FAST&AS	3	3/2.5	125	4.5	15/20	0.3
ECL	0.8	1.5/1.5	220	3	7/7	0.1
ABT240	3.5	1.5/1.5	220		6/200	.25
HC240	5	4/4	80		20/25	

*The straight TTL is an obsolete family, shown only for comparison with its successors. ABT240 and HC240 are driver series.

the worst condition correspond to low values of R_g, since this will be the current limiting parameter when, to the extreme, the gate will drive a heavy capacitive load, or the equivalent of a shorted line at resonant frequencies of the PCB trace layout. Therefore, even a shorted gate output (for the time of a transition) cannot deliver a current greater than V/R_g.

Fig..5.1 shows comparative plots of the spectral densities for several popular logic families, which are a direct measure of their propensity to EMI generation. The BB spectra are calculated for a single pulse of each family, all with a same 20ns pulsewidth, except for CMOS (200ns). The associated table shows the relevant parameters of the comparison, like spectral density $2A\text{x}\tau$, corner frequencies F_1, F_2, and the spectral density at F_2. We remark that in spite of slightly slower rise time, HC shows higher spectral density than FAST and AS, because of its larger dV/dt at the beginning. In spite of its high speed capability, ECL has the lowest spectrum

Therefore, even a shorted gate output (for the time of a transition) cannot deliver a current greater than V/R_g.

	A *(V)*	*2 xA.τ* *(V/MHz)*	*F_1=1/πτ* *(MHz)*	*F_2=1/πt* *(MHz)*	*Ampl. @ F_2* *(dBV/MHz)*
CMOS	5	2	1.6	6.4	- 6
LVMOS	3.3	0.13	16	110	- 34
LS	3	0.12	16	65	- 30
FAST,AS	3	0.12	"	125	- 37
HC	5	0.2	"	90	- 29
ABT	3.5	0.14	"	220	- 40
ACT	5	0.2	"	500	- 43
ECL	0.8	0.03	"	220	- 53

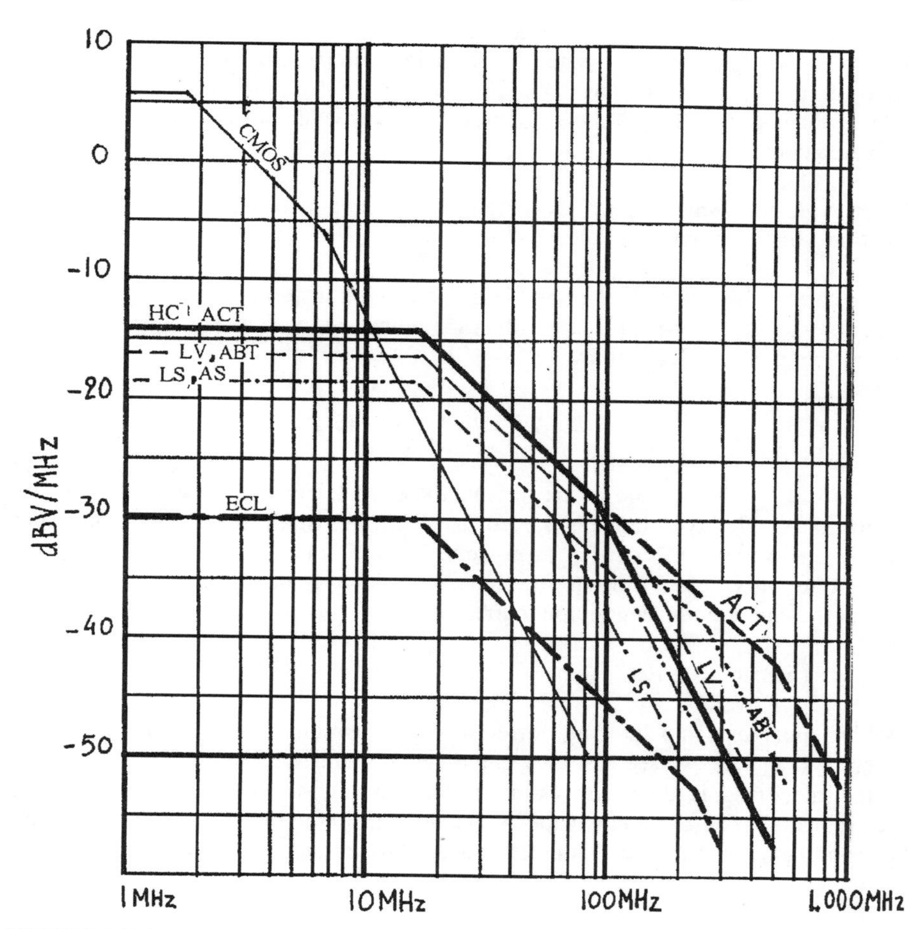

FIGURE 5.1 Voltage Spectra of Logic Devices, Assuming 20ns Bit Pulse Width for all Technologies except CMOS (200 ns)

The major players in noise generation mechanisms, some of them appearing Table 5.1 are reviewed hereafter.

1. Rise and fall times: the shorter they are, the wider the occupied spectrum of the corresponding logic pulses. Since EMI problems aggravate with F (for conducted EMI and crosstalk) and, often, F^2 (for radiated EMI), the consequence is obvious.

2. The power supply transition current: this is the instantaneous demand of the device alone during switching, independent of its loading. It can be very large, having nothing to do with the quiescent current during an established “1” or “0”. In TTL-derived families, and to some extent in the fast HCMOS technologies, this inrush current is due to the partial conduction overlap of the two output transistors arranged in a “totem pole”. During this overlap, the Vcc bus is virtually shorted to ground via two partially saturated transistors plus a limiting resistor. Recent designs have reduced this effect by using Schottky barrier diodes (SBDs) to prevent the output transistors from going into excessive saturation. Some manufacturers are also using “output edge control” circuits by replacing one large-output transistor with a group of smaller ones. However, the current peak is still significant and can pose PCB problems or even on-the-chip problems with gate arrays or other highly populated chips.

3. The voltage swing: this Lo -Hi or Hi -Lo excursion relates directly to electric radiation and capacitive crosstalk.

4. The current that the gate is forcing into (Lo-to-Hi transition), or pulling from (Hi-to-Lo) the driven gates is also larger than the quiescent current. For short lines, this load current can be calculated by:

$$I2 = CL\ \Delta V / \Delta t \qquad (5.1)$$

In Eq. (5.1), C_L is the sum of the driven trace capacitance to ground (0.1 to 0.3 pF/cm for single-layer boards, 0.3 to 1 pF/cm for multilayers) and the input capacitance of the driven gate(s), as given in Table 5.1. For instance, with a 3.5 V/3 ns rise time, driving a 5 cm long trace on a single-layer board with a fan-out of 5 gates at the end, the transient output current is:

$$I_2 = (5\text{cm} \times 0.3 \times 10^{-12}\ \text{F/cm} + 5 \times 5.10^{-12}\ \text{F/gate}) \times 3.5\ \text{V} / 3.10^{-9} = 30\text{mA}$$

Looking more in depth into the power supply transition current I_1 and the output current I_2 (Fig.5.2), we see that they combines in a non symmetrical way. For low-to-high transitions, load current I_2 adds up to I_1. For high-to-low, I_2 subtracts from I_1, since the gate is “sink” and the capacitive charge from the load has to discharge into the driving gate output, which appears as a short to ground.

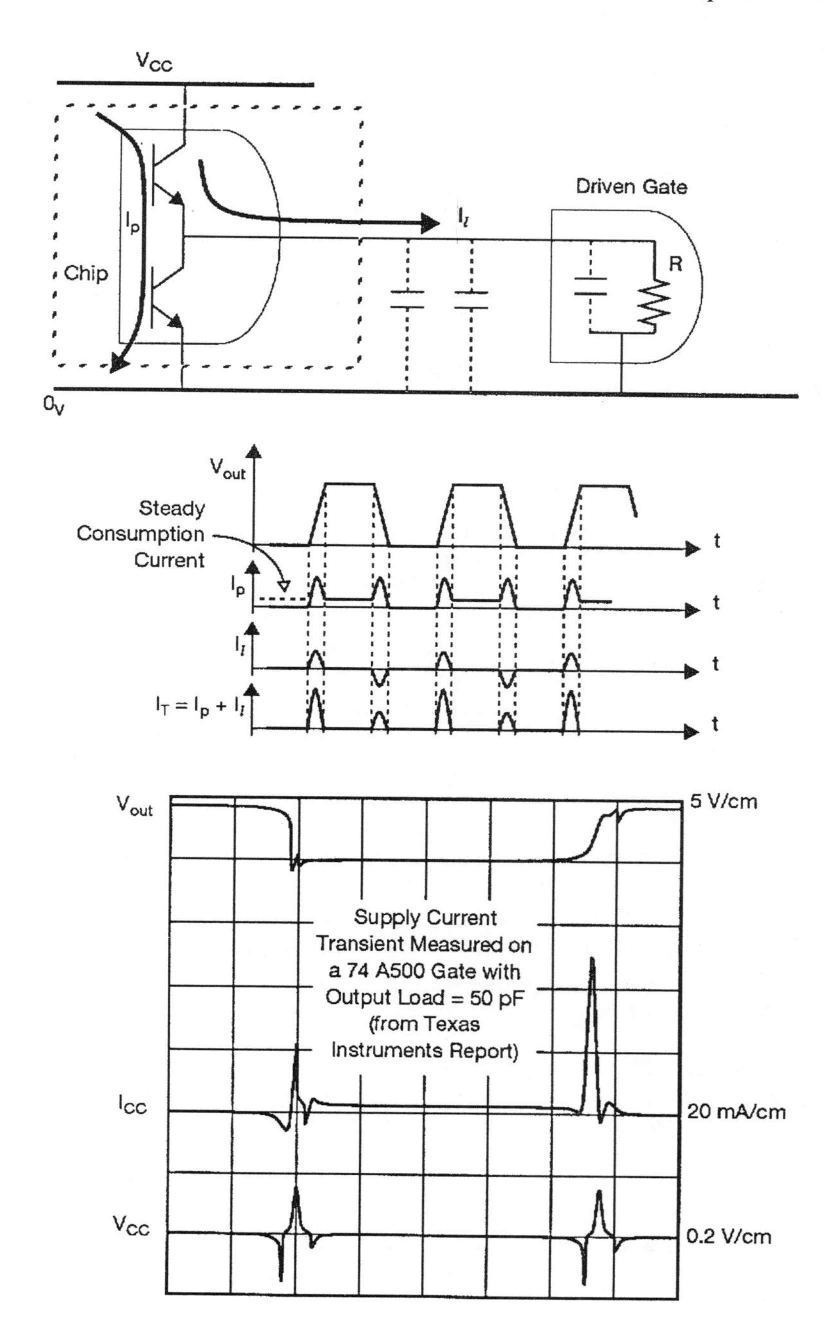

FIGURE 5.2 Transient Currents Drawn by a Gate off the Power Distribution Bus

When the driven line is electrically long, i.e. when its propagation delay is not small compared to the pulse rise times, the instantaneous output current I_2 becomes limited by Z_0, the characteristic impedance of the driven trace. In epoxy glass, propagation speed is approximately 14 cm /ns, corresponding to a propagation delay of 70ps /cm. Transmission line problems become serious when line delay T_d exceeds half of t_r, the pulse rise time. So, for estimating I_2, we can simply define a "long line condition" as follows:
if trace length l is $\geqslant$ 7cm x t_r(ns), the maximum possible output current during transition is equal to:

$$I2\ max = VL\text{-}H\ /Z0 \qquad (5.2)$$

where Z_0 is the characteristic impedance of the driven trace.
For instance if $Z_0 = 90\ \Omega$, a 3.5 V swing will produce:

$$I_2 = 3.5/\ 90 = 38\ mA$$

Ultimately, since very low line impedances would create very high currents, I_2 is limited by the output resistance of the gate.

Figures 5.3 through 5.5 show measured and computed values of conducted currents and radiated fields during the logic devices' operation. Since the basic building block of an LSI chip generally is still a gate, the radiation from an elementary gate in a DIP module can be calculated as shown in Fig 5.5. However, since thousands of gates can exist in a chip, the radiation from the whole chip cannot be the arithmetic sum of all the gates' radiation. Not all gates operate synchronously, and as the orientation of the radiating doublets in the chip is random, they can add or subtract as well.
Either some randomization is taken into account in math models or actual measurements must be made of complex chips. For microprocessors (since many operations with various data rates take place inside the chip), radiated and conducted profiles contain all frequencies corresponding to internal transactions between the arithmetic logic unit (ALU), the registers, buffers, and so forth. Their frequencies are the clock rate and its submultiples, plus all their harmonics.

Another aspect of radiated EMI from logic chips is the wide variance between manufacturers for the same IC device. The following differences between two vintages of a 74LS flip-flop have been reported (Ref. 9)

	E Field (dBµV/m) at 3 m	
Frequency	Vendor A	Vendor B
30 MHz	32.8	42.8
40 MHz	27.5	33.0
50 MHz	23.0	28.0

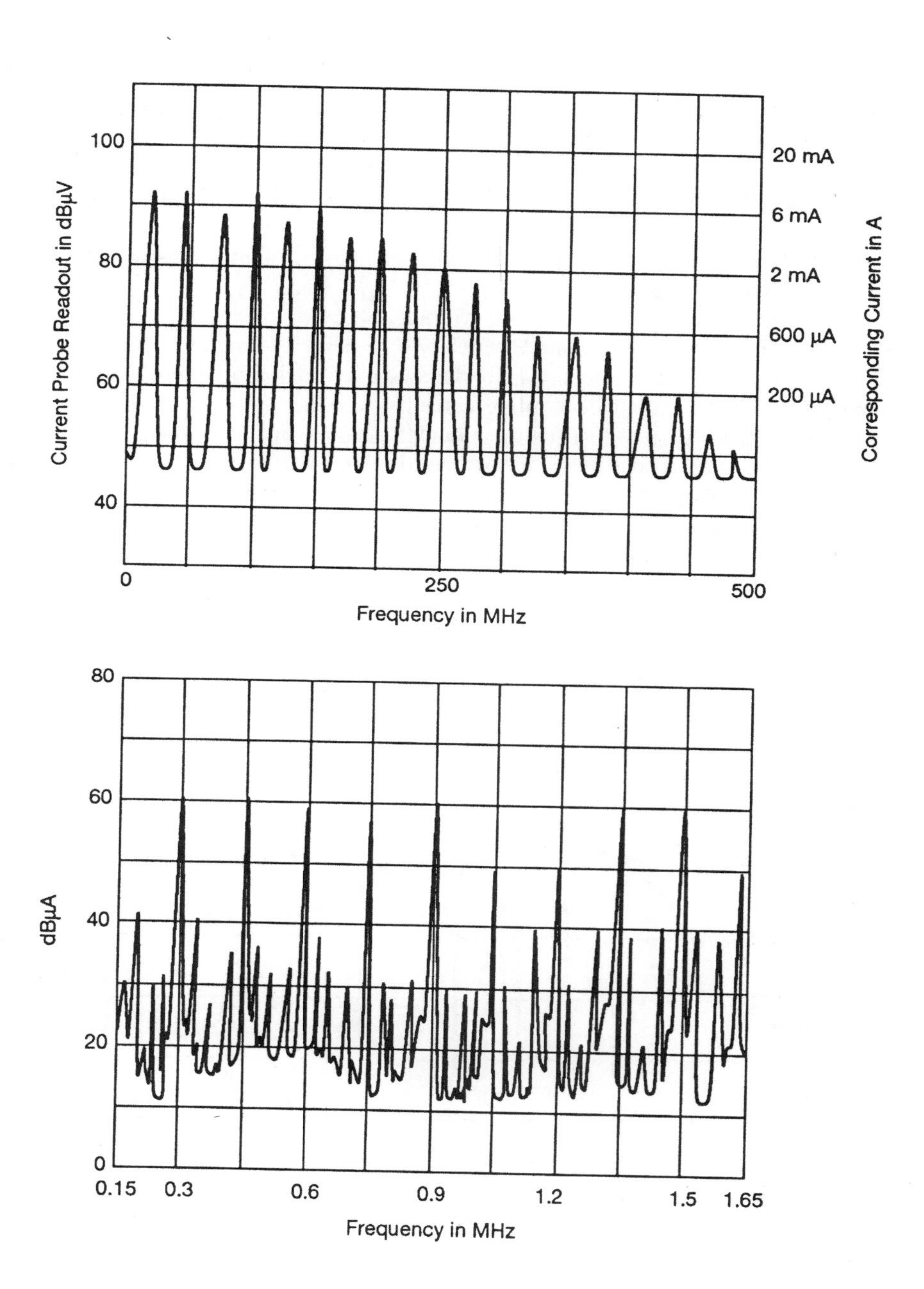

FIGURE 5.3 Frequency spectrum of Currents Demanded by Typical Digital Devices (Top, 74LS Gate; Bottom, 8420 Microprocessor)

With large IC modules, and especially fast processors with clock rates above 150-200MHz, a point is reached where the IC alone, even resting on a perfectly shielded PCB, may radiate above the authorized limit. Fig. 5.6 shows an example of the radiated field from a 270MHz Pentium®. The field value, translated at 3m, exceeds the FCC class B limit.

5.2 CALCULATING THE IDEAL BYPASS CAPACITOR

Two things are important in delivering the necessary instantaneous currents while the device is switching:

- the proper capacitor value, which we will address here
- the way this capacitor is mounted, which will be described in the next chapter

This capacitor and its associated conductors play an important role in radiation from PCB. The so-called decoupling capacitor can be regarded as a reservoir to provide the inrush current that the logic device needs to switch in the specified time. The reason for this is that by no means can the long wiring from the power supply regulator to the chip provide the peak current without excessive voltage drop. The value of the decoupling capacitor, C, close to the logic chips requiring a switching current, I, is:

$$C = I/(dV/dt) \qquad (5.3)$$

where dV = acceptable voltage drop at capacitor output (Vcc sag) caused by the demand of a current I during the time interval dt

dt = logic switching time $\approx t_r$

I = total transient current demand of the logic family = $I_1 + I_2$

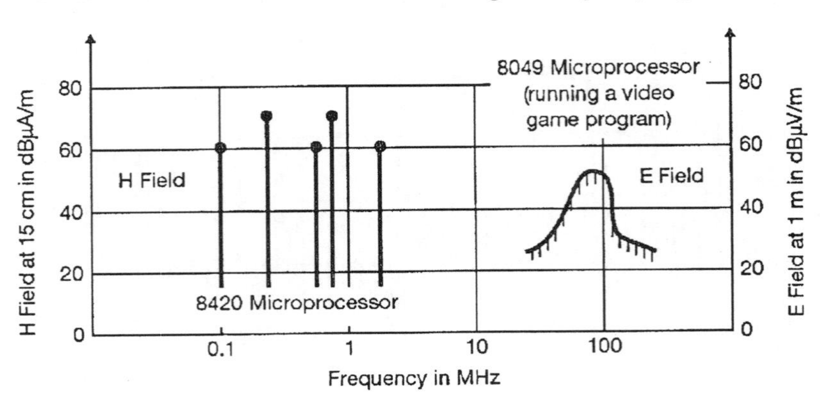

FIGURE 5.4 H Field at 15 cm from an 8420 Microprocessor and E Field at 1m from an 8049 Microprocessor

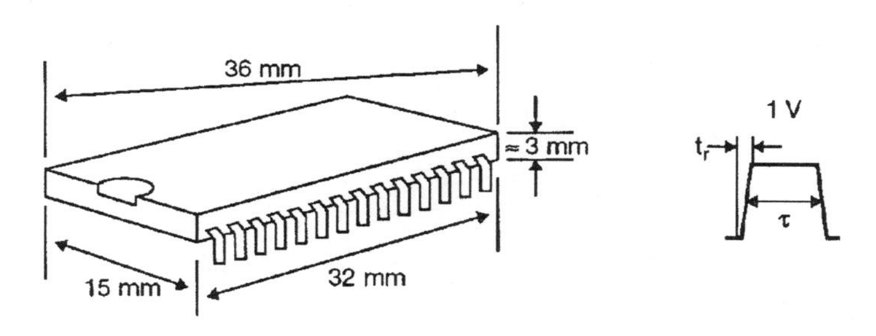

τ in ns	t_r in ns	E field @ 1 m in dBμV/m/MHz	F in MHz					
			3	10	30	100	300	1,000
100	10	A	0	−10	−18	−24	−24	−24
30	3	B	−10	−10	−18	−14	−14	−14
10	1	C	−20	−20	−18	−14	−4	−4

a. Calculated BB field for a 1 V pulse, one gate, and three different pulse characteristics. For different pulse voltages, fields will vary proportionally.

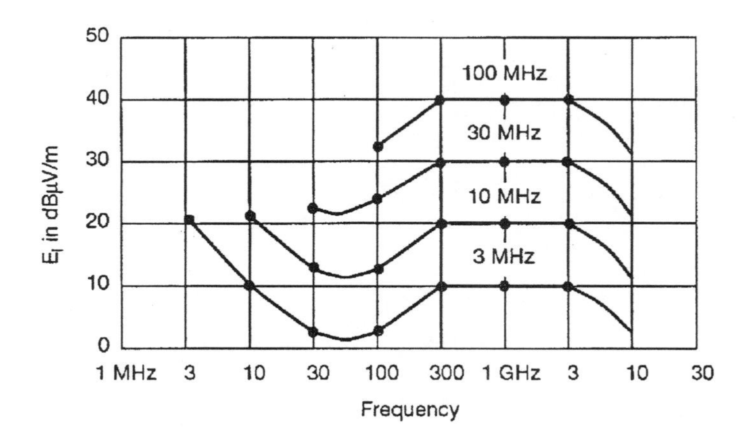

b. NB fields for fast 3.5 V/1 ns rise time and four different clock frequencies. Duty cycle = 50%, and there are 30 clock-triggered gates on the chip.

FIGURE 5.5 BB and NB Radiated Fields at 1 m from a Typical 28-Pin DIP Module

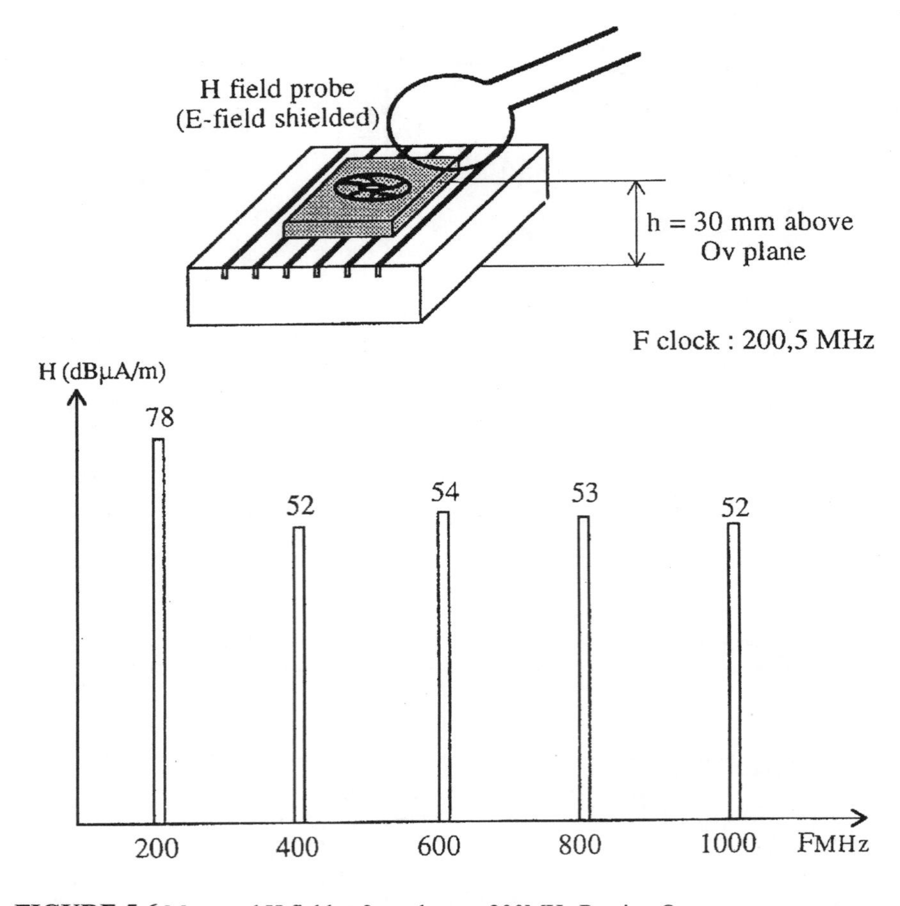

FIGURE 5.6 Measured H field at 3cm above a 200MHz Pentium®.

Table 5.2 lists values of C for some popular logic families, based on a maximum allowable Vcc drop equal to 25 percent of the noise immunity level. This is justified by an equal spread of the PCB noise budget over four contributors: power distribution noise, radiation pickup, crosstalk and mismatch reflections.

Calculated value of C is for only one active device switching at a time, and driving a fan-out of five gates. Several devices clocked in the same module will multiply the requirement for C by as many. Total current $I_1 + I_2$ has been checked against the maximum current imposed by the gate output resistances of table 5.1.

In the case of the standard (low-speed) CMOS, Table value is conservative because the assumption that the long wiring from the power supply cannot provide the peak current without excessive voltage drop is not true. For rise times in the 50 to 100 ns range, even with 1μH of supply lead inductance, the voltage drop would be acceptable. So unless severe ambient EMI is expected, this decoupling is not necessary, and one capacitor for every row of modules is enough. Notice also that although presumably "quieter" than bipolars, high-speed CMOS still exhibits significant switching currents due to its fast transition.

TABLE 5.2 Decoupling Capacitors Needed for Some Popular Logic

Logic Family	*Peak Transient current Requirements (mA)* 1 Gate Overcurrent I1 (mA)	1 Gate Drive I2 (mA)	*Decoupling Capacitor for a Fan-out of 5 gates + 10cm trace length* (pF)
CMOS	1	0.5	(700)
HC	15	6	1,100
ACT	40	20	1,300
LS	6	2	1,700
FAST	15	4.5	1,700
AS	30	4.5	2,400
ECL	1	7	850

Many chips are not so simple as a handful of gates. For instance, a dynamic 64 k RAM may need an extra current (above its normal consumption) of about 70 mA during a refresh cycle of 500 ns. To keep the supply rails within the desired voltage tolerance requires 220 to 270 nF of buffer capacitance. For a 256 kRAM, 330 to 390nF are required. The same is true for a microprocessor module which, too, requires about 220 nF (for a 16-bit device).

Another example is the high-density pin grid array (PGA) modules with 128 pins or more. In such packages, it is common to find simultaneous switching of 16 or 32 bus driver outputs capable of driving 100 mA each, with rise times of 2 or 3 ns. This, plus all the internal functions of the module, requires more than 150 nF per module.

From Eq. (5.3) and the number of gates that are switching synchronously, a proper value of C is selected for each module. Generally, rounding up to the next standard value gives sufficient size. More is not necessarily better, and excessive capacitance on cards could, in turn, draw too much current when the power is switched on or when the card is plugged into its socket.

5.3 GENERATION OF COMMON MODE BY IC GROUND BOUNCE

The so-called ground-bounce is a common impedance problem between the chip and the host PCB, (Ref.10) caused by the dI/dt effect in the IC ground lead (bonding wire plus pin). A single poorly designed oscillator, or drivers with too much peak current demand, can lift the whole chip 0V reference vs. the PCB ground. This transient voltage shift will be carried over by all the I/O signals pins, exporting this noise spike to the rest of the PCB. Fig.5.7 shows the basic mechanism for ground bounce, and Fig.5.8 shows comparative results for an ALS driver and an AC driver. In both cases, 7 out of 8 outputs are switched simultaneously while the last one is set to "0". The ground bounce on the non-switched pin reaches 1.5V for the AC family, a potential cause of EMI. Improved packaging styles (see sect. 5.5) exhibit less ground bounce, owing to the reduced parasitic inductance. For instance a 16 driver-SSOP module with 48 pins, including 8 distributed GND have only 0.4V of ground bounce during a 15/16 output simultaneous switching, i.e. three to four times less than a DIP package with corner GND pin.

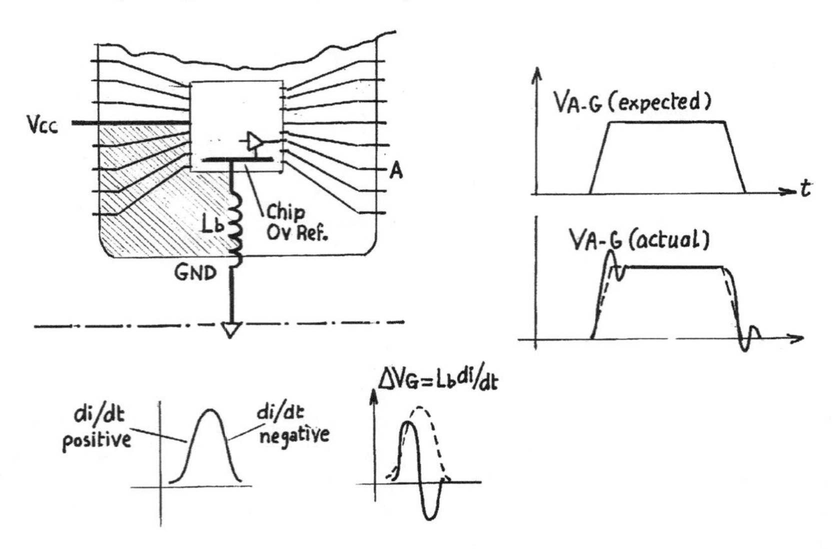

FIG.5.7 The Ground-Bounce phenomena at IC level. ΔV_g, the voltage drop in the bonding wire inductance L_b, is adding to the normal signal of the device being switched, but it also offset the chip 0V reference for all the other I/O pins. The overshoot/undershoot of the waveform can be easily mistaken for a mismatch problem. The heavied-up area, corresponds to the radiating loop of the IC alone.

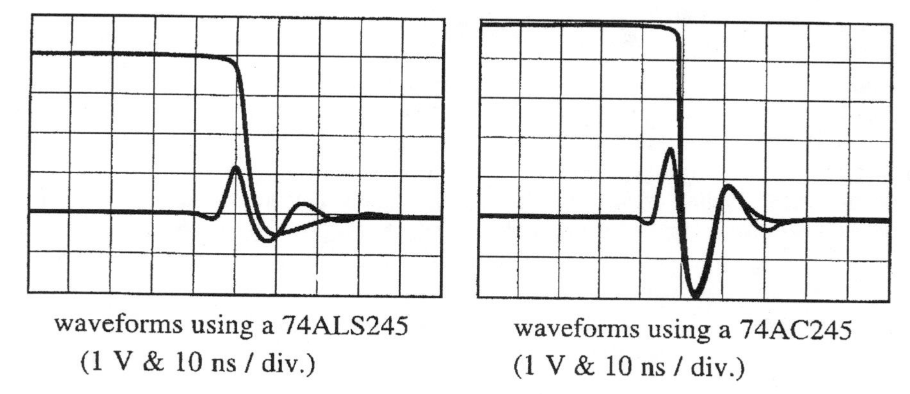

waveforms using a 74ALS245 (1 V & 10 ns / div.)

waveforms using a 74AC245 (1 V & 10 ns / div.)

FIG. 5.8. Ground Bounce with two logic drivers families

5.4 REDUCING EMI GENERATION AT THE IC ITSELF

Several IC manufacturers, aware of these problems have started since the early 90's to incorporate EMC in the design of their chips. This include geometrical improvements to reduce radiating loops areas and parasitic inductances, and voltage or current control to reduce the emission spectra. These efforts are supported and documented by now-standard methods for the characterization of chip emissions. This "EMC - on -the- chip" approach is somewhat remindful of those (not-so-old) days where we, EMC engineers had to fight to convince the designers to apply EMI analysis to their PCB.

5.4.1 EMI-Quiet MicroProcessors

Microprocessors, being the heart of a majority of functions and using the fastest clock frequency can often be regarded as one prime excitation source throughout the equipment. There may even be a point where all efforts in reducing P.C.board radiation will stumble on some uncompressible value: the field radiated by the processor module itself.
Many EMI reduction techniques are currently applied inside the IC :

- reducing the peak current demands from the internal signal bus buffers.
- improving the on-chip power supply distribution layout to the CPU core, ROM, RAM, Data & Address, Oscillator etc .. to reduce ripple on the internal DC rail. This is made by wide, fingered Vcc and Gnd (Ref. 11) metallizations.
- softening the 0-20% and 80-100% portions of the dV/dt transitions at the output stages, by using pre-drivers. This soft profile reduces the higher frequency contents.
- providing in-situ diffused capacitors of $\approx$ 1nF /mm^2.
- reducing loop areas in the die and the package.
- increasing the number of Vcc and Gnd pads, with a more even spread.
- die-down mounting on the substrate.
- using lower voltage excursions (Lo -to-Hi), like 1.5 or 2V.

5.4.2 EMI Contribution of Clock Oscillators

High frequency oscillators are serious players in radiated EMI, not just because they are the clocking master for all the fastest digital switching, but by the emission characteristics of the oscillator circuit itself (Ref. 12).
CMOS gates are often used as amplifiers in their linear region, to build inexpensive oscillators. This may create large loops, whereas the current in the L-C tank formed by the crystal with its two associated capacitors is highly non-sinusoidal, hence rich in harmonics. It is better to use elementary gates, like HC-U family, with a drive capability and bandwidth limited to what is strictly necessary. If the oscillator is made from a 4 or 8-gate module, do not assign unused gates to other functions than clock distribution, since the corresponding IC will tend to be heavily polluted by CM voltages at clock harmonic frequencies.

A substantial EMI reduction can be obtained by using microprocessors with integrated PLL (Ref. 13). While standard oscillators run typically at 2 or 4 times the base system clock rate, a PLL-based oscillator can run at the system clock frequency. For instance, to generate a 4MHz main clock, a 4MHz PLL integrated in the processor is sufficient, instead of the 16 MHz external resonator which one would normally use. On a Texas TMS370 microcontroller, the radiated EMI above 30MHz was typically 10dB below that of a standard oscillator version.

5.4.3 Standard methods for IC emissions measurements

With the growing needs for "quieter" ICs, standard methods of measurements become necessary for characterizing and comparing the EMI signatures of various devices, or similar devices from different sources. Even a minor change in the production masks or manufacturing process may cause significant changes in the emission level of a same P/N, from the same supplier.
Characterization methods belong to two general categories: conducted and radiated. Since one has to make sure that the measured emissions are coming from the sole IC and not from the test circuit, all methods are imposing a precise standard test jig:
4-layer PCB with integral ground plane, I/O lines treated as micro strip or striplines, coaxial fittings and impedance matching for the whole test gear, etc..

IC conducted Emissions measurements

Several techniques are proposed, which are generally studied, or ultimately endorsed by the IEC Technical Committee 47.A.

a) The RF current and voltage method (VDE)
The principle is shown on Fig. 5.9. The IC under test and its associated loads and controls are connected to the ground plane via 1Ω non-inductive resistors. The RF current waveforms a/o spectrum are derived from the voltage measured across each 1Ω shunt. The shunt in the IC ground common lead displays the instantaneous power supply "through current", when the entire device is operating. The 1Ω

resistors in each load ground lead(s) give the characteristics of each individual output current. These current measurements are completed by voltage waveforms and spectra, the I/O pins being loaded by 150Ω (or other value) resistors. The standard describes a set of templates with a 3-digit coding (for.inst.. B-4-F), corresponding to the three slopes - flat, 1/F and $1/F^2$ - of the overall spectrum envelope. The problem is that the templates scale correspond to ridiculously low amplitudes, totally unrealistic when toggle frequencies are exceeding few MHz. But the principle is valid.

b) The Workbench Faraday cage (Philips Co., Ref.14)
The IC, mounted on its test board is located in a small Faraday box. All I/O leads are fed through the box wall via coaxial connectors and loaded with 150Ω to ground. These 150Ω are deemed to represent the average CM impedance of the I/O cables in the final applications of the device.

IC radiated emission measurements *(Ref. 15, 16, 17)*

In its simplest form, the method is using a miniature H-field loop, for.inst.. a shielded 5 mm dia. loop, to probe just above the IC, in various orientations (see for.inst.. the measured results of Fig. 5.6).
The more elaborate method of SAE-J1752.13 places the test board on top of a miniature TEM cell, with the IC looking down through a square aperture in the top plate. The power measured at the cell output port can be converted in form of magnetic loop and electric dipole moments (μA x cm^2 and μA.x cm), such as the E and H fields at any distance can be derived. A coarse, but very practical correlation is that if we use V_{50}, the output port voltage, to compute the total IC radiated power, then translate it into an equivalent radiated field, we get:

$$E = 1/D \sqrt{30.Prad} \approx 1/D \sqrt{(30 \times 2 \times V_{50}^2 /50)} \qquad (5.4)$$

So, $E\ (dB\mu V/m)\ @\ 1m \approx V_{50}\ (dB\mu V)$

In many cases, the 30 - 300MHz correlation is rather close such as plots of V_{50}(dBμV) and actual one meter measurements of E(dBμV/m) are overlaying within +/- 3dB.
The method is sensitive to the IC placement vs. the cell axis, and directional effects are noted when the IC is rotated 180°, since magnetic and electric moments are combining differently (the highest reading must be retained).

Overview of IC /EMC test methods

Existing conducted methods provide a detailed profile of each pin contribution, allowing fine investigations of a specific device's output.
They have some basic limitations:

• the standard says that conducted measurements up to 1GHz are sufficient for characterizing emission properties, which is questionable (as of year 2000, processors with 500 - 1000MHz base clock are common place)
• they assume, a priori that direct radiation from the IC alone is not a problem, and that only its conducted emanations are of importance.

In contrast, the radiated method has the advantage of gathering the full radiated power of the IC. It does not permit detailed investigation of each pin, although it is always possible to enhance or inhibit a specific function from the test board interface.

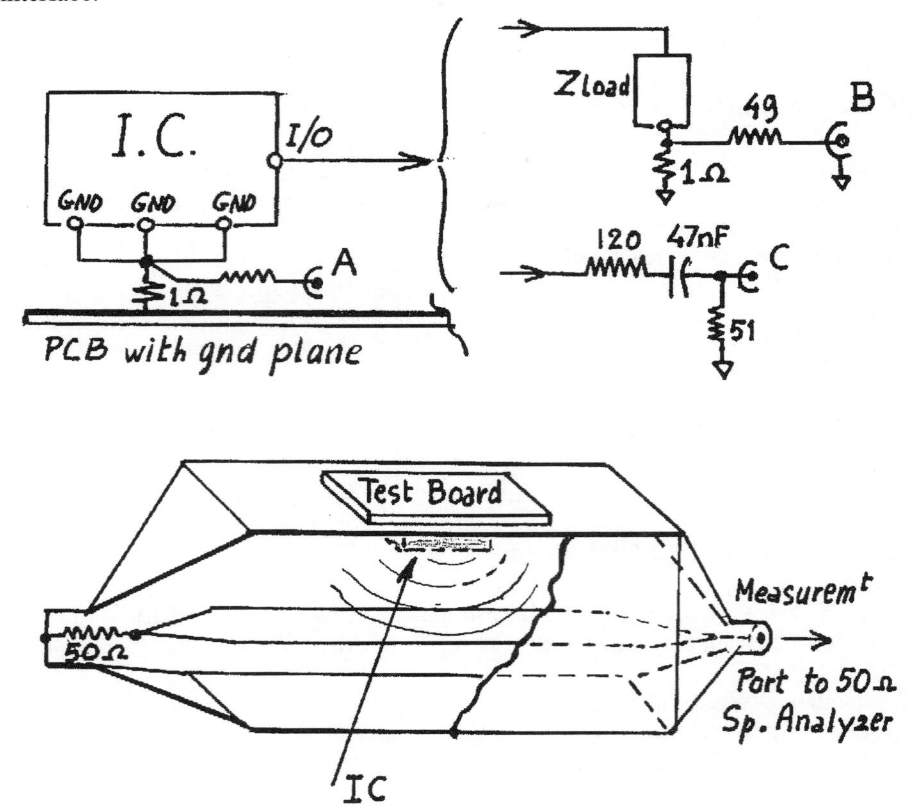

FIGURE 5.9 Standard Emission Test Configuration For ICs

5.5 INFLUENCE OF THE IC PACKAGE

Although it may seem minuscule, the loop surface made of the chip, its bonding wires and the module leads can become a significant EMI contributor in heavily populated cards, especially when multilayer boards are used. This is because, with multilayers, the trace radiating loops become so small that IC leads become the larger antennas. In this respect, the worst of all packages is the standard DIP, especially if mounted on an elevated socket.

In Fig.5.10 we see that the sole DIP's radiation, for instance, is only 20 to 12 dB below the traces' radiation. Flat packs with gull-wing leads represent a first step in loop area reduction. A further improvement is found with the center Vcc and ground pin layout provided by some IC manufacturers (see Fig. 5.11). Although this pin arrangement was designed primarily to reduce Vcc and ground voltage "bounce" by a lesser inductance, it also has a beneficial feature in reducing the worst-case loop area.

Finally, surface-mount components (SMCs), also commonly known as surface mount devices (SMDs) and surface-mount technology (SMT), achieve the best possible area reduction, as they exhibit about 40 percent reduction in component dimensions and, therefore, a 64 percent area reduction of the module radiating loop (see Fig. 5.12). In addition, they allow a significant reduction in board size and, therefore, in trace radiation. In Fig. 5.10, we see the decrease in radiation when changing (everything else remaining the same) to multilayer board and SMC. Common characteristics of the PCBs in Fig. 5.10 include eight 16-pin chips, six clocked gates per chip, clock frequency of 30 MHz, and a clock edge of 3 V/3 ns. Multilayer technology reduces trace-only radiation by 40 dB. SMC devices reduce chip radiation by 6 to 8 dB. Notice, though, that modules radiation now dominate trace radiation. An interesting side benefit of selecting low-profile, low-area packages is that the decoupling capacitor-to-chip loop area will also be reduced.

5.6 SHIELDING AT IC LEVEL

The most obvious application of shielding is by enclosing the entire equipment or P.C.board in a conductive envelope (see chap.10), since the printed traces and internal wiring are the dominant radiating antennas, inside the unit.

However, with the increasing use of large size ICs and MultiChip Modules, there are cases where a perfect EMI suppression at PCB traces will still leave intact the radiation of the IC itself, which alone may approach or exceed the limit objective.

If only one or few chips are responsible for the spec violation, or are causing internal EMI (self-jamming), while the rest of the packaging has been adequatly treated, it might be economically justified to shield the IC only (Fig. 5.13).

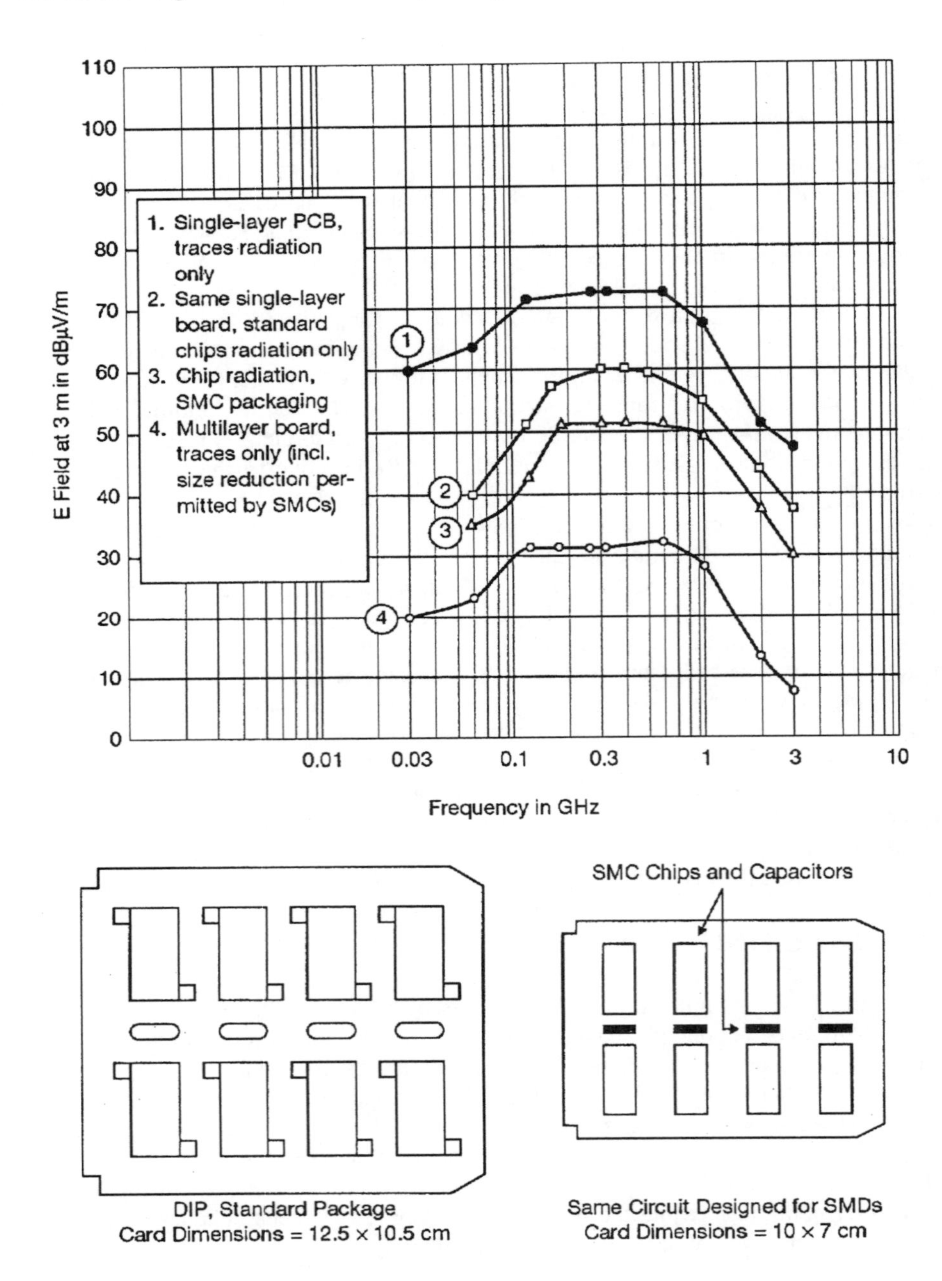

FIGURE 5.10 Radiated Field Comparison of a Single-Layer PCB (No Ground Plane) with Standard DIP Devices vs. the Same Card Redesigned as a Multilayer with Surface-Mount Components from Ref 4)

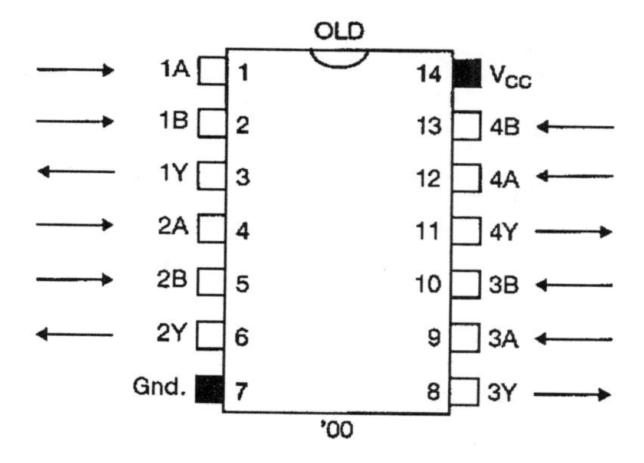

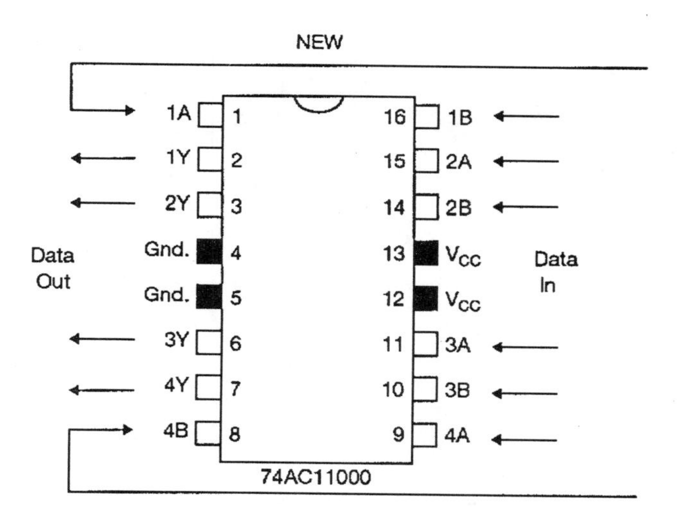

FIGURE 5.11 Texas Instruments' Advance CMOS uses a flow-through architecture that places the ground and Vcc pins in the center of the package rather than the traditional end-pin location. This allows simultaneous switching while keeping maximum performance from the device. It also ease board layout and save board real estate. For a typical high-speed logic transition, the noise glitches on the power distribution near the devices are reduced from about 2 V to 0.4 V. (Reprinted with permission from T. I. Tech Notice)

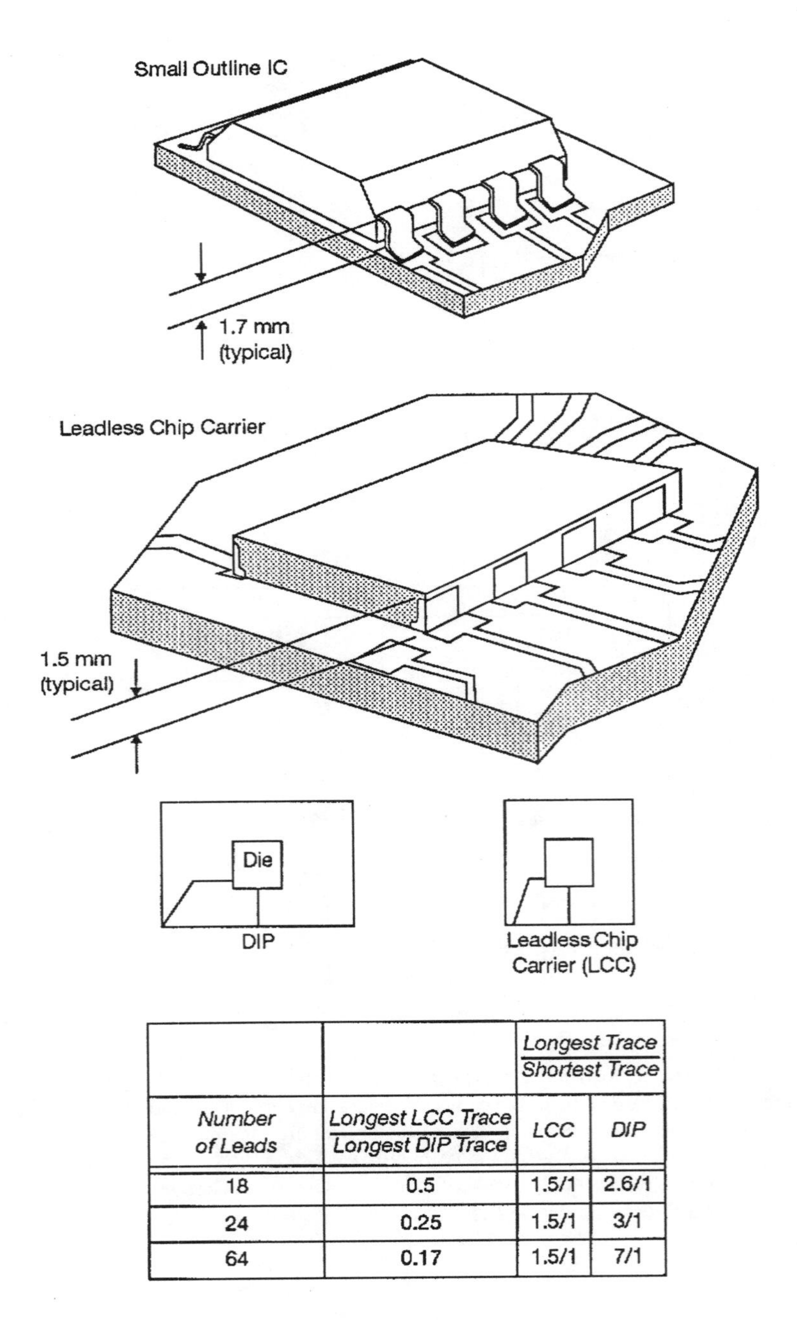

		Longest Trace / Shortest Trace	
Number of Leads	*Longest LCC Trace / Longest DIP Trace*	*LCC*	*DIP*
18	0.5	1.5/1	2.6/1
24	0.25	1.5/1	3/1
64	0.17	1.5/1	7/1

FIGURE 5.12 Loop Reduction by SMT Packaging. The 2:1 reduction in chip-to-chip trace area reduces radiated emissions and susceptibility.

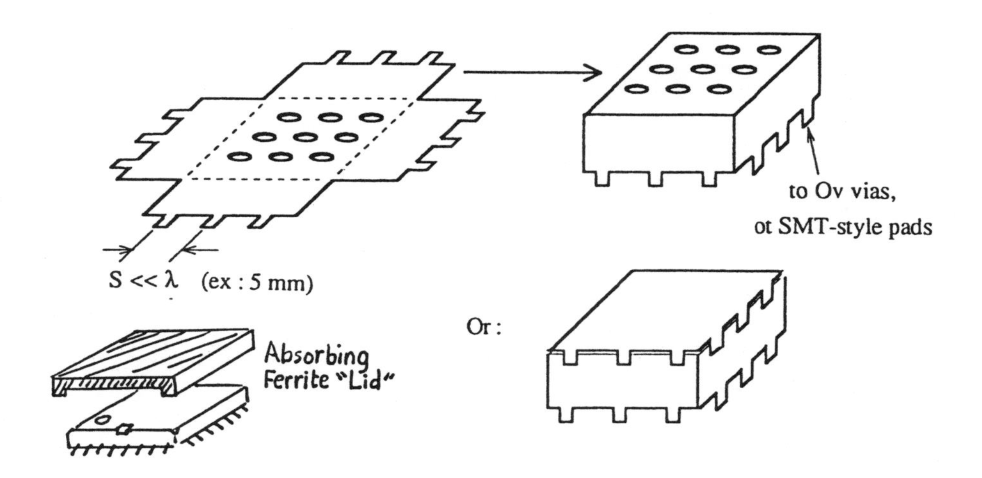

FIGURE 5.13 « Partial Shielding at the IC/Module level »

5.7 SUMMARY OF RADIATION CONTROL AT THE CHIP AND IC LEVEL

• Among the viable technologies to perform a given function, try to select the ones that are less "current hungry" at transitions; average power consumption is meaningless in this regard: transient current demand is what counts.
• Reserve fast switching technologies (tr, tf <10 ns) for functions where fast transitions are truly vital.
• Select package styles with minimal loop areas and optimized Vcc-Gnd pins location. Twenty-five DIP modules easily add up to 10 cm2 of radiating area.
• Use metal-enclosed modules (if cost permits), connect the can to the ground plane.
• When defining ASICs, specify output devices with the lowest driving capability suitable for the application: do not oversize.

Chapter 6

Printed Circuit Board Design

Other than choosing component technologies and packages that offer lower radiation levels, the designer has very few EMI-reduction options at the device level. In contrast, the PCB, as a building block, offers the first area in which a strong design action is possible.

A multitude of EMC "war stories" arrive at the inevitable conclusion that an in depth look at PCB design would have saved thousands of dollars in testing, last minute fixes and retesting, plus additional hundreds of dollars per unit, in hardware costs for shielding, gasketing and so forth. Not only is the PCB a radiating element by itself, but insufficient attention to board-level EMI can result in noise coupling to I/O lines and other external elements. These, in turn, carry the undesired signals away and radiate them. There are many cases where a handful of three-cent surface-mount capacitors and a cost-free rerouting of few traces will eliminate the need for expensive shields and filters.

Once the technical aspects of board radiation are understood, another stumbling block often appears: the computer-aided design (CAD) package for trace routing. Such programs usually ignore EMI problems and require manual data re-entry.

6.1 BOARD ZONING

Circuit boards, especially single-layer ones, should be laid out such that the higher frequency devices (e.g., fast logic, clock oscillators, bus drivers) are located as close as possible to the edge connector. This is illustrated in Fig. 6.1. Although other arrangements can be advocated with good results, this one is a trade-off for the lesser of several evils: radiating loops made by traces to edge connector pins, back and forth, are among the most offending ones in a PCB, so they are kept short for the faster logic. High dI/dt return currents, causing common-mode voltages along ground conductors, are also kept short.

The lower speed logic and memory, can be located farthest from the connector because they tolerate longer trace lengths without the proclivity for radiation, ground noise and crosstalk. Opto-isolators, signal isolation-transformers and signal filters should be located as close to the edge connector as possible to avoid crosstalk between undesired signals and the "clean" side of these isolators.

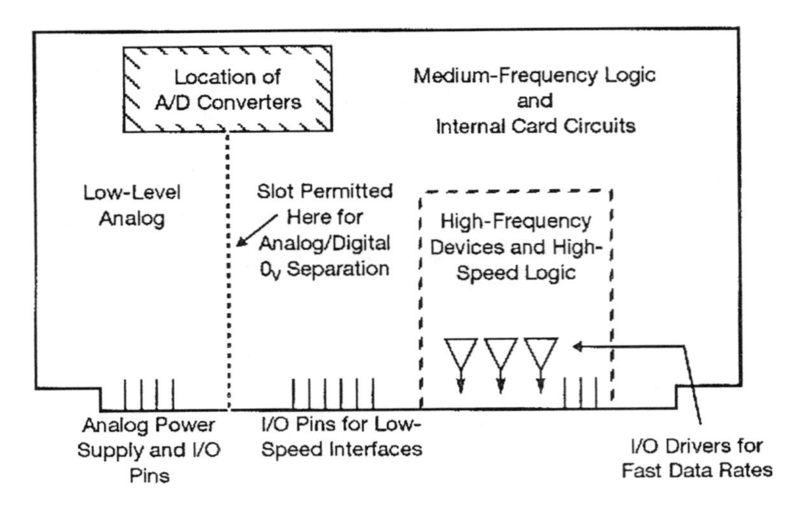

FIGURE 6.1 Applying Board Zoning Guidelines

6.2 BRUSH-UP ON SELF-INDUCTANCE OF CONDUCTORS.

Any conductor, in addition to its ohmic resistance, exhibits self-inductance. Although such blunt statement may seem trivial in a book written for electronic engineers, the author has seen so many instances where self-inductance was misunderstood, that a short brush-up may be useful.
The term self-induction defines the counter-electromotive force caused by magnetic field around a wire or trace, when the current is changing rapidly. Strictly speaking, self-inductance can only be calculated precisely if the two-way path of the current is defined. For such hair-pin geometry, the total self-inductance is:

$$L = 0.4 \text{ Ln } (2h/d) \ \mu H/m \quad (6.1)$$

with h = wire separation and d= wire diameter

If half of the total loop inductance is attributed to one of the wires, this "partial" inductance becomes:

$$L = 0.2 \text{ Ln } (2h/d) \ \mu H/m \quad (6.2)$$

Notice that this one-wire self-inductance can only be determined if one knows where the other wire is. However, to simplify calculations, it is often interesting to consider a practical worst-case for one wire. This is the value reached when the other wire is very far away; practically, an asymptotic value called "free-space" inductance is reached when the separation s becomes greater than the length *l* of the conductor pair. This worst-case value is:

$L_{(1\text{-wire})} \approx$ 1.5μH/m for a round wire
$\approx$ 1μH/m, or 10nH/cm, for a trace

This is the default value we will be using when a PCB trace is relatively far from the one carrying the opposite current. The table in Fig. 6.2 shows the corresponding impedances, including the slight variations brought by various trace widths.By comparison, we show also the impedance curve for a perfect, infinite copper plane. Such a ground plane has no self-inductance, since no magnetic field lines can close around the conductor: it only exhibits resistance, skin effect, and a very small internal inductance, within the metal. These increase only like the square root of frequency once the metal thickness exceeds the skin depth region. This compare very favourably with wire or trace impedance, which increases like frequency. For example, at 100MHz (the equivalent bandwidth for a 3.5 ns rise time), a wide plane shows only 3.7mΩ/sq: the switching of 30mA will cause only 100μV drop along the common ground, regardless the path length. Unfortunately, quasi-infinite ground (or Vcc) planes seldom exist in practice. Actual ground planes are finite, and current distribution across their width is not uniform. Cross section in Fig. 6.3 shows the return current spreading in the extended "shadow" of the overhead trace, a gross rule-of-thumb being that approximately 90% of the return current is found in a width equal to 10 x h. This, alone, corresponds to a strangled path which is no longer a surface impedance in Ω/sq but a linear impedance in Ω/cm.In addition to this, the last few % of the current reaching the edges correspond to the actual reclosure of the magnetic field lines around the finite plane, that is a self-inductance (Ref. 18,19). A close approximation to this partial inductance of a finite plane having a width "w" is given by:

$$L_{plane}\,(\mathrm{nH/cm}) \approx 5 \times h/w \qquad (6.3)$$

Notice how interesting is the concept of partial inductance: once the h,d and w geometry is set, one can calculate trace impedance regardless of the plane width, and plane impedance regardless of the trace width.

Summarizing these issues, Fig. 6.4 shows, for a same 10cm path length, a comparison between:

- 1mm wide, isolated trace, typical of a single layer board with Vcc and ground far apart
- the same trace, 1.6mm above a ground plane (Vcc trace for inst.)
- a gridded ground with 5mm mesh size ($\approx$ 3nH)
- a perforated ground plane with 1mm circular holes (1.25mm center to center)
- a plain copper plane 35μm (1oz) thick, 10cm wide (partial inductance $\approx$ 0.6nH)
- a theoretical infinite plane

a. Impedance of PCB Traces

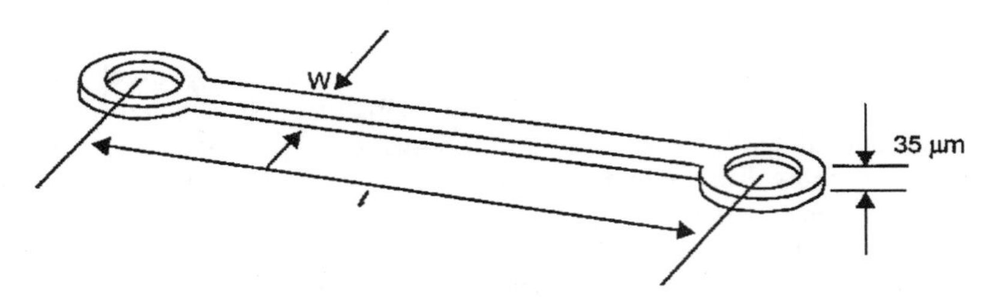

	Impedance						
	W = *1 mm*				*W* = *3 mm*		
	ℓ = *1 cm*	*ℓ* = *3 cm*	*ℓ* = *10 cm*	*ℓ* = *30 cm*	*ℓ* = *3 cm*	*ℓ* = *10 cm*	*ℓ* = *30 cm*
DC, 50 Hz to 1 kHz	5.7 mΩ	17 mΩ	57 mΩ	170 mΩ	5.7 mΩ	19 mΩ	57 mΩ
10 kHz	5.75 mΩ	17.3 mΩ	58 mΩ	175 mΩ	5.9 mΩ	20 mΩ	61 mΩ
100 kHz	7.2 mΩ	24 mΩ	92 mΩ	310 mΩ	14 mΩ	62 mΩ	225 mΩ
300 kHz	14.3 mΩ	54 mΩ	225 mΩ	800 mΩ	40 mΩ	175 mΩ	660 mΩ
1 MHz	44 mΩ	173 mΩ	730 mΩ	2.6 Ω	0.13 Ω	0.59 Ω	2.2 Ω
3 MHz	0.13 Ω	0.52 Ω	2.17 Ω	7.8 Ω	0.39 Ω	1.75 Ω	6.5 Ω
10 MHz	0.44 Ω	1.7 Ω	7.3 Ω	26 Ω	1.3 Ω	5.9 Ω	22 Ω
30 MHz	1.3 Ω	5.2 Ω	21.7 Ω	78 Ω	3.9 Ω	17.5 Ω	65 Ω
100 MHz	4.4 Ω	17 Ω	73 Ω	260 Ω	13 Ω	59 Ω	220 Ω
300 MHz	13 Ω	52 Ω	217 Ω		39 Ω	175 Ω	
1 GHz	44 Ω	170 Ω			130 Ω		

b. Impedance of Copper Planes

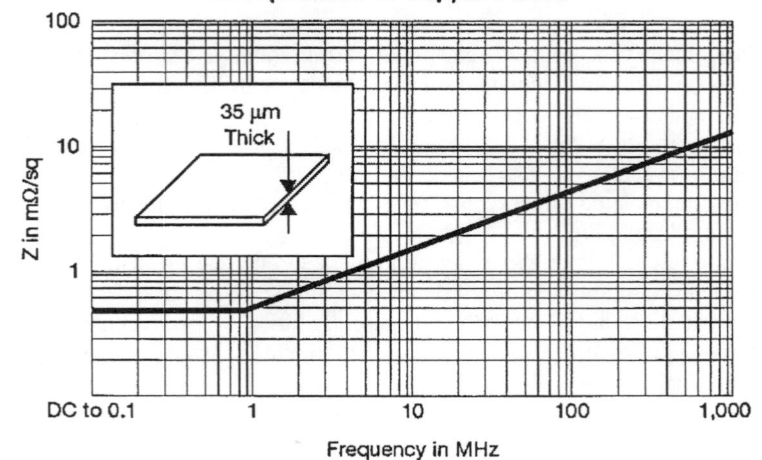

FIGURE 6.2 Impedance Comparison of Printed Circuit Power Distribution Traces and Planes

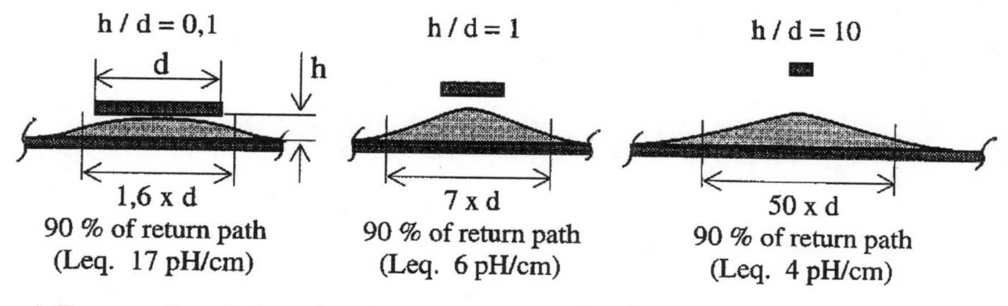

(Gross rule-of-thumb : 90 % of current is in ≈ 10 x h)

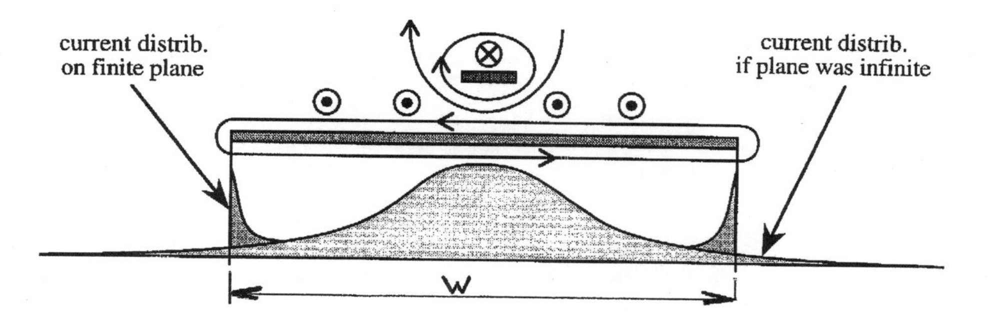

FIGURE 6.3 Actual Return Current Density with finite Planes

Example 6.1

From Fig. 6.4, calculate the ground longitudinal voltage noise for the following case and different ground options.We are looking at the harmonic #3 current from a 50MHz clock signal, 10 cm path length, assuming a fundamental current amplitude of 15mA :

Harm. #3 amplitude : 15 /3 = 5mA

	impedance @ 150MHz	ΔV
a) return path via thin ground trace	(curve A): 105Ω	≈ 0.5V
b) return path via a perfect ground plane	(curve F): 5mΩ	25μV
c) return path via perforated plane, 10cm wide	(curve D): 450mΩ	2.2mV

It is clear that, although hundred times more impedant than an ideal plane, the perforated finite plane is two hundred times less noisy than the isolated trace.

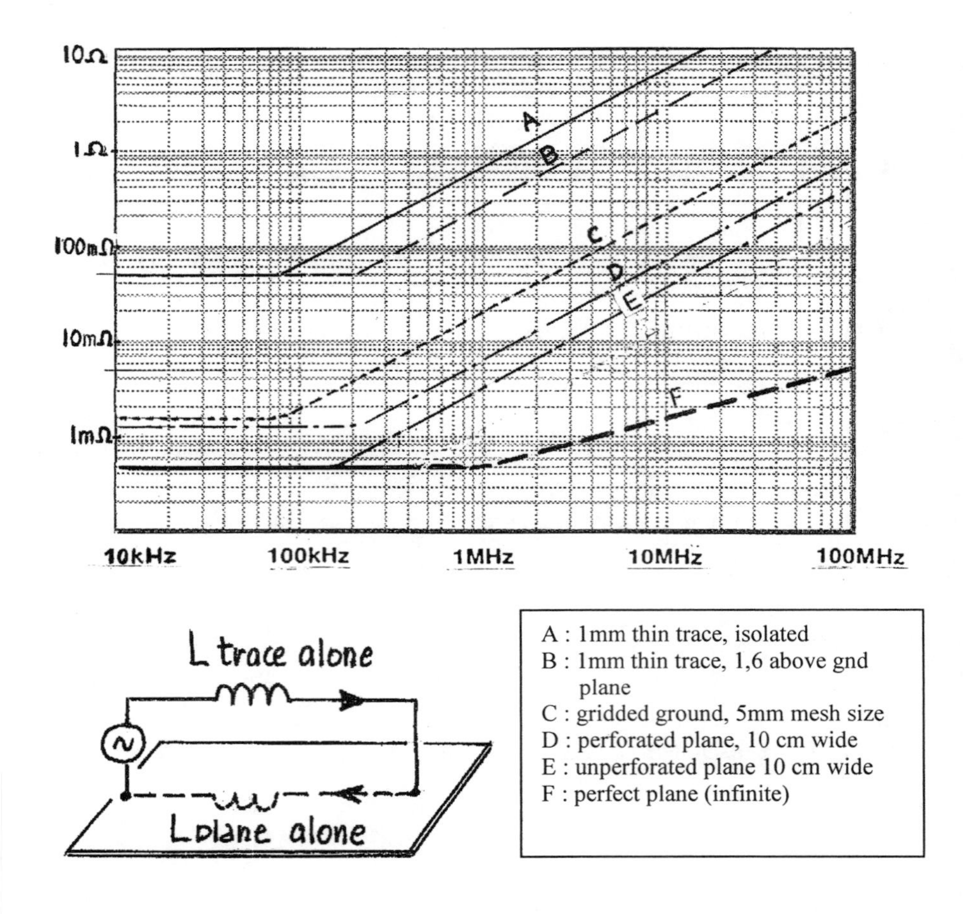

FIGURE. 6.4 Partial Impedance of Several Traces and Planes for 10 cm path Length

6.3 SINGLE-LAYER BOARDS

Single-layer PCBs, although cheaper, demand more precautions in controlling EMI (and radiated EMI in particular) because loop sizes are necessarily larger in all circumstances. This is true for the decoupling capacitor loop, the Vcc to Gnd trace separation and the signal trace to Gnd trace loops.

6.3.1 Power Supply Distributed Decoupling

Power distribution on single-layer cards is traditionally provided by supply and return traces. Their impedance (inductive reactance) is unimportant for slow-speed and/or low-power logic families such as ordinary CMOS (f.i. #4700 series). The need for capacitor decoupling is only one for every row of five or ten modules, plus one at the connector input. As the logic speed increases, considerably more care in layout is required due to increased trace impedance. High-frequency ceramic disc caps come to the rescue here, with one typically used to serve one or two modules.

Figure 6.5 shows a layout of power supply and return traces which are too far apart and therefore a poor design practice. Precautions must be taken to ensure that the capacitor, C, works properly. The problem, as shown in the equivalent circuit, results in an inductance of about 5nH for the capacitor leads (assuming they are cut very short) between the 5 V and ground traces and 2 x 5 nH for the DIP pin leads, outside the package itself. For a trace supply and return totalling 6 cm, the trace inductance is about 60 nH, for a total loop inductance of:

$$60 + 5 + 10 = 75 \text{ nH}$$

The IC internal inductance has not been considered because thay are already taken into account in manufacturer's data for intrinseque device characteristics : HF noise, ground bounce, etc...
A first, coarse approximation of the voltage drop from the capacitor to the IC, resulting from total loop inductance L, is:

$$\Delta V = L dI/dt \qquad (6.4)$$

Let us take a HCMOS gate having a transient supply current demand I_1 of 15mA/gate (from Table 5.2), plus a transient load demand I_2 of 35mA (for this device driving 5 gates). The situation of Fig. 6.5 results in a voltage drop:

$$\Delta V = 75 \times 10^{-9} \times (15 + 35).10^{-3} / 3.5\times10^{-9} \approx 1\text{V}, \text{ for one gate}$$

First of all, this is above the worst-case noise immunity level for this type of logic. Second, for emission aspects, this is a significant radiating loop, both by virtue of its size (about 4 cm^2) and by the current it carries (hundreds of mA when several gates are toggled at the same time). Finally, the unacceptably large inductance (and, hence, the voltage drop) also participates in Vcc and ground trace pollution, which contributes to the complete equipment radiation picture.

After such a gross estimation, let us look more in detail how this voltage drop distributes itself along the loop segments (Fig. 6.6). During a positive-going edge (low-to high), I_1 and I_2 are adding-up in the following traces: AB, BC, CO, EF. During a negative-going edge (see Chap.5, Fig.5.2), I_1 and I_2 are adding in segments OE, EF. This latter one EF is responsible for the ground-bounce of the IC, causing a CM impedance pollution of all the other IC leads.

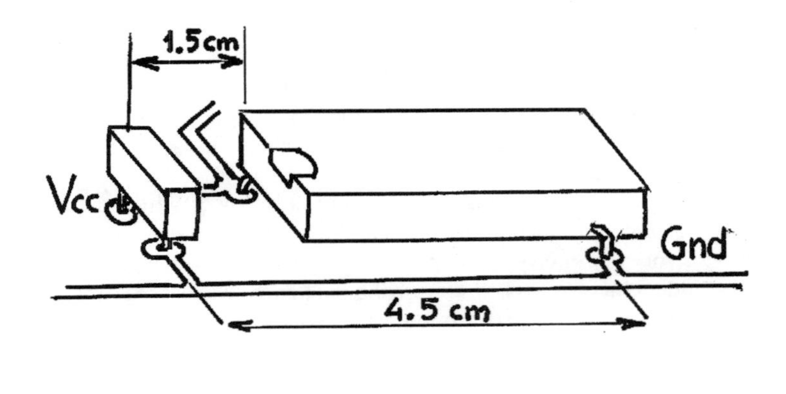

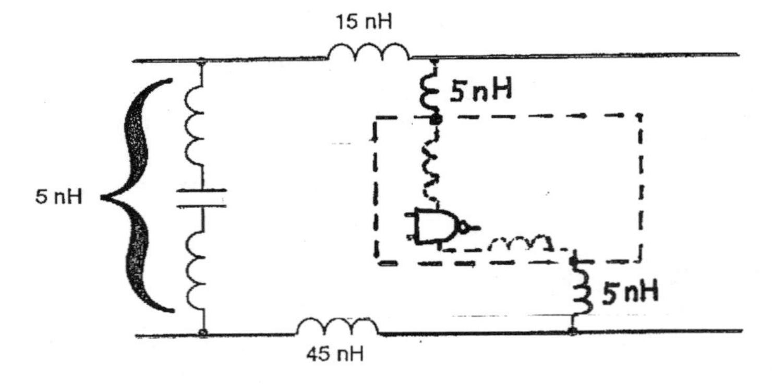

FIGURE 6.5 :Parasitic Inductances with DIP and Decoupling Capacitor

Voltage drops along the BC (for positive-going) and EF (for both transitions) segments are the most critical ones because:

- they do affect the whole dc distribution on this PCB branch, not just the gate or chip of concern
- this is where all the individual gates I_1 & I_2 currents are adding-up in a single Vcc or ground trace, causing longitudinal voltages which excite CM radiation of I/O cables.

Notice, though, that in terms of self-disturbance, $\Delta V_{(BC)}$ and $\Delta V_{(EF)}$ are not critical simultaneously: $\Delta V_{(AB)}$ relates to the Hi-state noise immunity, while $\Delta V_{(EF)}$ relates to Lo-state immunity which is generally more critical.

Figure 6.7 illustrates how Vcc and return traces should be routed close together to reduce capacitor-to-device loop area. This reduces the loop area by about 400 percent, self inductance also dropping to $\approx$ 8nH/cm.
An interesting concept (although as old as electromagnetism) is shown in Fig. 6.8. Whenever two identical circuits are simultaneously switching identical currents, one should consider whether the current paths can be arranged such that the magnetic

moments are subtractive. Doing so, the overall radiation seen at distances greater than loop separation is significantly reduced.
When a close routing of Vcc and Gnd traces is not possible, another option is to lay an additional ($\geqslant$ 1mm) trace for the capacitor's positive terminal to the module's Vcc pin (see Fig. 6.9). The negative capacitor terminal is located very close to the module Gnd pin. This reduces loop size and provides lesser voltage drop on the ground path (DE) which can be common to other users, hence more vulnerable. In this arrangement, the loop inductance of the Capacitor -Vcc pin-chip-Ground (ABCDE) loop must be regarded in 3-dimension. Its height above the PCB surface depends on the type of IC package, varying from 0.3mm (SMT) to $\approx$ 3mm (DIP). Its length and width depend on the traces layout, especially with single-layer.

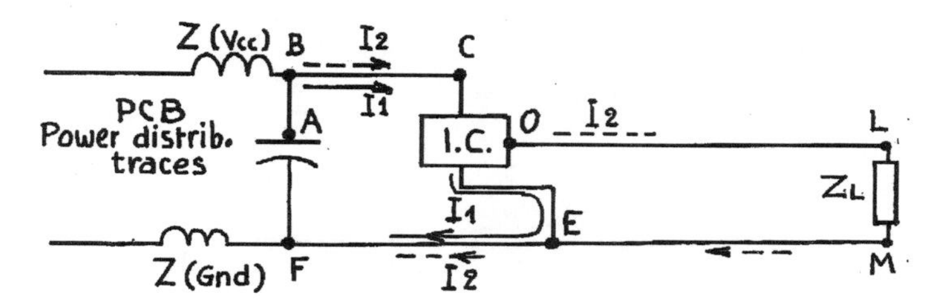

FIGURE 6.6 : Actual Current Paths During Logic Gate Switching

Other methods of power supply decoupling with minimal inductance and radiation are shown in Fig. 6.10. These include (1) leadless parallel plate capacitors located directly under the LCC or DIP packages and (2) minibus power distribution. The latter is especially useful when a ground plane cannot be incorporated.
Finally, an ultimate solution is to install on-site chip decoupling by including a multilayer ceramic pellet in the substrate. A promising technique (Ref. 20) consists of using multilayer IC substrates where supply voltages and ground returns are made by internal planes. Signals between the I/O pads and die are sandwiched between the Gnd and Vcc planes, as shown in Fig. 6.11.
When packaged in a surface-mount logic array, this mounting allows two voltage planes, two signal layers and several ground planes to be stacked with minimum intra-IC crosstalk supply and ground noises. Compared to the typical 5 nH of package inductance with a standard pin-grid array, the total inductance is reduced to less than 100 pH. For instance, with such a device having 224 I/O pads, 8 bus drivers can switch 100 mA each in 3 ns, with less than 100 mV of common ground noise at the chip level. Of these 100 mV, only 20 mV are due to inductance, the rest being caused by the bonding wires and contacts dc resistances. The need for an external decoupling capacitor is eliminated because it is built in.

Although the capacitor and IC loop may seem rather modest compared to the signal traces, the currents of all the internal circuits are exactly adding-up in this loop, and its contribution to total radiation can be significant (Fig.12A).

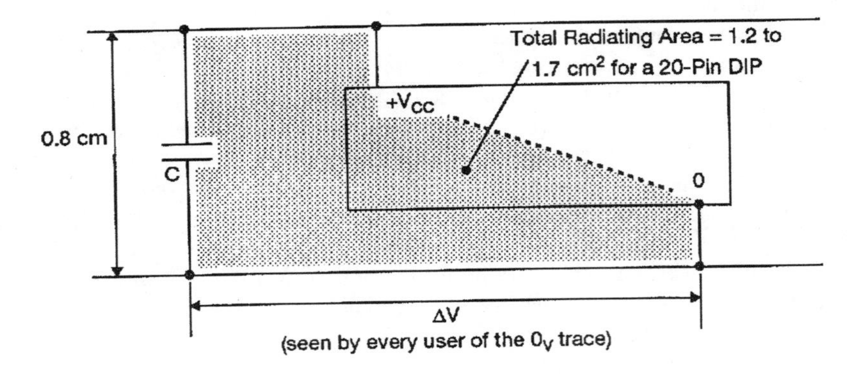

a. Initial Layout

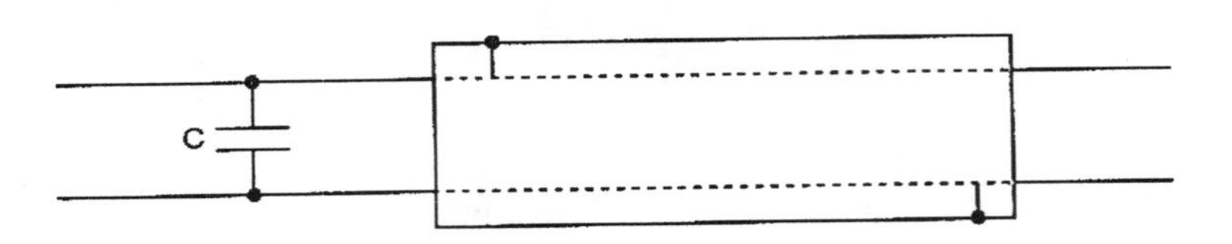

b. Loop Area Reduced via Closer Traces

FIGURE 6.7 Bringing Supply Traces Closer to Reduce Supply Current Loop (Single-Layer Board)

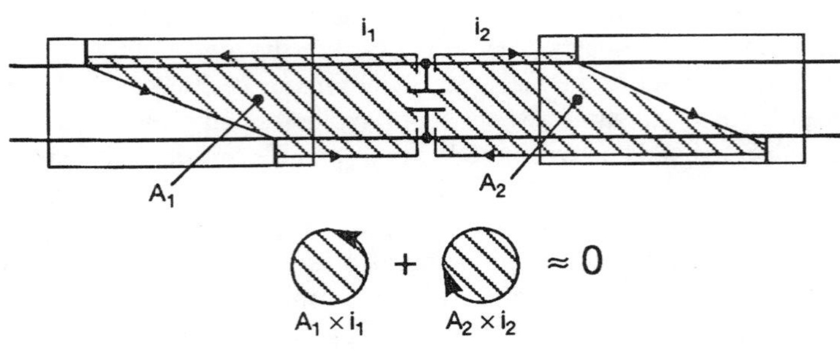

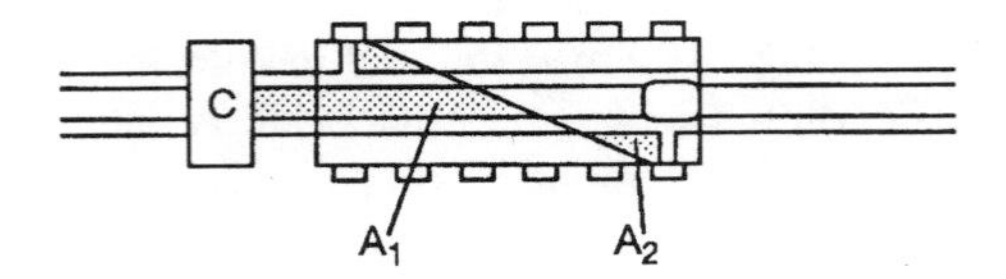

$A_{total} = A_1 - A_2$

FIGURE 6.8 Field Cancellation by Opposite Amp x cm^2 Magnetic Moments

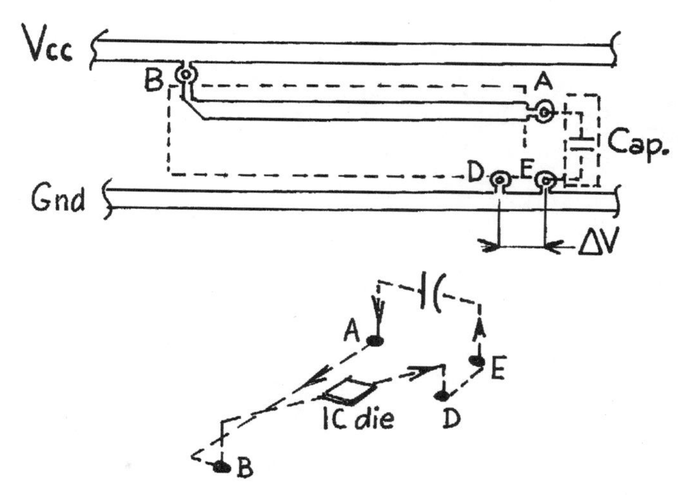

FIGURE 6.9 : Dedicated Capacitor Trace to Reduce Loop Size, and CM Impedance Pollution of the Ground Trace

6.3.2 The Problem with Bypass Capacitors in Parallel

With one decoupling capacitor per IC or pair of ICs, there are unavoidably many places where two or more capacitors will be in parallel on a same Vcc-Gnd traces pair, forming a resonant L-C tank. A self-resonance frequency will exist, with a Q factor limited only by the DC resistance of the traces and vias (Ref. 21).As a result, in a narrow frequency band, some of the switching current demand of IC1, while IC_2 is quiet (Fig. 6.12), may be in fact pumped out of C_2 instead of C_1. This causes a larger radiating loop than expected for the capacitive current. This problem may very well never hurt if there is no frequency to excite this mechanism. It could be predicted by a careful simulation with SPICE or other software, and if it occurs, it can be reduced by the following solutions:

a) enlarging the Vcc and Gnd traces, bringing them closer, or on top of each other, or better changing to a ground plane. By decreasing the self inductance L_t, the self resonance L_t // C_2 is shifted upwards in frequency. This may not seem very helpful, but in fact this new frequency corresponds to a lower impedance value for C_1, which will compete more efficiently with C_2 for feeding the current to IC_1, hence reducing the effective radiating loop size.

b) provide some resistive damping of the resonant circuit. This may prove less easy, technically and costwise. Adding few ohms to the DC distribution may create a permanent waste, and undesirable voltage drop for all the users on this particular board. For some critical devices like RF amplifiers, this can be done in the form of a RC filter decoupling instead of a simple C.

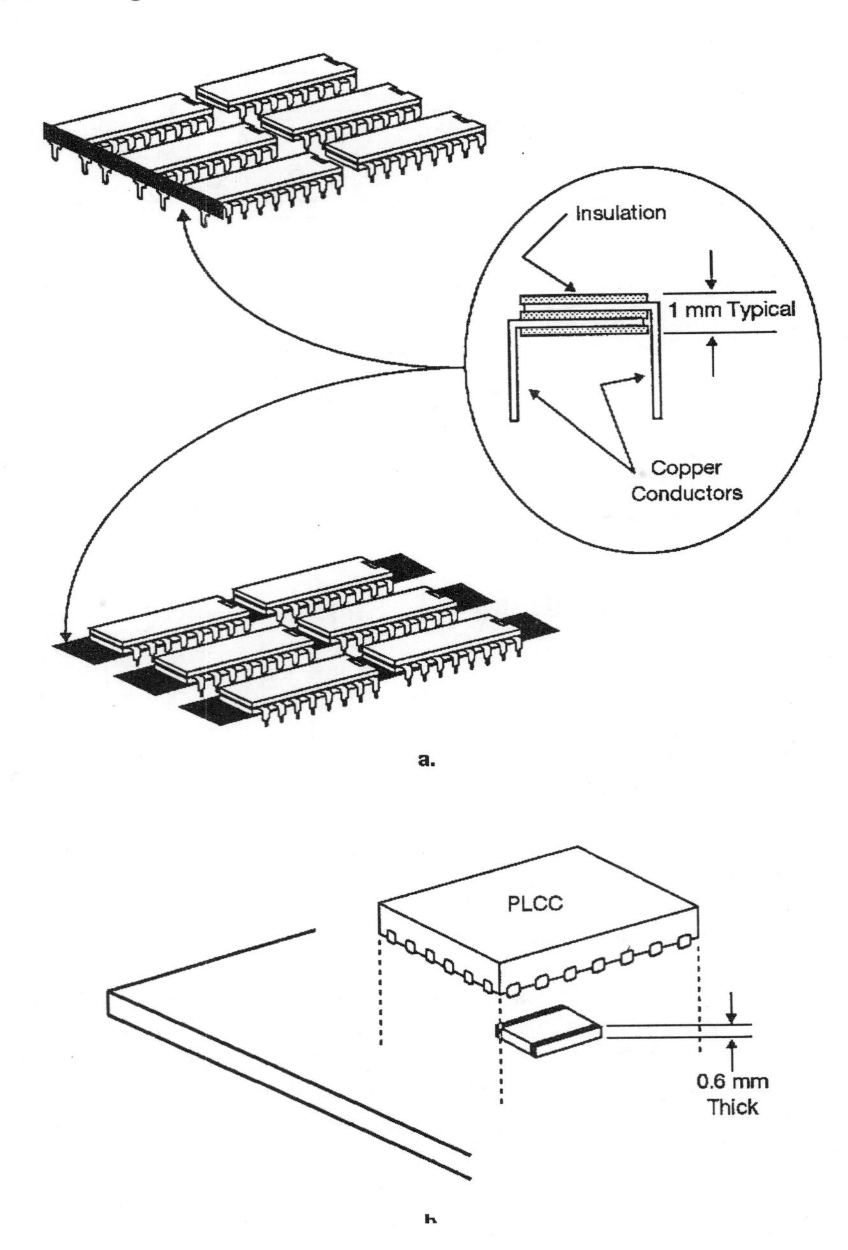

FIGURE 6.10 : Flat-Bus Distribution (a) and MLC Leadless Decoupling Capacitor (b). Both offer minimal parasitic inductance.

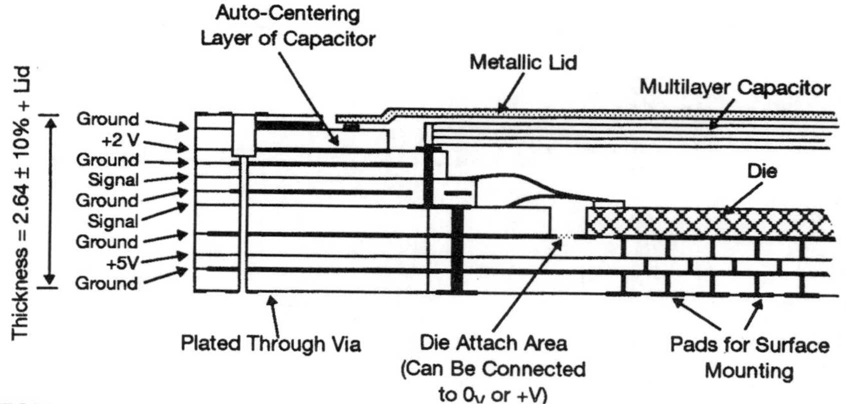

FIGURE 6.11 : Example of Integrated Decoupling Capacitor within a Large LCC Module. The ceramic substrate has, in addition, a multilayer power-0V-signal layout.

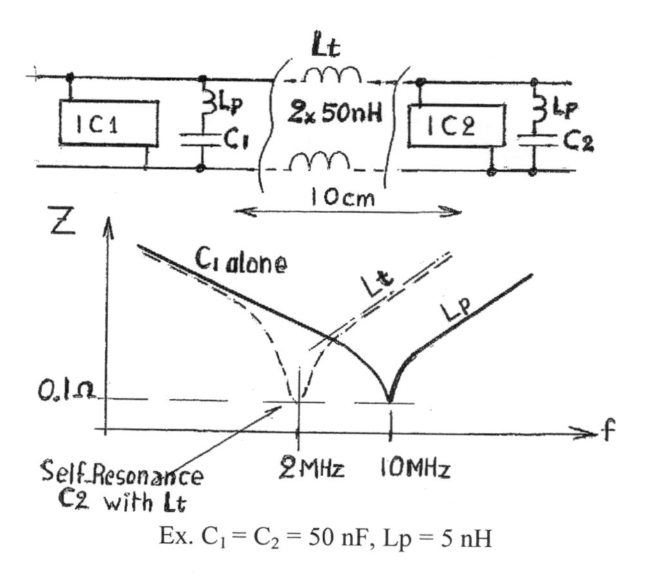

Ex. $C_1 = C_2 = 50$ nF, Lp = 5 nH

FIG. 6.12 Parasitic LC Resonances with Bypass Capacitors in Parallel

A better option is to insert a small, surface-mount, lossy ferrite between the Vcc rail and the capacitor lead. The ferrite must have a low inductance value but a high resistive term, to act as a frequency-dependent resistor. This both reduces the Q of the fortuitous L-C circuits, and forces the current demand of each IC to be taken off its dedicated capacitor.

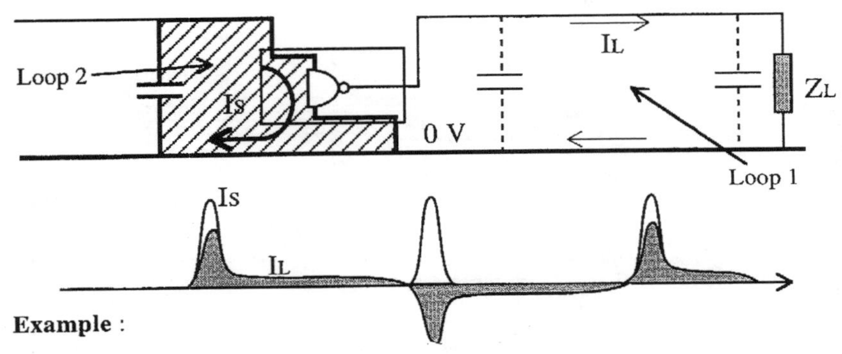

Example :

Single layer PCB, comparison based on 1 gate only.
30 MHz pulses, 20-pin module, TTL family
Loop # 1 : 10 cm x 1 cm, ZL ≥ 300 Ω, trace inductance = 15 nH/cm (two-way trip)
Loop # 2 : 1.5 cm^2

Spectrum og magnetic moments (I x S) for loops # 1 and 2, based on current spikes only

F (MHz)	30	90	150	330
Loop # 1 (15 mA pk), dB μA	73	67	58	44
Magn. moment, 10 cm^2	93	87	78	64
Loop # 2 (40mA peak), dB μA	82	76	67	53
Magn. moment, 1.5 cm^2	85	79	70	56

For 1 gate switching : the magnetic moment of Vcc loop # 2 is ≈ close to that of signal loop # 1, adding 3 dB to the radiated field.

FIG.6.12A Radiation contribution of the Bypass Capacitors loops

6.3.3 Decoupling at the Card Power Input

Because the many decoupling capacitors installed on a card need to be recharged after each logic transition, the dV/dt imposed on the edge connector, and subsequently on the rest of the power distribution array, may be too high, even though the corresponding time constant is greater. This, in turn, would "export" the daughter card noise emissions into the rest of the power distribution circuit, mother board and the like. To avoid this, it is a good precaution to install a large ceramic (MLC) or aluminium capacitor right at the edge connector Vcc and ground leads. The value of this capacitor, which becomes part of the "bucket brigade", should approximate (but not exceed) the sum of the individual capacitors installed near the ICs.

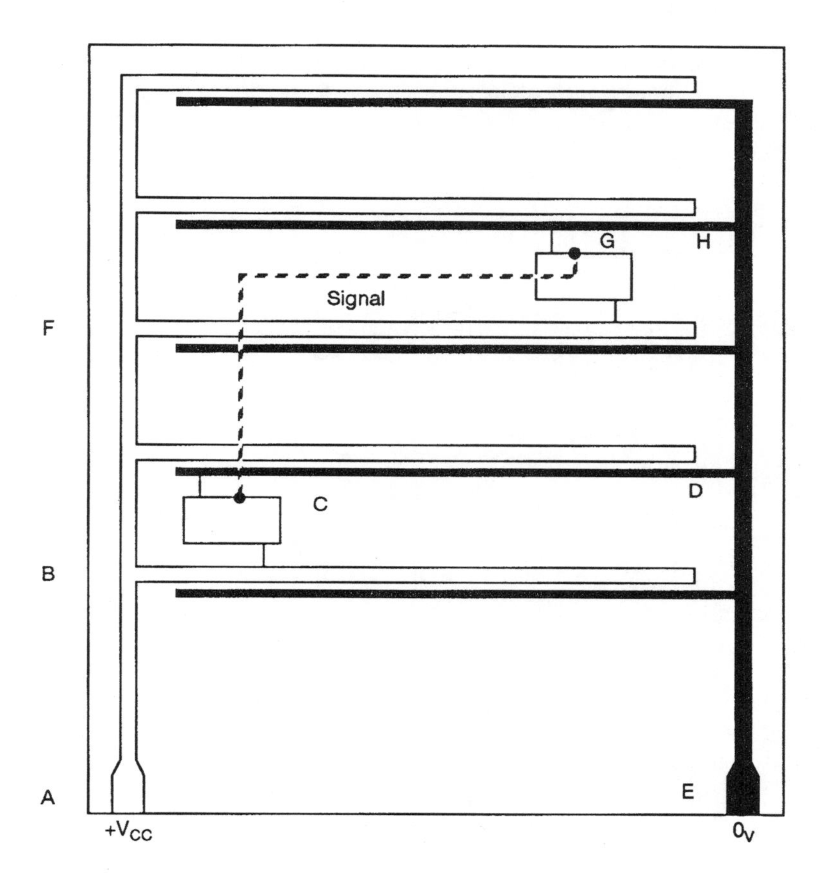

FIGURE 6.13 A Poor Layout (Single Layer) Creating Huge Radiating Loops

6.3.4 Vcc and Ground Trace Layout (Single-Layer Board)

With regard to power distribution, one consequence of too much separation between the power supply and return rails was discussed in Section 6.2.1. This is further illustrated in Fig. 6.13, where a poor PCB layout is used with the main supply trace on one side of the board an the return trace laid along the opposite side. This, along with excessive inductance, creates huge radiating loops such as ABCDE and AFGHE for transient power supply currents. Similarly, signal currents between modules are forced to return by some very sneaky pattern.

A better layout is shown in Fig 6.14. Here the Vcc and Gnd traces are laid close together in horizontal parallel runs. They form a low-impedance, low-area transmission line while allowing signals to run horizontally unobstructed on the solder side of the board. However, signal traces to ground return still can form significant loops.

Imposing an additional constraint of running most horizontal signals on the solder side and vertical signals on the component side, an even better layout is suggested in

Fig. 6.15. (Ref. 22, 23) Although it cannot match the performance of a multilayer board, this design offers a grid pattern for Vcc and Gnd traces; all intersections are connected by via holes. At the edge connector, at least every tenth pin is assigned for ground, unless signal line impedance matching requires more ground pins. (This requirement is discussed in Section 6.6.)

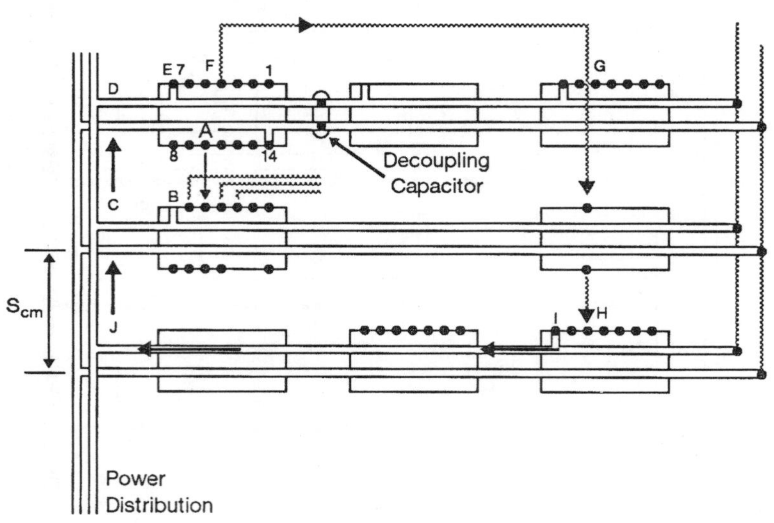

FIGURE 6.14 A Somewhat Better Layout, Reducing Power Distribution Loops. However, signal traces-to-Gnd loops are still large, with typical spacing, h, of 2 to 2.5 cm. Areas such as ABCDEA represent $\approx$ 3 cm^2; areas such as FGHIJDEA represent 20 to 30 cm^2.

6.3.5 The Need for Ground Planes and Ground Areas (Single-Layer Board)

Single-layer boards do not easily lend themselves to hosting a ground plane. Except for very sparsely populated boards, all of the available real estate is needed for signal traces, with not much room left for power distribution traces, let alone for a whole ground plane. However, even if an entire side of a double-faced, single layer PCB cannot be sacrificed for a copper foil, several simple arrangements can produce comparable performance.

As seen in the Fig. 6.3 comparison, when changing from thin or very long traces to a copper plane, the benefit comes from the drastic decrease in inductance. By leaving as much copper as possible rather than etching it away, a low impedance is achieved for both the supply and ground since they are close together. This preferred practice can be expressed by a simple guideline: make the PCB as optically opaque as possible by extending supply and ground return traces into large areas.

If the board is a single-layer, single-sided type, the Vcc and signal lines will have a better opportunity to run near a ground copper land (see Fig. 6.16). Above few hundred kilohertz, it is true that returning currents will not use the whole copper surface and will concentrate on the edge, so the effective reactance will be higher

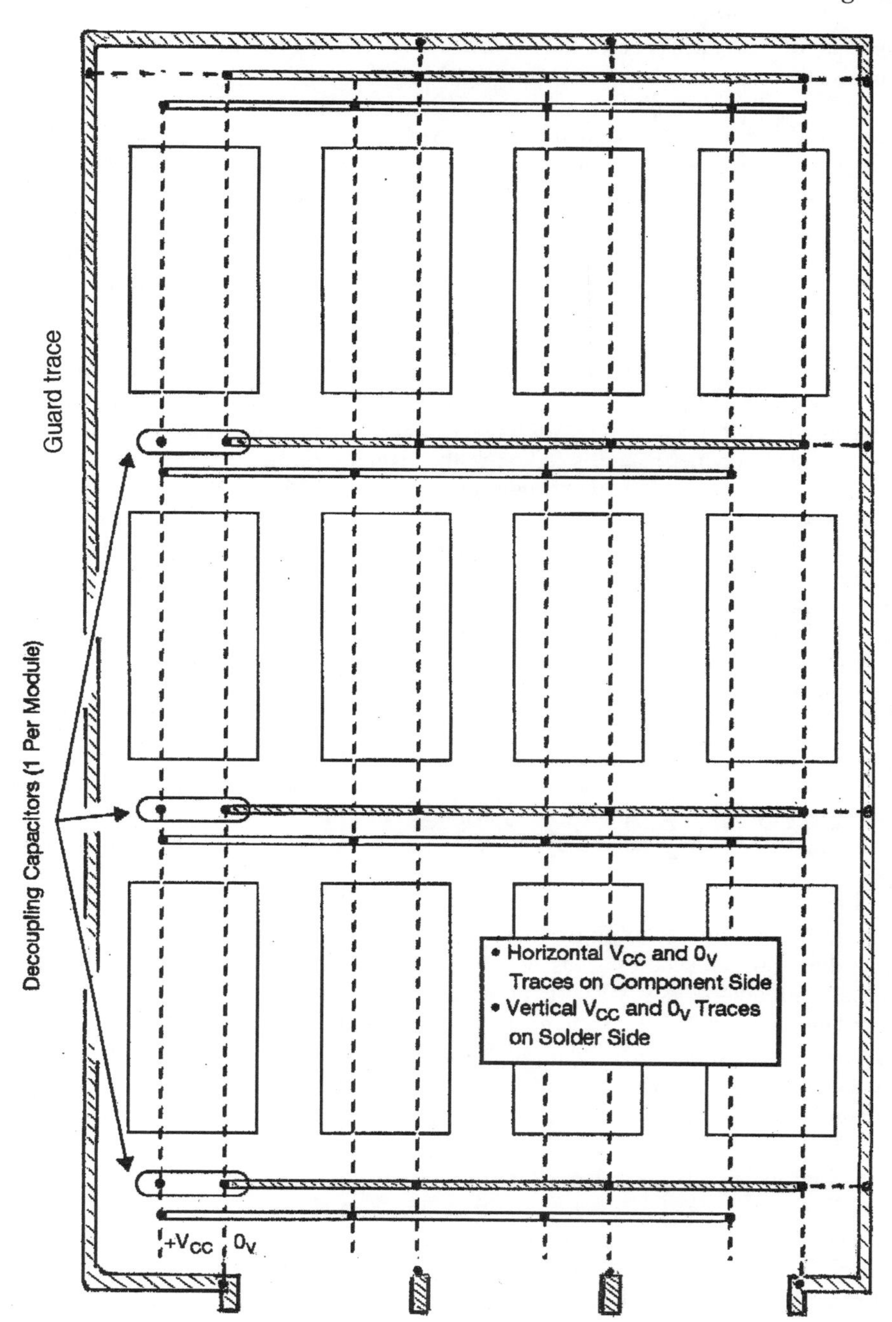

FIGURE 6.15 An Even Better Single-Layer Design, Reducing Both Power Distribution and Signal-to Gnd Trace Loops

than that of the ideal plane, but nevertheless most loop sizes will be reduced significantly.

If the board is a single-layer, double-sided type, the Vcc and signal lines often will be above a ground copper land, and less current will circulate on the copper plane edges. Radiating loop size and power distribution noise will be further reduced.

It has been often reported that such islands created by copper land-fill areas are useless, or even counter-effective. When examined carefully, these critiques have several foundations:

a) the PCB designer (or the computer auto-router) may forget to connect some of these islands to the rest of the ground network. A floating copper spot is worse than no copper, because it may aggravate capacitive crosstalk and radiation.
b) sometimes a copper island is linked to the ground grid at only one point, reason advanced being the famous "single point myth". Unless it is used simply as an electrostatic screen, the landfill's essential purpose is to reduce the whole ground grid inductance, which means there must be current in it. With a one point connection, there will be no current.
c) sometimes a copper island is correctly tied with the ground grid, at several points, but there is no critical (high-speed, or sensitive) trace running atop or nearby, therefore current paths are taking place elsewhere, and indeed, the island birngs no improvement at all.

Answer to a) and b) is simple: land-fills must be connected to the general ground patch at every possible occasion, and especially at places where a signal trace is running nearby or above. Answer to c) is that, even if the copper island is useless for the moment, a future change may move or add traces which will benefit from its proximity.

Frequently, for manufacturing reasons (etching and tin flow), large copper lands on the card top are replaced by grids. Although this creates a slight increase in ground impedance, it does not significantly affect the concept, provided there are no flagrant discontinuities and the grid elements remain large (see Fig. 6.17). Notice that bottlenecks like those associated to ground vias have been kept to minimum inductance in caption (A). In contrast, narrow grid bars and long bottleneck (caption B) severely degrade the grounding impedance.

6.3.6 Trace-to-Chassis Parasitic Coupling

One should avoid placing fast signal traces next to the edge of a card, which in most cases also places them close to a chassis plate, causing capacitive leakages. There should be at least a ground trace or plane edge extending far enough to ensure that HF currents will return by the PCB ground rather than by some uncontrolled path over the chassis length.
This might seem contradictory with something we will recommend later on, i.e., decoupling I/O lines to chassis. However, this will be done in a controlled manner,

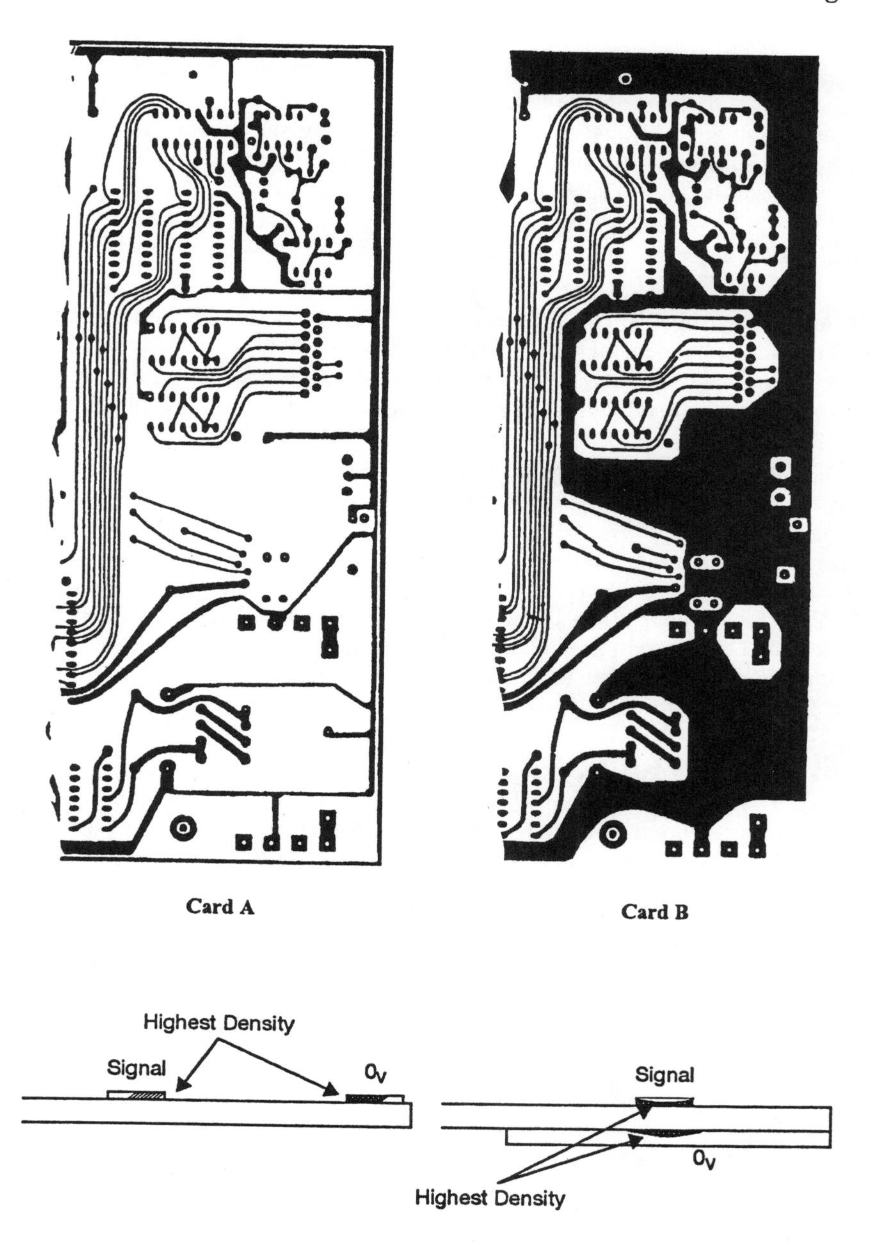

FIGURE 6.16 Advantage of Ground Copper Lands. By not etching copper for Vcc and Gnd, as shown in card B, the cost does not exceed that of card A, but radiating loops are smaller, as the current density distribution is more favorable owing to a nearby ground.

at the I/O connector zone, where HF current circulation will be confined to a well-defined portion of the chassis: the connector plate. Having uncontrolled HF currents returning from a "hot" trace to the source via a long run across the chassis may exacerbate radiation through the enclosure, unless this enclosure has been fully treated to act as a Faraday cage.

6.3.7 Clock traces

With single layer boards, clock traces are especially prone to radiation and crosstalk coupling, due to the general lack of a continuous ground plane as areturn. Any trace carrying clocks ⩾ 10 MHz must have a dedicated ground trace running parallel from end to end, either on the same side at a distance < 1mm, or underneath.

If clocks ⩾ 50 MHz are run on a single layer board (not a very safe practice), the clock trace must be bound by two ground traces, at equal distances. This creates two identical loops, each one carrying half the return current, but with opposite magnetic moments hence less radiation.
This can prove necessary even for a single layer board with ground plane on one side. Take for instance a thin clock trace like 0.16 mm width (6 mils) and a board thickness of 1.6 mm: two grounded traces at 0.16mm on each side (center-to-center 0.32mm) are more efficient for radiation and crosstalk reduction, than the ground plane which, (in terms of h/w ratio), is farther.

6.4 MULTILAYER BOARDS

Multilayer boards are the ultimate answer to PCB noise in general, and especially to radiated EMI. At a cost of 2 to 2.5 times that of a single-layer board, they often eliminate the need for expensive overall box shielding. In general, it is difficult to operate high speed logic successfully on a single-layer board because of common impedance coupling. While many articles have described such single-layer board achievements, they require considerably more attention to fabrication details, making quality control and repeatability difficult during mass production. To avoid a range of problems, including excessive radiation, multilayer boards are recommended wherein the power supply and signal ground reference are realized on separate one-ounce, copper foil planes.

6.4.1 Multilayer stacking

In the commercial-type multilayer, there are usually n + 1 levels for an n-layer board, as shown in Fig. 6.18. Here, a four-level board configuration is illustrated. Level A is the component side, and it includes the interconnect traces. Combined with the upper face of level B, they form a microstrip line. The upper and lower faces of ground return level C are, in theory, electrically isolated by three skin depths (i.e., about 25 dB) around 30 MHz. Thus, currents returning from D and flowing on the lower face of C do not appear on the upper face, and vice versa. This

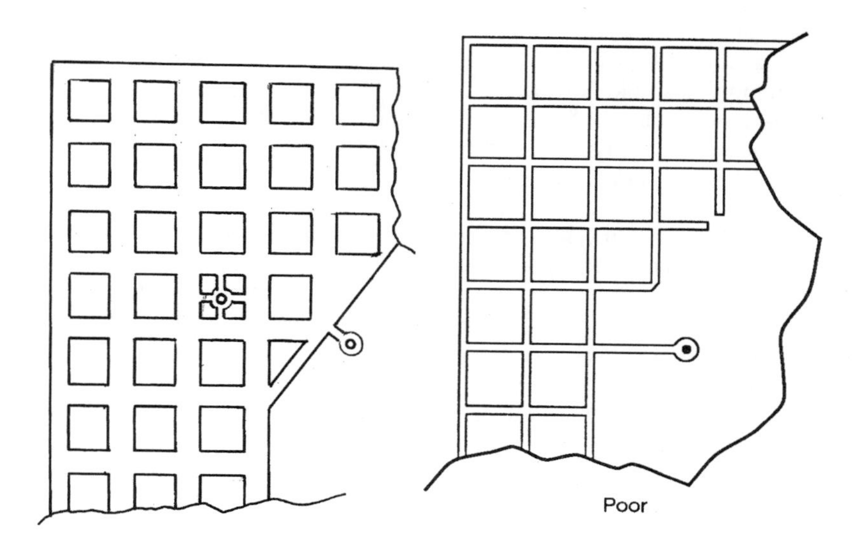

FIGURE 6.17 Acceptable vs. Poor Grids in Copper Planes

is more academic than applicable to the real world, as ground planes generally are heavily perforated. Signal level A forms microstrip lines with B and C, which are strongly coupled in the case of high frequencies. Likewise, signal level D forms microstrip lines with 0V plane C.

Compared to a typical single-layer board where the distances between signal and ground traces are in the range of a few centimeters, the 0.2 to 0.5 mm layer separation makes radiating loops about 30 to 100 times smaller with multilayers.

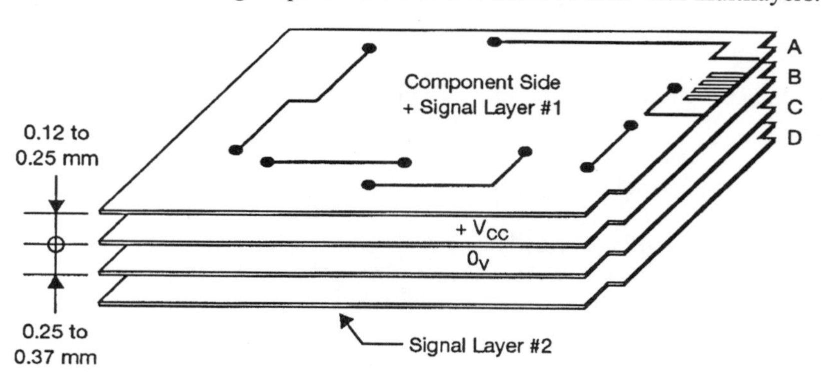

FIGURE 6.18 Typical Multilayer Board with Signal Traces Outside

Other advantages of the Fig. 6.18 arrangement are:
• easy access to signal traces for repairs and temporary wiring changes
• minimum crosstalk between the two signal layers
• minimum spacing between the Vcc and Gnd planes (only one layer thickness). This makes the two planes behave as a very low characteristic impedance line (typically a few ohms).

Figure 6.19 shows the inverse arrangement, where Vcc and Gnd are the external planes. This arrangement provides some shielding by "sandwiching" all signal lines between two copper foils. It also provides a more even transmission line configuration (true stripline) than the asymetrical microstrip of Fig. 6.18. The corresponding drawbacks are:
• It is difficult to access the signal layers.
• It becomes mandatory to run signal levels 1 and 2 at 90° to avoid heavy crosstalk between layers.
• It is impossible to use leadless modules without etching large holes in the copper planes (a very bad practice).
• It provides lower characteristic impedance for the signal traces (typically 1.4 times less than the Fig. 6.18 arrangement), which causes an increase in dynamic loading for digital ICs outputs.

For very dense boards and military or aerospace applications, multilayers with 6, 8 and even up to 14 levels can be used. In terms of radiated emissions, the aspects discussed above are applicable to any number of layers. A recommended arrangement is shown in Fig. 6.20 for a 6-layer board:

• Layer #1: low-speed or steady-type signals. Eventually high-speed signals but only for the shortest runs (no more than few cm).

• Layer #2: high-speed signals. These must be run at 90° angles with layer #1, or staggered as shown.

• Layer #3 & 4: Gnd and Vcc. Eventually, the dielectric thickness between 3 and 4 can be thinner, and made of a material with a higher ε_r, for increasing the buried capacitance.

• Layer #5 and 6: signal layers, treated like #2 and #1, respectively.

When changing from layer 1 to 2, or 5 to 6, one can use buried (blind) vias, rather than drilling holes in power and ground planes. In certain application, however, blind vias are suspected of causing long-term reliability problems under harsh environment.
As an alternative, layers #2 and 3 can be interchanged, which releases the constraint for the 90° runs, and keep the critical traces buried in a true stripline mode. This is to the expense of doubling the height between the Vcc and Gnd layer, i.e. half the buried capacitance.

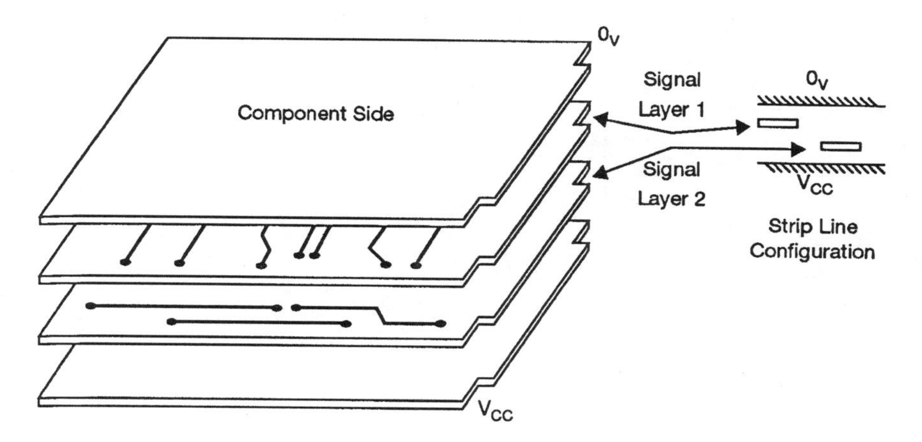

FIGURE 6.19 Multilayer Board with Buried Signal Traces

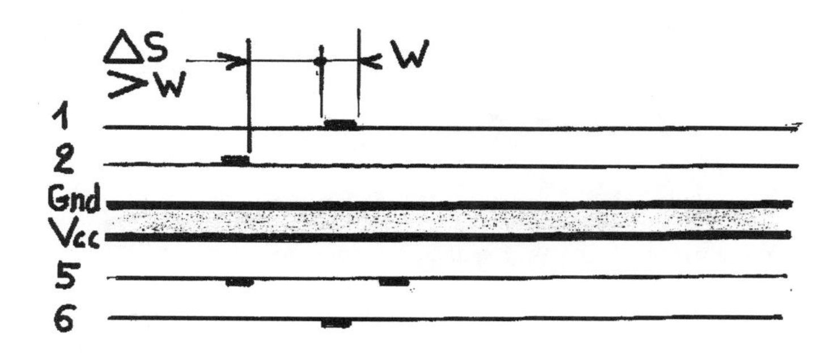

FIG. 6.20 Recommended arrangement for 6-layer

Before concluding this discussion, a word of caution is appropriate. In PC boards that mix digital, analog (slow), high-level RF analog (transmit), low-level RF analog (receive) and so forth, there is often a need for separate power and ground planes. This is to avoid the possibility of return currents from one circuit polluting the others and, eventually, reradiating at unexpected places. When such plane segregation is needed, the different planes should be laid side by side—never on top of each other. Due to the large capacitance, stacking planes from different functions will create a heavy coupling effect, causing a fraction of the high level currents to circulate in low level ground plane.

6.4.2 Decoupling Capacitor Requirements with Multilayers

A question often arises: with the Vcc and Gnd planes being so close, don't they create huge in-situ capacitance, which reduces or eliminates the need for discrete decoupling capacitors? In fact, except for very specialized boards where a ceramic

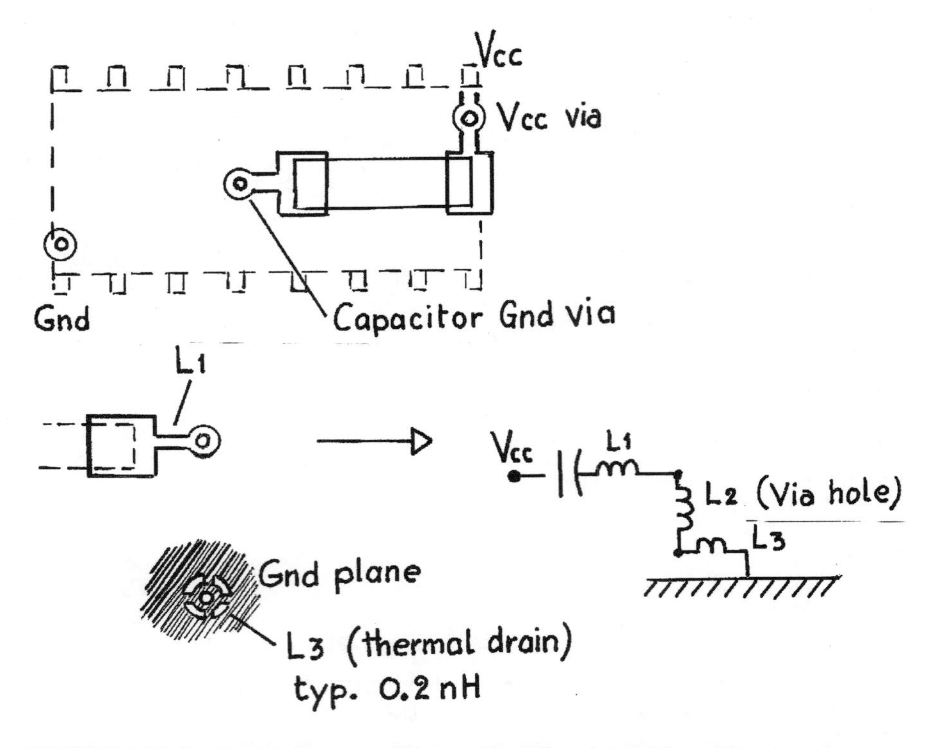

FIGURE 6.21 Residual Inductance of Bypass Capacitors in Multilayer Boards.

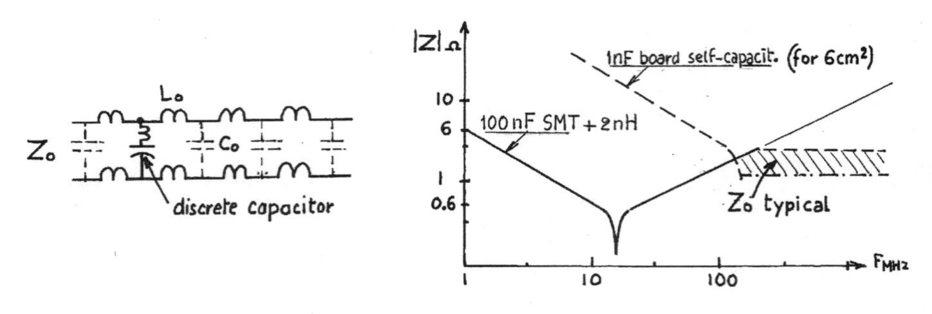

FIGURE 6.22 Combined Impedances vs Frequency of discrete and buried Capacitors

compound is laid between the two power planes, a regular multilayer offers only 30 to 300 pF/cm^2 of "free" capacitance. For a typical 20 or 24-pin package occupying approximately 6 to 8 cm^2 of board real estate, this is totally insufficient and discrete capacitor requirements remain. Nevertheless, as frequency increases to the VHF region, a point is reached where the unavoidable parasitic inductance of discrete power supply capacitors will make them progressively useless. A typical manufacturing practice is shown in Fig. 6.21 , with the soldering lands, via hole and

thermal drains concurring to a parasitic inductance which spoils some of the zero-inductance concept of leadless capacitors.
Fig.6.22 shows such configuration where the 100nF SMT capacitor, although close to perfection, will resonate around 12MHz with the 2nH inductance of its via hole and soldering pad. By contrast, the 1000pF of board self-capacitance (for ≈ 6cm^2 real estate) is competing efficiently with the discrete component above 100MHz. Soon after, the Vcc and Gnd planes stack is seen as a parallel plate line whose distributed L,C parameters are providing a characteristic impedance $Z_0 = \sqrt{L/C}$, with typical values of 1 to 3Ω.

6.4.3 Perforated Planes: The Swiss Cheese Syndrome

With densely populated boards, the ideal plane concept that is advanced when advocating multilayers becomes something like the illustration in Fig. 6. 23. Not only is the copper area reduced (in fact, by no more than 10 to 15 percent, which would be tolerable) but many IC zones are so perforated that some holes overlap, the string of holes becoming a slot (Ref. 25). This leads to three observations:

a) A slot in and of itself would be of no consequence; the problem arises when a signal (especially a clock trace) crosses over the slot. The return current density is disturbed by the slot, and current lines will concentrate on the slot edges (following a loop route equivalent to an inductance of ≈ 1 nH/cm). This is exactly the same as exciting a slot antenna if one were to transmit intentional radio signals, which is exactly what we want to avoid.
b) The problem is reduced if a copper rib is maintained between holes. But this still turns the plane into an array of bottlenecks, especially in high current density areas. For instance, in Fig. 6.24, the decoupling capacitor has to serve the PGA modules through narrow, inductive paths. A better decoupling, and therefore less radiation, is achieved by putting one capacitor on each side. SMT packages have another plus here: they do not degrade internal planes.
c). One should never make a slot in a ground plane for burying-in a clock trace, hoping that it will reduce EMI. This may effectively reduce this particular trace radiation but increase seriously the radiation of many other signals that will cross this slot.

6.4.4 Allowable Slots

With mixed-function cards, particularly those combining analog and digital circuits, a gapped plane is sometimes used to avoid a certain amount of logic current returning by the analog ground land. With such a layout, absolutely no signal trace should be permitted to cross over the slot area. Signals that must go from one zone to the other should pass only above the small copper land (see Fig. 6.25, top of the card) where the A/D converter is located.

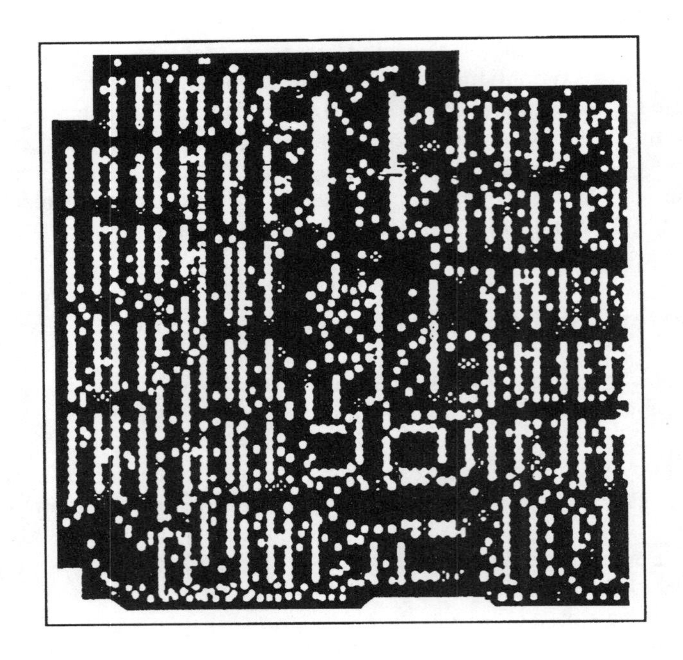

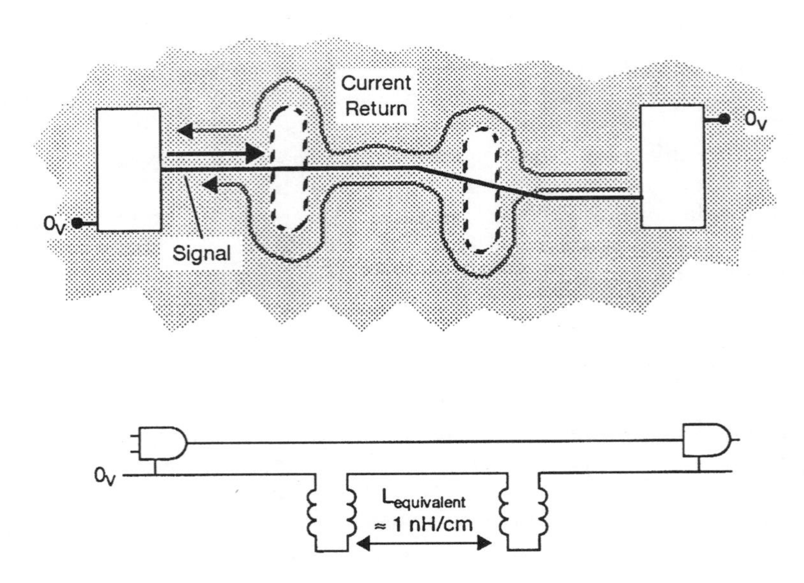

FIGURE 6.23 What a Ground Plane Can Become

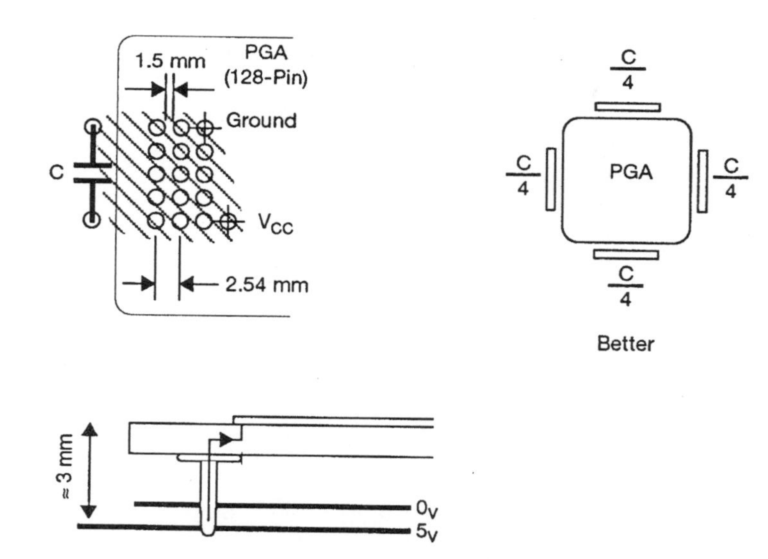

FIGURE 6.24 Perforated Plane Disturbing Capacitor Discharge Path

6.4.5 Placement of High-Speed Traces

Some traces must be routed very carefully, due to their higher propensity to radiate at discrete, stable frequencies. Such traces include those for clocks, crystals, least significant bit (LSB) (especially with address bus), RF oscillators and video circuits.

The following guidelines must be observed for proper high-speed trace placement:

- Above perforated plane areas, do not run critical traces across a row of holes. Try to run them parallel instead.
- Do not run critical traces at the extreme edge of the card. This will nullify the ground plane advantage and pull return current to the edge, increasing radiation (see Fig. 6.26). Beware of a frequent manufacturing practice of etching approximately 3 to 4 mm of copper from around the entire PCB edge.

6.4.6 Analog Digital Mix

A PCB hosting both low-level analog and digital circuit is always a designer's nightmare. Analog circuits are often operating with sensitivity thresholds in the mV range for audio applications, down to the μV for the input section of RF modules (mobile phones, radio or TV receivers, RF modems etc..). This is 60 to 120 dB below the amplitude of digital switched levels, which can be located within a cm range on the same card. The compatibility between the two portions is first a matter of functionality, not in the scope of this book. However, there are few typical implications with radiated interference.

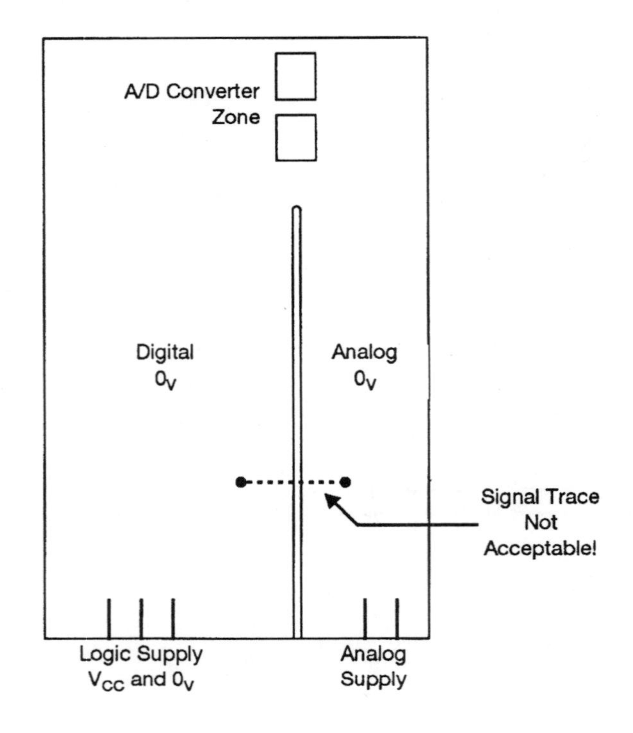

FIGURE 6.25 Allowable Slot with Analog/Digital Mix on PCBs

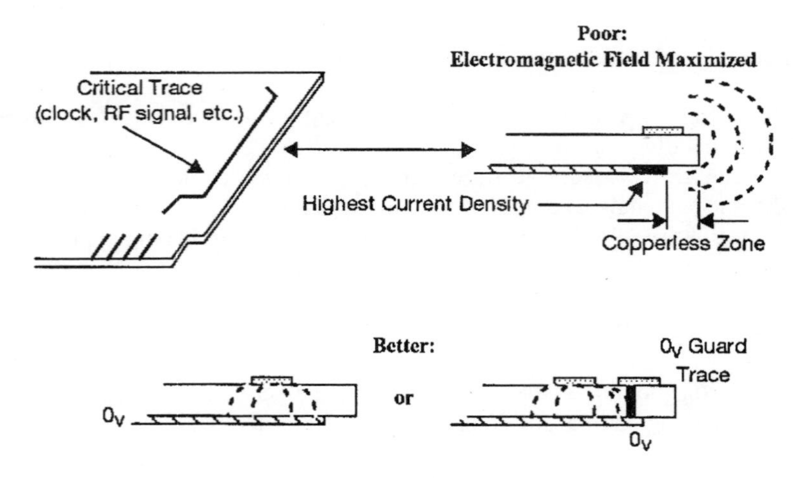

FIGURE 6.26 Problems with High-Speed Traces on Card Edges

- residues of clock harmonics with few tens of mV can appear on analog traces, out-of-band of the analog circuits, therefore not of concern for their performance. Nevertheless, the polluted analog signals can turn into fortuitous, efficient RFI radiators (see sec. 2.5).

- sensitive RF receiving circuits can be in the near field radiated by the digital portion of the PCB.

Example 6.2
Assume a μProcessor module incorporating a 300 MHz clock, radiating just 10dB below FCC class B, at 900 MHz. This gives:

46 dBμV/m - 10 dB = 36 dBμV/m at 3 meters

The Near /Far Field transition for 900 MHz is D(NF -FF) = 48 /F = 0.05m
Therefore, the 36 dBμV/m field @ 3m can be transposed at 5cm as :

E(5cm) = E(3m) + 20Log (3/0.05) = 36dBμV/m + 36dB = 72dBμV/m

On this same PCB, a cellular telephone receiving section is located, with its receiving antenna trace. The typical sensitivity of a 900MHz mobile phone is -100dBm (7dBμV), which can be translated into an equivalent field strength by using a typical antenna factor, ending in 34dBμV/m of minimum discernable field. Comparing with the field radiated by the processor section nearby:

72dBμV/m - 34dBμV/m = 38dB

The field received at 900MHz from the μP circuit is almost 100 times stronger than the RF signal sensitivity. This require drastic protection of the RF input traces.

Because of these risks, the PCB zoning (Fig.6.1) and DC distribution clean-up at the digital-to-analog frontier must be carefully applied. Since, typically, 40-50dB of decoupling are needed, simple capacitive filtering is generally not enough and 3-stage filter like the feed-thru or SMT “T” style must be provided (see Chap. 9). If an IC module (ASIC, DSP etc) performs both analog and digital functions, the dilemna arises as to where it should be located: if placed in the digital section there is risk of having few analog input/output traces in the middle of noisy logic circuits, and placed in the analog section, the risk is just the opposite. The best would be to have a module where all critical analog pins, including DC voltages and AN.Gnd would be on the same side, and all noisiest logic pins on the other side, allowing the module to be mounted exactly over the AN-DIG. frontier of the board. Unfortunately, this is not always possible. The solution in this case (Fig. 6.27) is to locate the AN-Dig module in the analog area, with its digital I/O traces, including Vcc, immersed between two wide digital ground traces, such as the digital portion of the IC becomes an “extension” of the PCB digital section.

6.4.7 Thin foil PCBs

The growing trend to miniaturized equipment (laptop PCs, notepads, cameras, cellular phones) has called for printed circuits which are thinner and offer more density than the traditional 1 - 1.6mm epoxy glass. This is the case of the RCC (resin-coated copper) where a 5μm (0.25 mil) copper foil is laminated with a 50μm resin. This allows for traces that are thinner than with the highest density class-5 PCB. Trace widths of 50-75μm (2-3 mils) are obtainable, with 50μm dia. micro-vias. Ultra-thin multilayers are made by stacking several RCC foils. The small thickness and absence of glass fibers make the precision drilling of vias by laser economically feasible, and blind vias are obtainable.
Considering the scaling factor, thin foil PCBs are not much different from regular multilayers, since the width /height ratios for traces above ground plane are comparable. Thus, self-inductance and crosstalk coefficients are similar. Major difference is with trace dc resistance, reaching 0.7Ω/cm for a 50μm wide, 5μm thick trace, such as resistive voltage drop can be an element of concern.

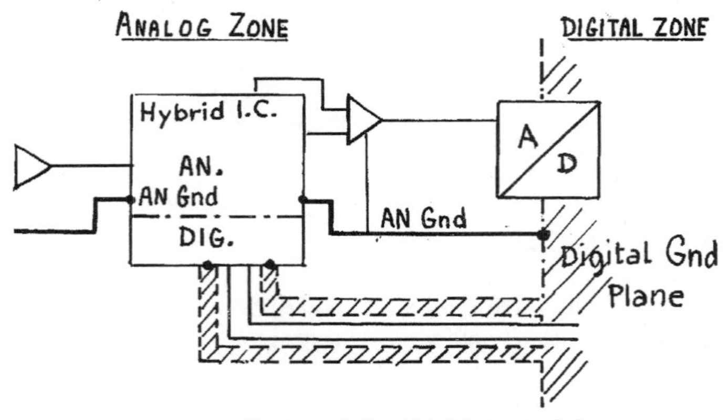

FIGURE. 6.27 Zoning trade-off with a Mix AN-DIG. Module

6.5 CROSSTALK

Crosstalk occurs when a wire or trace carrying fast signals runs parallel to another conductor. By mutual capacitance and inductance, the *culprit* conductor induces a certain percentage of its voltage into the *victim* conductor (see Fig. 6.28). Crosstalk increases with the proximity of the culprit and victim wires, increasing frequency (or decreasing rise times) and higher victim impedance. It also increases when the culprit and victim conductors are more distant from their return wire or plane. Crosstalk is defined in dB, as follows :

$$\text{Xtalk (dB)} = 20 \text{ Log } (V_{vict.} / V_{culp.}) \qquad (6.5)$$

Therefore, -20 dB of crosstalk means that for 1V of culprit voltage, a 0.1V amplitude will appear on the victim circuit. A more detailed analysis of crosstalk can be found in Ref. 2 and 10.

At first glance, crosstalk is primarily an internal, or functional, EMC concern. The designer will worry mostly about crosstalk causing a self-disturbance in his equipment. But in fact, crosstalk is an insidious and significant player in the generation of radiated EMI. High-speed clocks and HF circuits that are used only for internal functions may unintentionally couple into I/O lines by crosstalk, then radiate. As such, crosstalk reduction is a radiated EMI problem as well as an internal concern.

It has been calculated that to have a reasonable probability of staying below FCC (or CISPR) radiated levels, and using the I_{cm} criteria described in Section 4.3, the undesired harmonics induced internally on traces that later exit as I/O cables should not exceed certain levels. These levels are:

- for FCC Class B: 10 mV per harmonic, at frequencies > 30 MHz
- for FCC Class A: 30 mV per harmonic, at frequencies > 30 MHz
- for MIL-STD-461: RE02: 20 mV per harmonic, at frequencies > 30MHz

This is assuming a "bare bones" situation with:

- unshielded cables
- unfiltered I/O ports
- poorly balanced (10%) I/O links

To achieve this, with typical digital pulses of 3 to 5 V peak amplitudes, the internal couplings between culprit and victim must not exceed -40 dB (i.e., 1 percent) above 30 MHz. This is much more restrictive than what it would take to simply avoid self jamming.

6.5.1 Capacitive Crosstalk

Although both magnetic and capacitive crosstalk phenomena exist, the latter generally predominates in PCBs because of the high dielectric constant of epoxy. The simplified prediction table of Table. 6.1 give average values of capacitive crosstalk per cm length for few typical trace-to-trace configurations. For longer runs, the crosstalk increases accordingly.

For the equivalent circuit shown in Fig. 6.28, crosstalk expresses by:

$$\text{Xtalk (dB)} = 20 \text{ Log } \{(\omega R_v (C_{1\text{-}2}) / \sqrt{[\omega R_v (C_2+C_{1\text{-}2})^2 + 1]}\} \qquad (6.6)$$

The table 6.1 has been scaled in function the s/h ratio which we found more practical than d/h (center-to-center), which is sometimes used. Anyway, this d/h ratio has been shown in parenthesis, for reference. The value of C_2, trace to ground capacitance has been shown as it governs the maximum possible X_{talk}, given by :

$$\text{Xtalk (max)} = 20 \text{ Log } [C_{1\text{-}2} /(C_{1\text{-}2} + C_2)] \qquad (6.\text{-}7)$$

The last column on the right corresponds to a PCB without ground plane, such as no firm value can be given to C_2, other than an approximate default value of 0.2pF/cm,

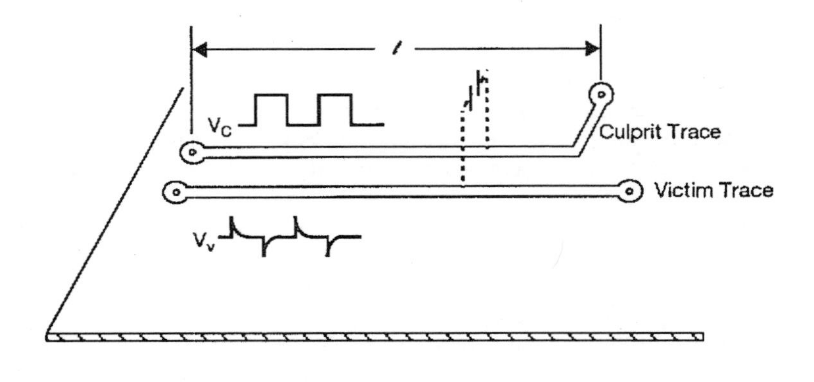

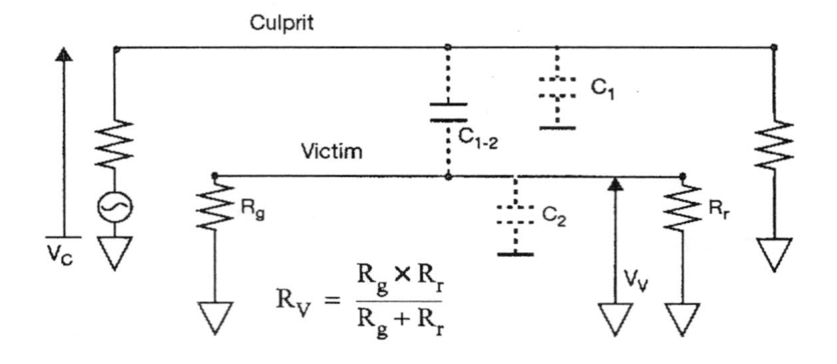

FIGURE 6.28 Basic Mechanism for Capacitive Crosstalk

assuming that the corresponding ground trace is farther away than the traces separation itself.

Crosstalk increases linearly with the length of the parallel run, as long as this one is small compared to the culprit wavelength. As length, or frequency, increase, a point is reached where crosstalk will not increase anymore. Since the victim voltage is of a general form V_c x R x C_{1-2} ω. cosωt, this maximum value at a given frequency is the sum of the cos. terms over a segment equal to half-wavelength. This corresponds to a weighted length of 0.7λ/2, corrected by $\sqrt{\varepsilon_r}$ for the wavespeed in epoxy. The limit is translated into a maximum coupling length of 5000cm /F_{MHz}.

Notice how C_{1-2} values are falling off rapidly with the separation distance, when there is a ground plane, and very weakly when there is no ground plane.

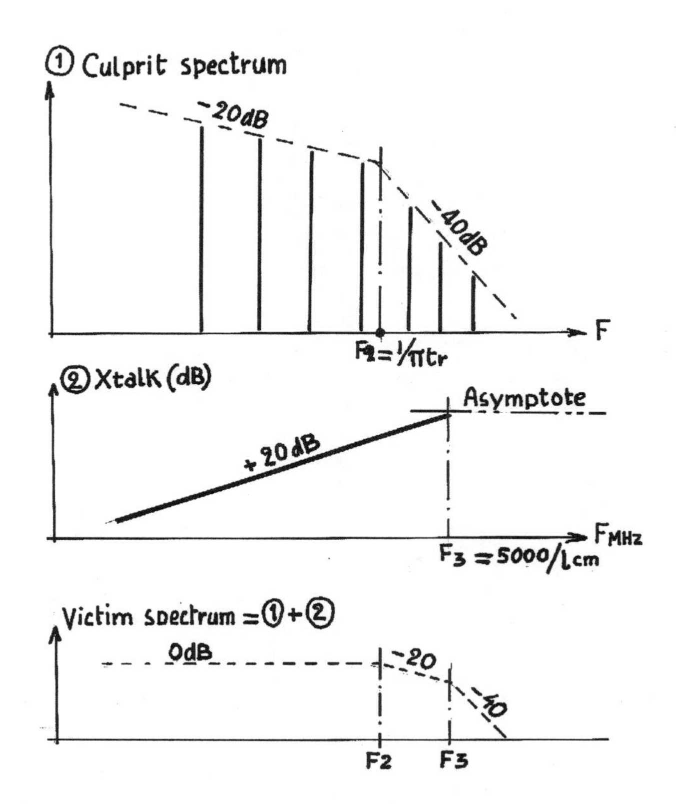

FIGURE 6.28 (Continued) Basic Mechanism for Capacitive Crosstalk

TABLE 6.1 Capacitive Crosstalk between PCB Traces, for Rv(total) = 100 Ω and 1 cm Length. For other values of length and Rv, apply the correction: 20 log [1 cm x Rv /100]. The clamp in dB on bottom line is the maximum possible crosstalk, ever. For buried traces, clamp is 10dB lower.

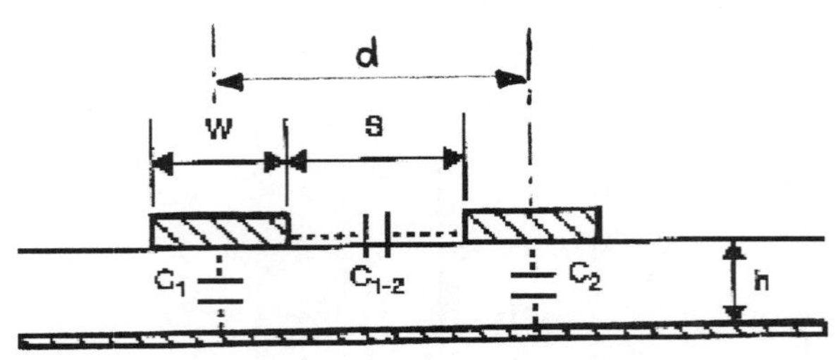

C2 Z0	w/h=3 1.6pF/cm 40Ω			w/h=1 0.8pF/cm 70Ω				w/h=0.3 0.5pF/cm 120Ω				w/h=0.3 - 0.6 No Gnd Plane			
s/h	10	3	1	10	3	1	0.3	10	3	1	0.3	10	3	1	0.3
corresp. d/h	(13)	(6)	(4)	(11)	(4)	(2)	(1.3)	(10.3)	(3.3)	(1.3)	(.6)	(10.4)	(3.4)	(1.4)	(.9)
C1-2 (pF/cm)	.016	.06	.16	.007	.04	.14	.25	.003	.03	.1	.2	.18	.22	.35	.5
Freq.															
1kHz	-160	-148	-140	-167	-152	-141	-136	-174	-154	-144	-138	-139	-137	-133	-130
3kHz	-150	-138	-130	-157	-142	-131	-126	-164	-144	-134	-128	-129	-127	-123	-120
10kHz	-140	-128	-120	-147	-132	-121	-116	-154	-134	-124	-118	-119	-117	-113	-110
30kHz	-130	-118	-110	-137	-122	-111	-106	-144	-124	-114	-108	-109	-107	-103	-100
100kHz	-120	-108	-100	-127	-112	-101	-96	-134	-114	-104	-98	-99	-97	-93	-90
300kHz	-110	-98	-90	-117	-102	-91	-86	-124	-104	-94	-88	-89	-87	-83	-80
1MHz	-100	-88	-80	-107	-92	-81	-76	-114	-94	-84	-78	-79	-77	-73	-70
3MHz	-90	-78	-70	-97	-82	-71	-66	-104	-84	-74	-68	-69	-67	-63	-60
10MHz	-80	-68	-60	-87	-72	-61	-66	-94	-74	-64	-58	-59	-57	-53	-50
30MHz	-70	-58	-50	-77	-62	-51	-56	-84	-64	-54	-48	-49	-47	-43	-40
100MHz	-60	-48	-40	-67	-52	-41	-46	-74	-54	-44	-38	-39	-37	-33	-30
300MHz	-50	-38	-30	-57	-42	-31	-36	-64	-44	-34	-28	-29	-27	-23	-20
1GHz	-40	-28	-20	-47	-32	-21	-26	-54	-34	-24	-18	-19	-17	-13	-10
Asymp.	-40	-29	-21	-40	-26	-16	-12	-44	-25	-16	-11	-7	-6	-4	-3

The procedure in using this simplified table and model is as follows:

1. Select (or interpolate) the cross-section geometry of the culprit/victim traces.
2. Define the culprit frequency (or frequencies). This may require a quick Fourier analysis of the culprit signal, based on the repetition frequency and rise time. Time-domain calculation of crosstalk is not recommended here, as we need to determine the harmonics of the coupled voltage.
3. Find the corresponding crosstalk per cm length.
4. Apply length correction = 20 log l_{cm} . Do not use above l_{max} = 5000 cm/F_{MHz}, corresponding to $\lambda/2$ maximum coupling length in epoxy dielectric.
5. Apply the following impedance correction if Z_{victim} (i.e., victim source and load resistances in parallel) is $\neq 100\ \Omega$:

   $$20 \log (Z_{victim} / 100\Omega)$$

   If the $\lambda/2$ limit has been passed, replace Z_v by the victim trace characteristic impedance Z_0, shown in the table.
6. At any frequency, whatever is the result of length + impedance corrections, the maximum possible crosstalk cannot exceed the clamp value shown at the bottom of the table.
7. If the victim trace is sided by two identical culprit traces, the two crosstalks are adding-up, i.e 6dB are added to the calculated value.
8. Although table is for surface traces, it can be extended to buried traces. The $C_{1\text{-}2}$ capacitance is approximately 0.6 - 1 time that of surface traces (≈ -3dB to 0dB difference). C_2 is exactly double, and Z_0 about $\sqrt{2}$ times lower, therefore the asymptote of maximum Xtalk is ≈ 10dB lower.

Example 6.3

Two traces have a 10cm parallel run on a single layer board (Fig. 6.21). The pertinent data are:

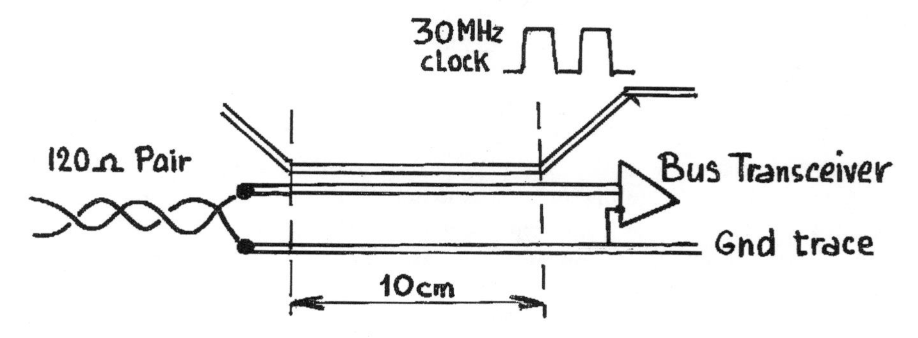

FIGURE 6.29 Configuration as in Crosstalk Example 6.

• w = 15 mils (0.38 mm)
• s = 15 mils (0.38 mm)
• Board thickness, h = 1.2mm
• All ground traces or ground areas are more than 1.2 mm (3 x w) away.
• The culprit is a 30 MHz clock with t_r = 2ns (i.e 2nd corner freq. is 160MHz).
• The receiving trace impedance consist in the input of a data line receiver (>> 1k Ω) in parallel with the characteristic impedance of an I/O pair, Z_0 = 125 Ω, so $Z_{victim} \approx$ 125 Ω.

Estimate the crosstalk at harmonic #3, 5, 9 frequencies, with respect to RF pollution of I/O cable.

The calculated values are as follows, for w/h = 0.3, s/h = 0.3 and no ground plane

Xtalk at:	90MHz	150MHz	270MHz
1. Table value, dB/cm	- 31	- 26	- 21
2. Length corr. for 10cm (10 cm is < λ/2)**	+ 20	+ 20	+ 20
3. Impedance corr.	+ 2	+ 2	+ 2
4. Total Xtalk	- 9	- 4	(+ 1)* clamp to - 3
5. V_{cuprit} (from Fourier spectrum)	0 dBV	- 6 dBV	- 16 dBV
6. V_{victim} = #4 + #5 :	- 9 dBV (0.35V)	- 10 dBV (0.3V)	- 19 dBV (0.11V)

Remarks: * Crosstalk cannot be positive nor exceed the maximum clamp value in the table.
** The 10cm length correction will stand up to F= 5000/10 = 500MHz. Above this frequency,the Xtalk coefficient will not increase any more.

Taking into account a DM-to-CM conversion of -10 dB for unbalanced link (Ref. Sect. 2.5 .) this is about 40 dB higher than the permissible voltage excitation for long I/O cables, if FCC Class B compliance is desired (30 dB too high if Class A is the aim). Therefore, a crosstalk reduction is in order, as discussed next. Otherwise, shielded I/O cables or filtered I/O connectors will be needed.
Notice that, as frequency increases, the harmonic amplitude decreases but crosstalk coefficient increases, resulting in approximately the same victim voltage. With periodic rectangular pulses, crosstalk-coupled voltages are ≈ constant with frequency and do not depend on the basic repetition rate, until the second-corner frequency, $1/\pi t_r$, is reached.

There are several ways to decrease capacitive crosstalk in PCBs:

1. Increase culprit-to-victim spacing; there is a limited latitude to do this. Without ground plane, it takes a tenfold increase in distance (changing from s/h = 1 to s/h = 10) to lose ≈ 6 dB of crosstalk. Most of the time, there is not enough empty room on a PCB to allow generous spacing. A 3 x increase in spacing will save about 3 dB of coupling, which is always good to get, but seldom sufficient. A variation of this consists of increasing s by simply reducing the width w (if possible), while

keeping the same center-to-center spacing, but this, too, will not bring a drastic reduction.

2. Decrease the length of the parallel run; the crosstalk will decrease in proportion.

3. Preferably, run the culprit and/or victim traces above the return plane or traces. The same ten-fold increase in separation will reduce crosstalk by 20 to 30dB, decreasing C_{1-2} and increasing C_1 or C_2. Multilayers, in this respect, offer more possibilities for crosstalk reduction.

4. Insert a grounded trace (guard trace) between the culprit and victim traces. This guard trace should be grounded at least at each end, and more frequently if possible. A grounded trace inserted at the half-way point reduces crosstalk, on typical 10 cm runs, by at least 20 dB up to 100 MHz. As an extension of the above, high speed culprit traces could be "buried" in a ground land at the same level. But never slot an existing ground plane to do this (see the previous Section 6.4).
   In lieu of a grounded lead, any trace that remains on the same PCB (does not exit) and interconnects low-impedance ($<<100\ \Omega$) circuits such as a DC line, bias voltage, status line at "0" level, etc.. can be used as a substitute guard trace.

In summary, the best way to avoid I/O line pollution from internal high-frequency circuits is to strictly ban any close parallel runs that do not have at least a ground trace between them.

6.5.2 Magnetic Crosstalk

For traces carrying larger currents, like those with characteistic impedance of 50Ω or less driven by bus transceivers, the magnetic contribution to crosstalk may not be negligible. (Ref. 10) At worst, magnetic and capacitive crosstalk may combine additively at the victim's near-end (the side which is close to the culrit generator side). Table 6.2 shows the mutual inductance value, M_{1-2}, in nH/cm between two adjacent traces. The victim coupled voltage is calculated by:

$$V_{vict} = M_{1-2}\ \omega\ I_{culp}$$

Table 6.2 Mutual inductance M_{1-2} (nH/cm) between PCB traces, vs trace separation s

Ratio s /h	10	3	1	0.3
A. Microstrip • • h -----------				
M(nH/cm) = $Ln[1 +(2s/h)^2]$	0.04	0.4	1.6	3.8
B. Stripline ------------ • • h ------------				
M(nH /cm = $Ln[1 +(h/s)^2]$	0.01	0.1	0.7	2.5

Notice, in the table, that buried traces (striplines) produce significantly less magnetic crosstalk than surface traces (microstrip). By contrast with trace-to-trace capacitances of Table 6.1 , it is not possible to give a typical value when there is no ground plane underneath, since M_{1-2} will depend enormously on the distance to the nearest ground trace. Mutual inductance is approximately independant of trace width. But the trace width plays a role in the trace self-inductance, L, which governs the characteristic impedance (hence, the maximum culprit current) and the maximum possible crosstalk: M_{1-2}/L.

Although a full treatment of magnetic crosstalk would outgrow the purpose of this book, we will describe one practical example.

Example 6.4: Magnetic Crosstalk in Multilayer PCB
Culprit circuit: internal bus, operating at 5V / 30MHz. The maximum load, for a long trace, is given by trace characteristic impedance of 50Ω.
Culprit - victim separation: 0.5mm Height above ground: 0.4mm
Microstrip configuration. Parallel length: 6cm.
Calculate magnetic crosstalk at the fundamental, Harm. #3 and #5 frequencies, since the victim trace is an I/O line, candidate to EMI radiation.

F:	30MHz	90MHz	150MHz
V_{culp} (from Fourier spectr.)	3V	1V	0.6V
$I = V/50\Omega$	60mA	20mA	12mA
M_{1-2}, nH/cm	1.3	1.3	1.3
M_{1-2} for 6cm	8	8	8
$V_{vict} = M \times 2\pi F \times I =$	90mV	90mV	90mV

Notice that, because the large value of I_{culp}, and in spite of the existence of a ground plane, this is a significant coupling, causing as much noise on victim trace than capacitive crosstalk in a similar geometry. Our limit objective for maximum crosstalk is exceeded.

This last example was assuming a very low impedance for the culprit trace, with a corresponding large current. While such condition is not very frequent in ordinary PCBs, it is extremely common in backplanes with high-speed parallel bus (see Chap.7)

6.6 IMPEDANCE MATCHING

With increasing frequencies (or shorter rise times) traces tend to become electrically "long", i.e. the propagation delay along this line is no longer small compared to the wavefront rise time. At this juncture, it becomes imperative to terminate the line in a matched load to avoid pulse ringing.
Here again, we have a case where something that could be a purely functional concern becomes an EMI radiation issue. Unterminated lines will exhibit oscillations that, in addition to causing serious functional problems, can as much as double emissions and crosstalk levels (see Fig. 6.30). There is an abundant and accurate supply of literature on the problem of transmission line matching, and the subject is beyond the scope of this book. Therefore, we will simply provide some basic guidelines.

An electromagnetic wave propagates at a speed:

$C = 3 \times 10^8$ m/s or 30 cm/ns in free space

$= 30\sqrt{\varepsilon_r}$ cm/ns in a medium with dielectric constant ε_r

For instance, with standard PCB, ε_r is 4 to 4.5, so propagation delay over a length, *l*, will be:

$$Td = l \times \sqrt{\varepsilon r} / 30 \qquad (6.8)$$

with T_d in ns, for *l* in cm.

This would be for lines entirely buried in epoxy. If the signal and return traces are both on the same surface (one side, single-layer board), part of the electromagnetic wave propagates in air instead, and the dielectric constant to use is approximately:

$$\varepsilon' r = (1 + \varepsilon r) / 2 \qquad (6.9)$$

When T_d exceeds $t_r/2$ (half the signal rise time) that is approximately when, in PCBs:

$$tr\ (ns) < 0.14 \times l(cm) \qquad (6.10)$$

the line must terminate in a resistance equal or closest as possible to the line characteristic impedance.

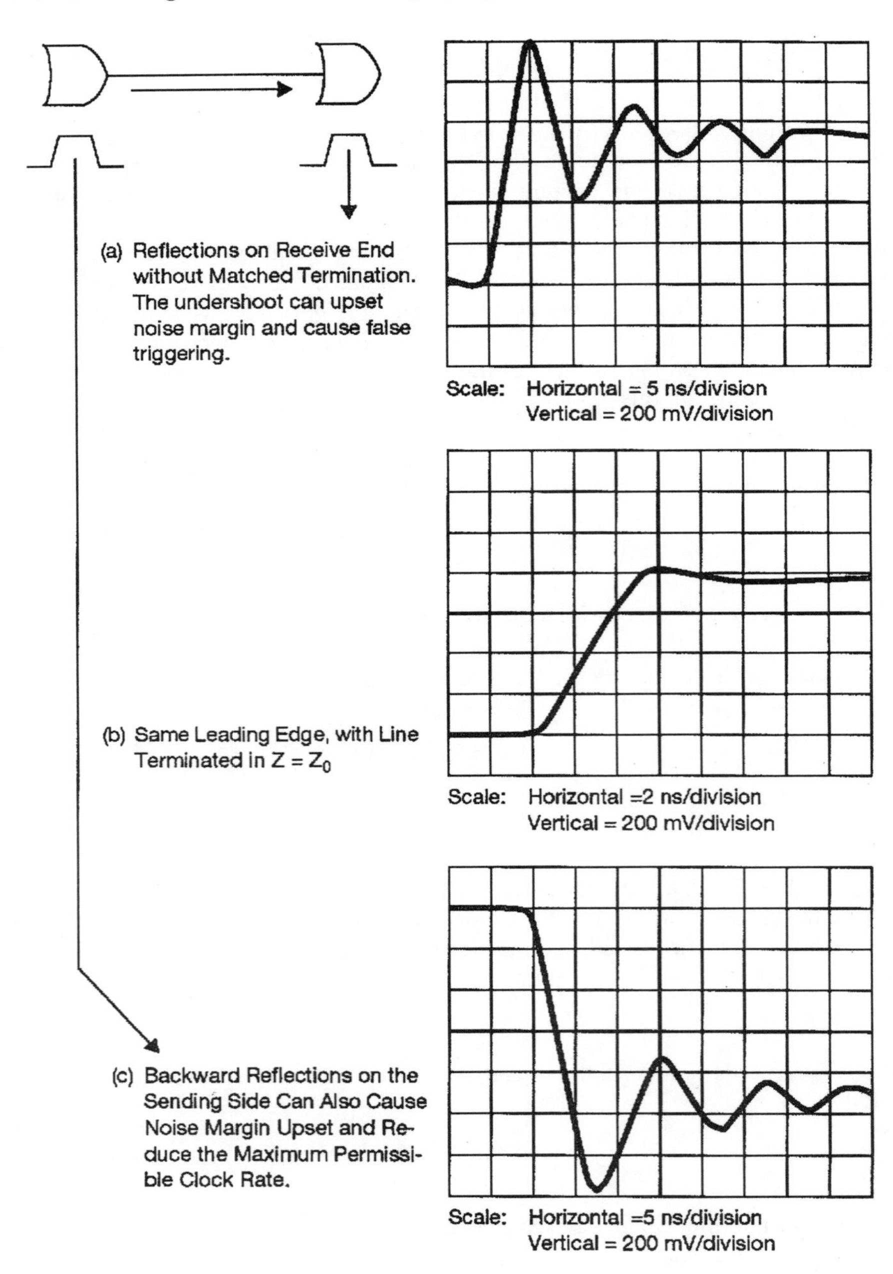

FIGURE 6.30 Example of Mismatch Problems with Fast Rise Times, Shown for ECL-10k and 20 cm Line (Source: Motorola ECL Application Notes, reprinted with permission)

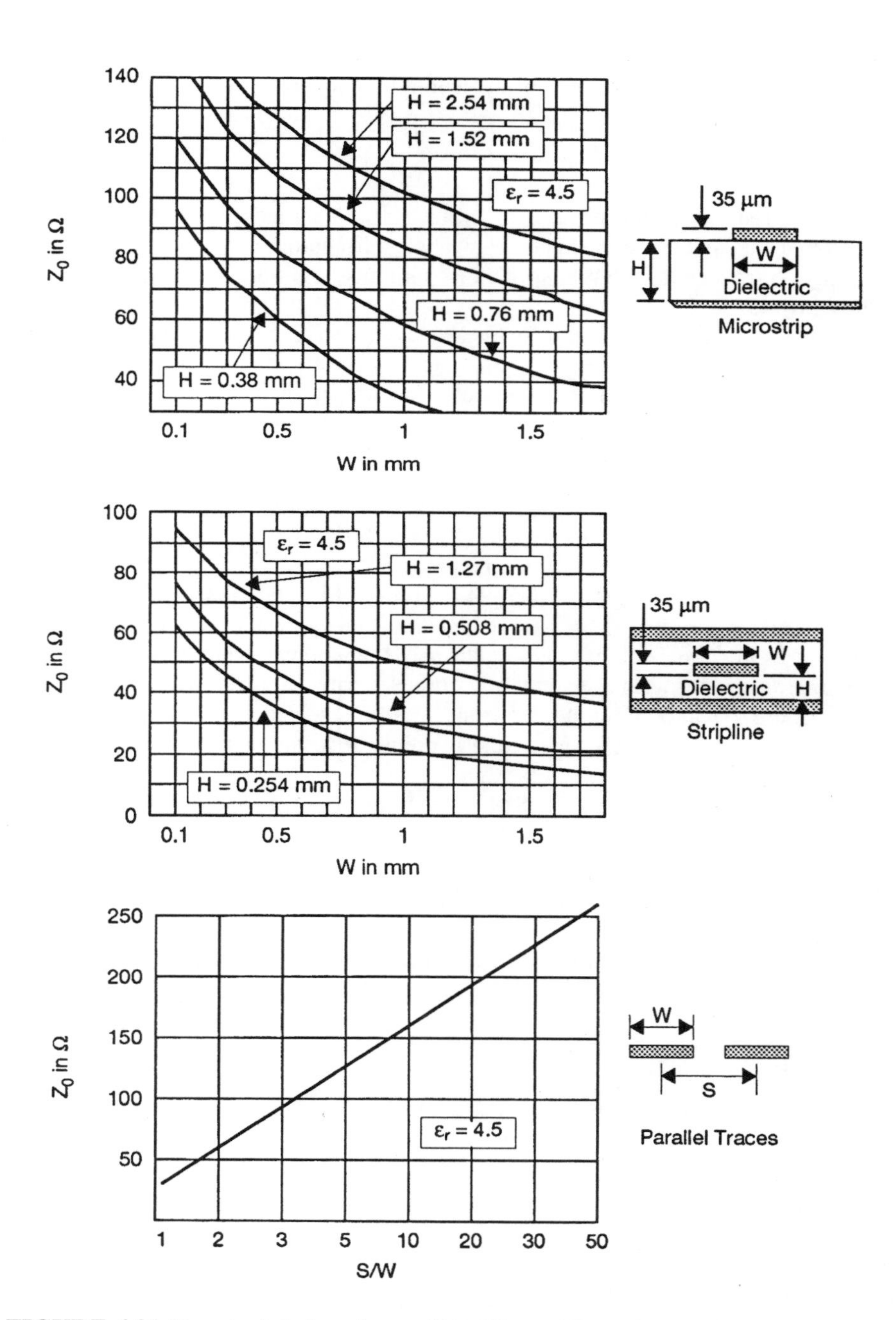

FIGURE 6.31 Characteristic Impedances (Z_0) of Several Trace Geometries

Correct termination can consist of:

It is easy to calculate the ringing overshoot with improper terminations, given that the reflection coefficient at the end of the length l loaded by Z_L is:

$$\rho = \frac{Z_L - Z_0}{Z_L + Z_0} \quad (6.11)$$

The characteristic impedances, Z_0, for typical trace geometries are shown in Fig. 6.23. Notice also that 90° corners create an abrupt discontinuity (locally, L and C are modified). To reduce the VSWR at this discontinuity, right-angle corners should be truncated by 45°.

- a resistance equal to Z_0 (with the drawback of wasting power during the DC plateau of digital pulses)
- a set of pull-up/pull-down resistor pairs, such that their parallel combination approximates Z_0, but only half the power is wasted since digital pulses will be either high or low
- a series RC network across the line end, such that termination R is only seen during the propagation delay, T_d (the only time where matching is needed). Compact sets of such RC networks in single in-line (SIP) packages are available, with values of C ranging from 10 to 500 pF.
- a series resistor on the source side. This allows one reflection to occur, but the reflected wave terminates on a matched end and does not bounce back. This solution has the drawback of affecting the drive capability of the source device.
- as an alternate to the above, a small few-100nH ferrite can take you out of the need for matching, by smoothing the rise front. The time constant (63% crossing point) is ≈ unchanged, but the 90% rise time is doubled, so the new rise time might no longer compete with T_d.
- a clamping diode at gate input. Several vendors incorporate this diode to limit signal overshoot.

6.7 CARD CONNECTOR PIN ASSIGNMENT

As far as EMI emissions are concerned, the card-level connector features that interest us are:

- crosstalk
- characteristic impedance
- contact impedance

Card connectors may become limiting factors in circuits operating at high frequencies.

6.7.1 Crosstalk

Although the coupling length may seem minuscule in a connector, the proximity of contacts embedded in a dielectric material can easily cause a contamination of I/O lines (our permanent concern) via internal high-frequency signals. For a 2.54 mm contact spacing, the 25 mm coupling length (typical of a male + female team)

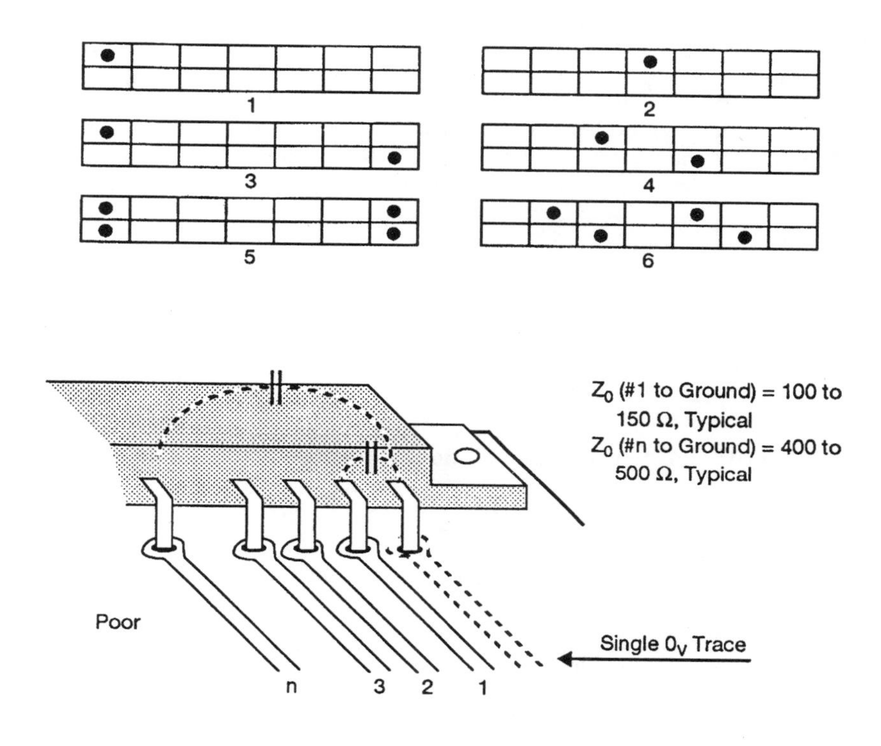

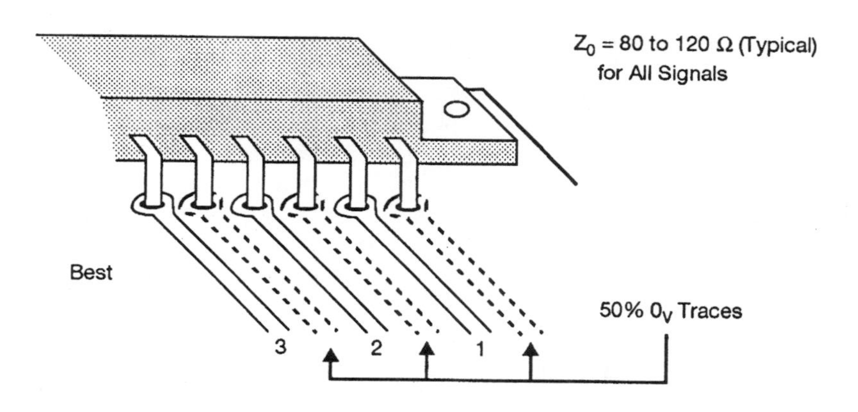

FIGURE 6.32 Reducing Crosstalk and Impedance Mismatch at Connector Crossing. At the top, performance improves from # 1 to # 6. The single 0v assignment labeled "poor" creates more crosstalk between the remote pins ("n"), and the odd spread of L and C parameters creates different Z0 and propagation delays.

typically creates -40 dB of capacitive crosstalk between adjacent pins at 100 MHz. For digital signals, this is already eating-up our entire allocation for such coupling (see Section 6.5).
If the culprit pins carry more power, such as an RF signal that should not leak outside the box, an even more substantial isolation may be needed. For instance if a video signal of several watts is leaving the card (before being driven into a coaxial line) by the same connector as ordinary, low-speed I/O lines, 60 or 80 dB isolation is required.

Crosstalk in connectors can be reduced by:

- careful segregation of culprit and victim pins (simply by continuing the segregation that was done at the trace level)
- interposition of ground pins (see Fig. 6.32)

6.7.2 Characteristic Impedance and Mismatch

When a line enters or leaves the PCB, the mating connector should, ideally, have the same characteristic impedance, Z_0, to avoid mismatch. Up to about 100 MHz, this is of little importance, but when the connector length starts to represent a non negligible fraction of the signal wavelength, connector mismatch will create reflections and ringing. Taking a connector pair whose total pin and body length represents about 3 cm, and given a velocity in connector insulation of 15 cm/ns, the connector trip represents a 0.2 ns delay. Taking as a rule of thumb $\approx$ 1/10 of the trip delay for an indiscernible mismatch, this connector must be matched when signal rise times get shorter than 2 ns (or frequencies above 150 MHz).
Part of the mismatch problem is the fact that different distances to the ground pin, as in Fig. 6.32, not only create a different Z_0 but also different propagation delays for each contact (speed and Z_0 are both related through the distributed L and C in the connector).
Here again, just like for crosstalk, a good precaution is the regular insertion of ground pins or, even better, alternating pins in a signal-0v-signal configuration for bandwidths exceeding 30 MHz. In some modern card connectors, a ground plane is built in the connector so that, except at the soldering pin level, all contacts are laid above or below a ground plane (see Fig. 6.34).

6.7.3 Contact Impedance

When daughter cards communicate via backplanes, or several PCBs are linked by a flat ribbon cable or flexprint, unexplained EMI problems may still occur, eventhough the designer has used ground planes for all of these circuits. These problems are usually traceable to the card's deck. The cause is that an insufficient number of Vcc or ground pins have been allocated at the edge connector. This creates an inductive bottleneck at each current transition, specially with high current returns from I/O drivers that share a single Gnd pin (see Fig. 6.33). Therefore, in terms of high frequencies, the card becomes "hot" with respect to the backplane. Here again, the answer is: allocate more, evenly spaced ground pins.

Example 6.5
Assume a PCB with eight simultaneous drivers at 50 mA each, with a 5 ns rise time, sharing only one ground pin. Taking 2.5 cm of average length for the edge connector pins plus the lead-in trace, we get:

$$\Delta V = L \times \Sigma\, dI/dt$$

$$= 2.5\text{cm} \times 10\ \text{nH/cm} \times (8 \times 50\ \text{mA}) / 5.10^{-9}\text{s} = 2\text{V}$$

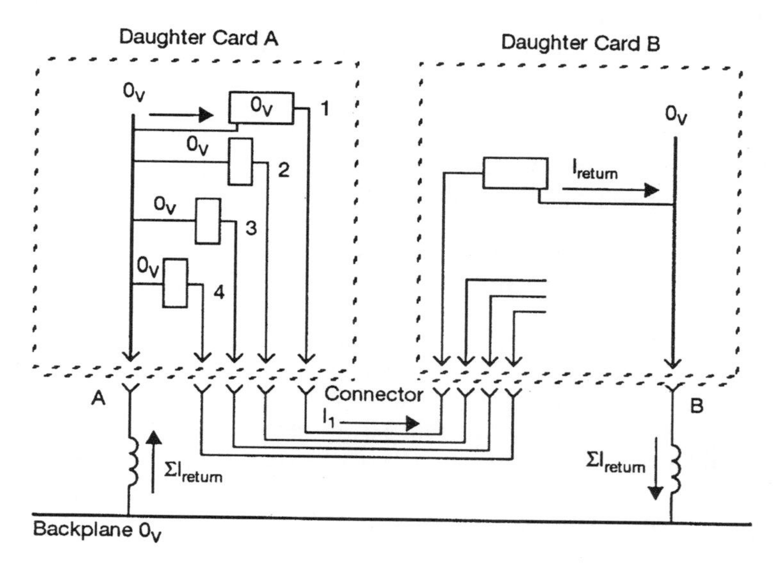

FIGURE 6.33 Common Impedance Noise at Connector Crossings

The problems described in Sections 6.6.1, 6.6.2 and 6.6.3 become critical above a few tens of megahertz. Therefore, with cards exchanging signals below this frequency range, the assignment scheme in Fig. 6.15 is generally sufficient. Notice, however, that the 10 percent of ground pins are evenly spread over the connector width. This spread is as important as the number of pins: the same number of ground pins stacked on one side would reduce contact DC resistance but would not reduce contact inductance to any practical extent. For higher speeds, the "one signal, one Gnd" scheme becomes mandatory. In some modern card connectors, at the soldering pin level, all contacts are running above or below a ground plane (see Fig 6.34).

6.8 GROUNDING OF 0V REFERENCE TO CHASSIS

The decision of whether to connect the PCB ground to the chassis must be a part of the general grounding strategy in the equipment, and furthermore, in the entire system, including its installation. Traditional low-frequency analog and audio practices have for long dictated that signal reference should be isolated from chassis (then from earthed structures) everywhere but at one connection point: the center of the star grounding system.
Toward the opposite end of the frequency spectrum, RF designers, especially in the VHF/UHF and microwave domains, have always practiced multipoint grounding. With this approach, the signal reference and box skin are connected as many times as possible.
Although with a single PCB one could conceivably handle signal speeds from DC to daylight, within the scope of this book, "radiated EMI" generally implies a high-frequency problem. Therefore, even if the equipment of concern handles both low-frequency analog and high-frequency RF or digital signals, the grounding strategy should be:

- For the low-frequency analog circuits, keep the Gnd isolated from the chassis to avoid ground loops, except at one point, which is generally the DC power supply 0V terminal.
- For the high-frequency circuits, connect the Gnd to chassis at the PCB level, as close as possible to the I/O cable entry points. This works because floating signal references become meaningless above a few megahertz; the PCB to chassis capacitance (typically few tens to hundreds of picofarads) tends to close the loop anyway and also creates parasitic resonances between the whole PCB and chassis.

Therefore, we may have several packaging situations:

a) All the electronics are mounted on one main PCB, and there is no backplane or mother board. In this situation, the logic ground is connected to the chassis within the I/O connector zone. This should be made via a short and wide strap (no wire), a set of spring contacts or with several screws. There is an interaction here with the grounding of the power supply module, as addressed in Chap. 9.

b) The active PCBs are pluggable daughter cards, mounted on a mother board. Because of the need to remove the cards easily, no permanent ground-to-chassis bond can be made at the daughter cards. In this case, the daughter card's ground plane is continued by the mother board ground plane through the connectors ground pins. Mother board ground plane is, in turn, grounded to chassis as in situation a). An other option exists, if card guides are used ; ground the daughter cards via metallic card guides, using spring contacts pressing on the card edge traces.

c) For various and sundry reasons, the end user or the purchasing authority explicitly requires that the signal reference be totally floated from the chassis. In such a case, the mother board should still have a chassis ground copper land, isolated from the 0V, to serve as common area for the HF decoupling capacitors (see Chapter 9, "I/O Connector Area").

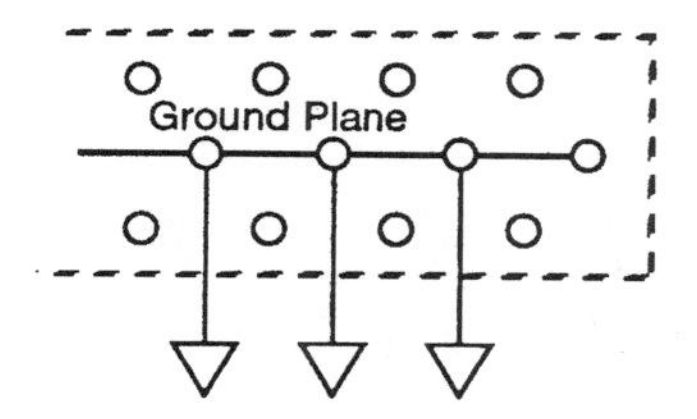

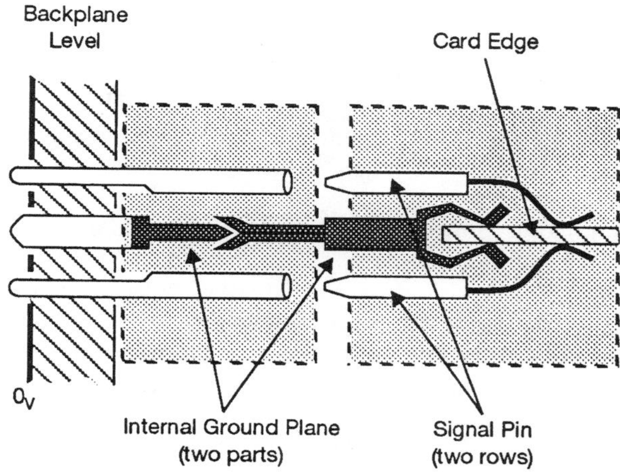

FIGURE 6.34 Special High-Speed Card Connector with Internal Ground Plane (Amphenol/Socapex High Density Connector)

6.9 EMC SOFTWARE TOOLS FOR PCB DESIGNS

With the endless trend for reducing the time-to-market, product engineers need a final version of their PCB which is both functionnally perfect and EMC compliant, within shorter time, and less and less delays for prototyping and iterations.
Ordinary design habits like placement and spacing rules, experience and flair may not be sufficient. We have already shown that a deterministic approach for predicting radiated EMI is achievable. When a PCB reaches a certain complexity, with more than several hundred nets, whose 25% are critical, a manual calculation becomes monumental. Computer simulation may be needed, and to avoid the tedious entry of all the layout dimensions and component parameters, it is best to have the simulation software teamed with the PCB routing software.

As early as 1983, a CAD program (#5300 software by Don White Consultants Inc.), was commercialy available. Using sets of friendly, interactive entries, and coarse trade-offs for averaging trace lengths, number of clock-carrying traces etc.. this precursor was displaying the radiated emission spectrum from a complete PCB. As of the year 2000, software tools are available, which are interfaçable with PCB autorouters like Mentor Graphics, Orcad, Visula, Cadence etc..

Several philosophies are used:

a) the EMC program is a rules-checker which verifies, after the layout is made, that a precise set of rules for traces separation, guard-traces, line resonances and mismatch overshoot, delays, X-Y tracking, Vcc decoupling, isolated copper lands etc... are not violated (e.g. “Design Adviser” by Zuken-Redac).

b) or, the EMC program is a field equations - solver, calculating the radiated field for each trace, combining them into a global field at the prescribed test distance.

c) instead of using typical waveforms derived from the specific logic technologies, the EMI program may extract the results from a signal analysis software which calculates first the exact signal waveforms, including the influence of Xtalk, mismatch etc. Then these actual signal characteristics are used to calculate the radiated field (ex: “Presto” by HDT, or “Quantic EMC”).

In all cases, the program provides a display of rules violation (a) or radiated field spectrum (b), with a list of the nets which are responsible.

As of the year 2000, the major critique regarding these software tools is that they are essentially post-layout analyzers, not interactive expert systems. Once the culprit traces have been spotted, one has to re-route them differently, which means a new run of the auto-router will have to move dozens of other traces to accomodate the relocation. A new PCB layout will result, which may solve the original problem but uncover a few new culprit traces etc.. The process is usually a converging one, but is extremely hungry in CPU time. For instance, a typical field-solver will take one hour on a Pentium 450 MHz dual-processor to plot the radiation from a circuit with one single IC, at one frequency.
Therefore, the next step toward a really helpful software would be an EMC package which automatically guides the auto-router into a successful “virtual PCB prototype”, at the first pass.

6.10 SUMMARY OF RADIATION CONTROL AT PCB LEVEL

Reducing emissions really starts at the board level. Briefly, the proper approaches are:

Common rules for ALL Boards:

1. Decouple Vcc for every module (or every two modules) with 10 nF to 47 nF capacitors. Use ceramic and, if possible, leadless components. For RAM, microprocessors and pin grid arrays, use 220 to 470 nF. Decouple again Vcc at the edge connector with 1 to 10µF capacitor.

2. Keep critical traces (master clocks, divided clocks with the same fast rise etc.) away from board edges. For clocks >10MHz, border them with ground traces alongside.

3. At edge connectors, allocate about every tenth pin to Gnd. Increase to every four or every other pin if impedance matching is required.

4. Check for crosstalk on long parallel runs. If I/O traces get more than 10 to 20 mV per discrete harmonic, increase trace spacing or add a guard trace.

5. Locate I/O drivers & receivers away from the highest frequency sources, closest to their respective I/O connector pins.

6. Provide a guard ring around board edge, tied frequently to the ground net, or plane.

7. Where HC or AC is used, and unless load-end matching is required, add a small resistor (47 to 330Ω) at the gates output.

8. Force necessary changes in auto-router layout, to clear EMC rules violations.

Addition for single-layer (1 or 2-sided) boards:

1. Apply one of the following practices (listed from "fair" to "best")
a) Distribute Vcc and Gnd by large traces > 1 mm, running side by side or one above the other. Install ICs above Vcc-Gnd pairs, with their longest dimension in the trace direction.
b) Landfill open areas with ground. Viewed in transparency, the board should be as opaque as possible.
c) Implement a grid pattern of Vcc & Gnd traces.
d) Consider adding distribution by flat busses.
e) Devote one side of the board to ground plane.

Addition, for multilayer boards:

1. Multilayers reduce emissions by reducing signal-to-ground loop size. Do not demolish the concept by allowing slots or overlapping holes.
2. Beware of crosstalk when using inside signal layers : stagger, or force perpendicular routings.
3. Decouple modules as we did for single-layer boards.
4. Do not neglect the "guard ring and guard traces rule", eventhough there is a ground plane. When guard traces are run along a hot trace which is changing level (e.g. switching from layer 1 to layer 4), provide a ground via hole for the guard trace, close to the signal through-hole.

Chapter 7

Emission Control in Mother Boards and Backplanes

Unless the equipment is of small dimensions, with all components housed on a single card, the design will probably include a mother board. Although some mother boards contain active devices (bulky discrete components, power supplies, etc.), their major function is to effect interconnections between the daughter cards and with the I/O lines.

One problem with mother boards is that dimensions are fairly large; therefore all noise mechanisms are aggravated by one order of magnitude. For instance:

1. Because long parallel runs (highways) exist from one card location to another, they enhance crosstalk, which is a lesser problem with smaller PCBs.
2. Lines become electrically long and may require impedance matching and other EMI reduction techniques.
3. Because backplanes may carry hundreds of interconnect lines that are in a switching state during any given strobe gate, the propensity for radiation is severe.

A good place to start, before all PCB layouts are frozen, is to complete the connector pin assignment and trace routes at the mother board. By doing this first, undesireable proximities will be avoided, and the stage will be naturally set for a good layout of the daughter boards. For example, the designer should organize the runs by families:

- Internal-only digital runs (daughter to daughter)
- Digital I/O runs
- Analog (low level)
- Analog (high level) and Power or Analog Video

  etc ..

Each family is separated from the next ones by ground traces or land, or run on different layers with a Gnd or Vcc plane in between.

Completing these rules,

1. No high-speed traces (e.g., clock, LSB, video) may run close to sensitive traces (analog, alarm, reset) or to wires running to I/O connectors (interfacing to the outside world). This will prevent a single clock wire to contaminate many other wires via crosstalk.
2. No high-speed clocks or data wires should run without a ground trace alongside (see Fig. 7.1).
3. Approximately every tenth connector pin should be a ground pin. For lines which must respect impedance matching, this rule is re-inforced to one ground pin every four signal pins, or ultimately one for each signal pin (for.inst. for parallel bus with rise times $\leqslant$ 3ns).
4. The +Vcc distribution must run close to the ground traces or plane.

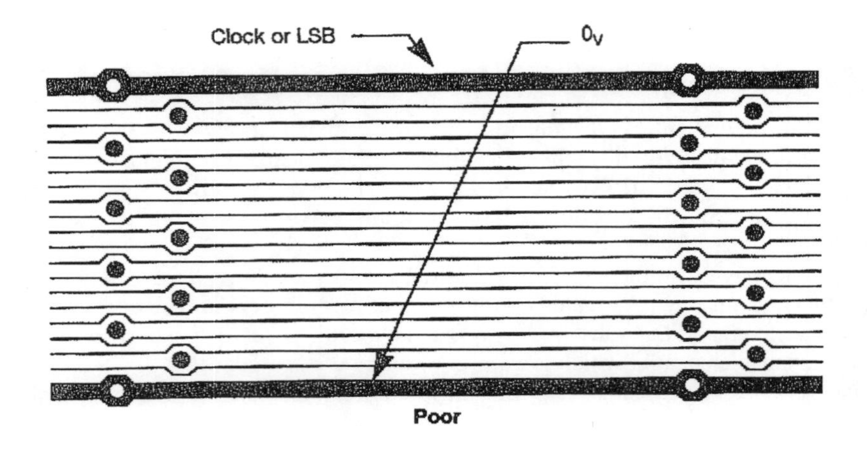

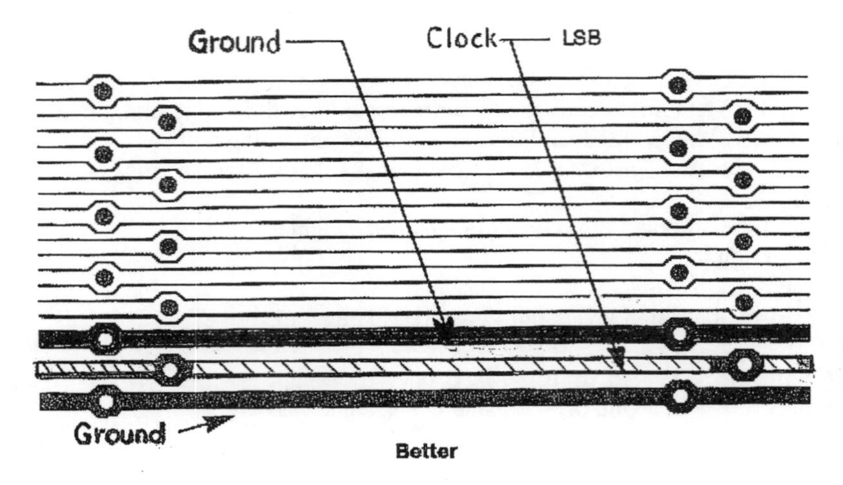

FIGURE 7.1 Reducing High-Speed Line Loop Areas in Mother Boards

7.1 WIRE-WRAPPED BACKPLANES

Although wire-wrapped designs are being phased out, particularly for active PC boards, this is still a popular technique for backplanes in such cases as (1) limited production runs and (2) equipment that must be custom tailored to the buyer's requirements. We do not consider here the cases of prototypes and breadboarding because such equipment does not need to comply with EMI radiation limits.

Wire wrapping is evidently more prone to HF radiation because of the larger size of the wire-to-ground loops.

To minimize this problem, the following procedures are recommended:

- Organize the wiring list so that the longer wires are mounted first. That way, they have a good chance to be close to the ground plane. (We assume that there is a ground plane.)
- Connect the shorter wires last.
- Do not try to lay the wires in an X-Y pattern; rather, run them via the shortest path. These random directions will reduce radiation and crosstalk.
- Over the signal wires' wrapping, install an X-Y grid of Gnd wires, interconnecting the 0V pins of the backplane/daughter card connectors. (As explained in Chapter 6, these should include one Gnd wire for every tenth pin.)

7.2 SINGLE OR MULTILAYER MOTHER BOARDS WITH Vcc AND GROUND PLANES

In backplanes, the general flow of signal traces being parallel, it is generally easy to provide full copper planes without obstructing signal routing. In this case, accessibility to traces is not crucial, so it is feasible to have the Vcc and Gnd planes on the exterior and the signal layers in between, thereby reducing radiation. The constraint mentioned with regard to multilayers PCBs remains, i.e., signal layers, should not be stacked on top of each other without a ground or power plane in between. If this is not feasible, the signal traces in two stacked layers should be staggered such as they never run on top of each other.

7.3 CROSSTALK AND IMPEDANCE MATCHING

One reason for advising the designer to make connector pin assignments at the mother board first was that crosstalk, radiation, and other EMI aspects are exacerbated by the "highway" nature of the signal flow. If this is done properly, the clean distribution of mother board traces will naturally continue via the connectors at the daughter card entries. The designer must consider crosstalk and impedance matching at both levels, as described below.

7.3.1 Crosstalk

Crosstalk is a higher risk due to the long parallel runs. All the crosstalk aspects discussed in Section 6.5 are aggravated here, so the crosstalk budget must incorporate the noise picked up by I/O lines during their trip to and from the mother board. However, chances are good that the wavelength limitation will be reached and, therefore, that the full trace length need not be entered in crosstalk estimation. With a 2ns rise time, for instance, maximum crosstalk is reached after about 30 to 35 cm of trace length, and this must also incorporate the corresponding daughter card and connector length.

When the culprit or victim is a differential signal (using two traces), an interesting remedy to crosstalk in the backplane would be to twist the traces (culprit or victim, but not both). Provided an even number of loops are created, it takes only a few twists to create a significant reduction in crosstalk and radiation. Twisting requires the traces to cross and a jumper /via-hole at each crossing, so it complicates board fabrication and is seldom used. The solution is valid, though, and efficient.

The remarks made in Chap. 6 regarding capacitive crosstalk being the dominant mode in PCB may not stand here. Backplanes are often forcing low characteristic impedances (see next section), causing significant current loops, hence magnetic crosstalk, in addition to the capacitive one. Inductive and capacitive crosstalk are additive at the near-end side (the end which is near the culprit signal source), and subtractive at the far-end. Thus, the near-end crosstalk, sometimes termed "backward" can be twice what would be due to capacitive or magnetic couplings taken separately.

7.3.2 Impedance Matching

Impedance matching also is a more frequent requirement for backplanes than for daughter cards, due to the longer line length. An additional problem exists in mother boards, caused by the addition of several, periodic lumped capacitances (see Fig. 7.2) corresponding to:

• the trace enlargement at the connector via holes, where the trace enlarges into a circle

• the signal-to-ground capacitance at each connector level

The result is a decrease in the actual value of Z_0, as well as the signal speed. It not rare in multilayer backplanes to see the calculated Z_0, typically 50 to 70 Ω, actually dropping to 25 or 40 Ω. This means more current, hence more radiation.

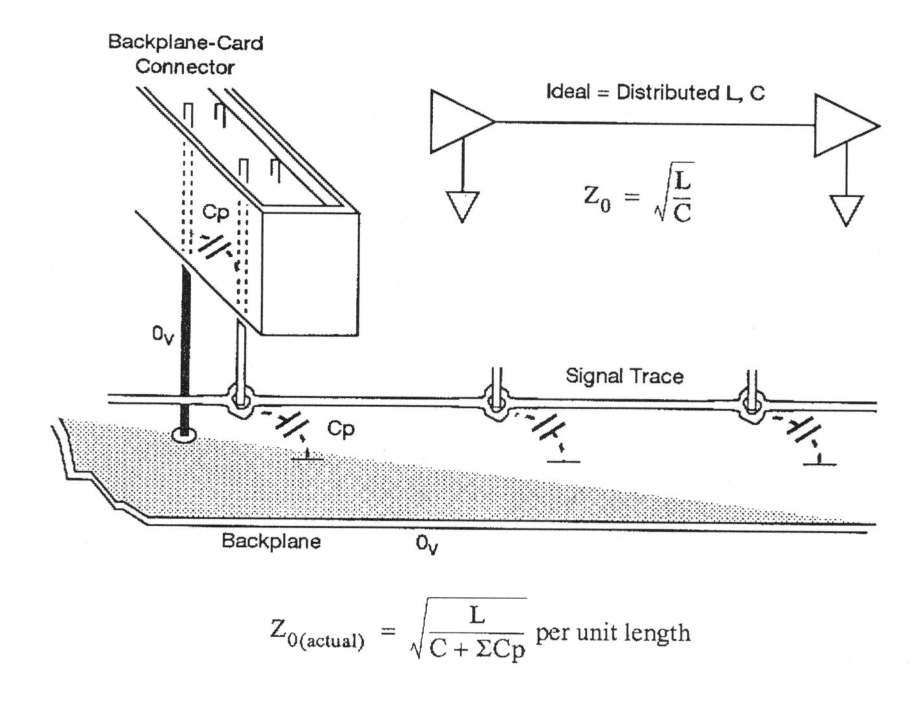

$$Z_{0(actual)} = \sqrt{\frac{L}{C + \Sigma Cp}} \text{ per unit length}$$

FIGURE 7.2 Actual vs. Ideal Characteristic Impedance in Mother Boards

7.4 CONNECTOR AREAS AT BACKPLANE INTERFACES

In a large equipment, the mother board typically is fitted with end connectors for interfacing with other motherboards or the rest of the machine. This is accomplished via flat ribbon cables or other means. The I/O connector area must continue the PCB-to-motherboard, noise-free concept. If high speeds are involved, the connector areas should respect impedance matching considerations, as do the printed traces and cables. This means that alternate signal-Gnd pins may have to be provided at the end connector to avoid discontinuities in characteristic impedance.

In any case, an extension of the board ground traces or plane underneath the connector area is recommended (see Fig. 7.3). This allows the most direct connection of all the I/O signal ground returns and makes it easier to decouple directly the noisy lines at the connector level, using discrete or planar capacitors.

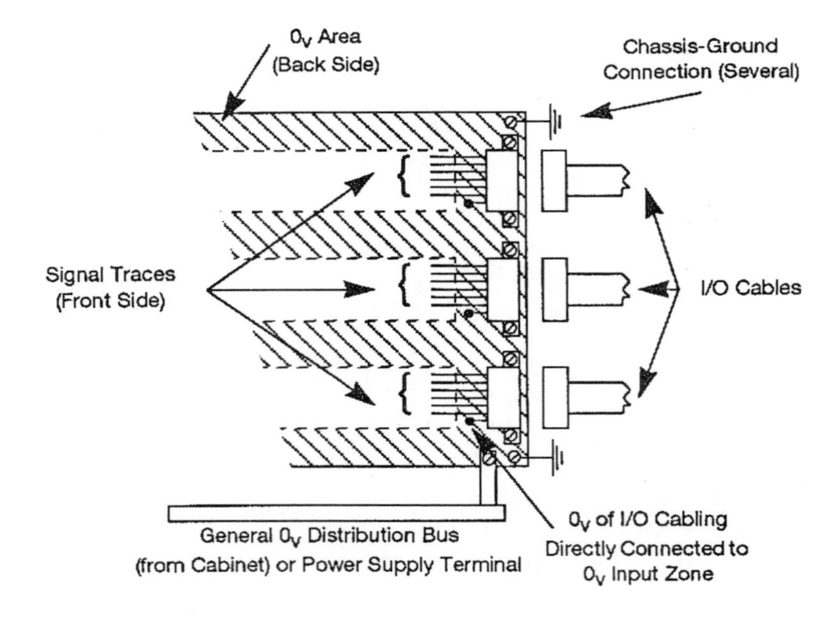

FIGURE 7.3 Connector Area at the Mother Board (or Single Board) Interface with External Cables

7.5 INCREASED RADIATION IN CONNECTORS AREAS

With high-density daughter card connectors, like the DIN #41612 with 4 or 5 rows of contacts, Fig. 7.4 shows the risk of increased radiation simply caused by the connector un-masked loops. For the signal going out on pos. #5 and returning via pos.#1, the radiating loop A-B-C-D can represent easily an area of 4cm^2, which requires only 3mA of loop current, at 100MHz, to exceed class B limit. Such loop exist regardless of the fact that both the daughter and mother boards may have a ground plane. The only way to reduce it is to use a male-female connector system incorporating an internal ground plane (see Fig. 6.32), or add more ground pins close to each signal pin.

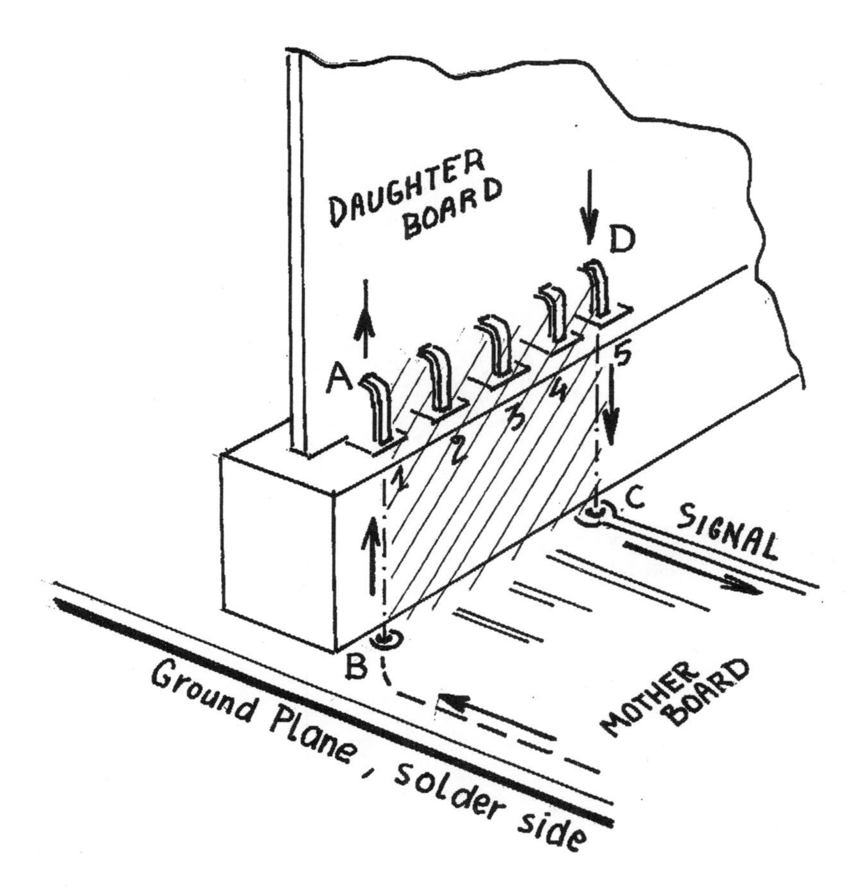

FIGURE 7.4 Unmasked Radiating Loops caused by Backplane Connectors

Chapter 8

Controlling Radiated Fields from Switch-Mode Power Supplies

Since their introduction at the end of the 1960s, switch-mode power supplies (SMPS) have become progressively popular, such as the vast majority of today's electronic equipment uses this type of regulator. With regard to EMI, SMPS have always been a serious concern, primarily because of their conducted emissions but also, to a lesser extent, radiated emissions. The first aspect has been covered extensively in Ref. 26, by J. Fluke. Therefore, we will concentrate on radiated field generation and suppression.

There is no official regulation covering power supply radiated EMI levels, assuming that the device is not sold as a stand-alone item. However, more and more power supply vendors incorporate filtering and shielding in their product to lighten the burden at the host machine level. In addition to the issue of specification compliance, SMPSs can be a source of internal EMI if sensitive circuitry is located nearby. So, depending on the design strategy, an equipment designer will have to deal with a noncompliant, homemade power supply, or a compliant, commercially built one. Assuming the worst, this chapter will address radiation control from the earliest level.

8.1 BASIC RADIATING SOURCES

Figure 8.1 shows the essential radiating sources or circuits in a simplified, one transistor SMPS:

1. The primary loop comprises the transformer (or storage inductor, in an SMPS with no isolation), the switching transistor(s) and the primary capacitor. This loop carries pulsed HF current and is generally the major radiator. Its radiated field can be computed quite accurately from its current spectrum, and the formulas in Chapter 2 for differential-mode loops.

2. The secondary loop comprises the transformer secondary, the rectifiers and filter capacitors (and, preferably, inductors). This loop carries the rectified (but not yet smoothed) current. Current is generally higher. The radiation can be calculated as in 1.

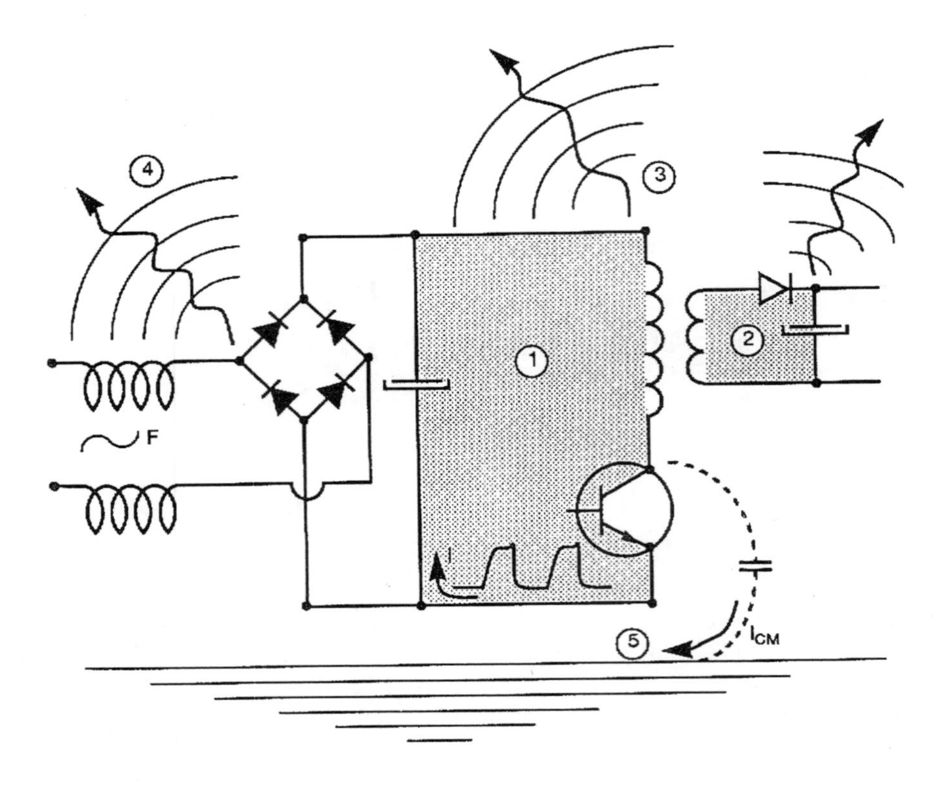

FIGURE 8.1 Principal Radiating Elements in a Switch-Mode Power Supply

3.The transformer (or switched inductor) itself tends to radiate during the currenι peaks where it is brought to saturation. The leakage field can be significant in close proximity to the transformer.
4. The filter inductors, which ironically have the mission of suppressing conducted EMI, in turn tend to convert some of the reactive power into radiated EMI unless packaged carefully.

To a lesser extent, the CM loop due to the stray capacitance to ground of switching transistors, GTOs and diodes is also a candidate radiator. Such a loop can carry CM currents in large areas of the SMPS and host machine. However, this circuit is so strongly filtered to control conducted EMI up to 30 MHz (civilian) or 50 MHz (military) that its radiated field generally is not a problem (exceptions acknowledged).

Example 8.1. Radiation from an Elementary One-Transistor SMPS

For the simplified 1kW /25 kHz SMPS of Fig. 8.2, calculate the radiated H field,
a) at 1 m distance, against the VDE 871 class B limits. Although German VDE 871 is no longer legally required since the EC Directive has replaced all national

requirements, there are still procurement specs which require VDE 871magnetic limit. The reason is that it is the only widespread civilian limit addressing H field.

b) repeat for a 7 cm distance against MIL-Std 461C, RE0 1.

Solution:

First, we need to calculate the peak current in the primary switching loop, per the following formula:

$$I_p = \frac{P_{secondary}}{V_{primary}\sqrt{2} \times \text{efficiency}} \times \frac{T}{\tau} = \frac{10^3 w}{120\ \sqrt{2} \times 0.8} \times \frac{40\ \mu s}{18\ \mu s} = 16\ A$$

Then, we calculate the characteristic frequencies of the Fourier envelope:

$F_1 = 1/\pi\tau = 17$ kHz

$F_2 = 1/\pi t_r = 2$ MHz

The fundamental amplitude will be:

$2A/\pi = 0.64 \times 16\ A \approx 10\ A$

At 1 m distance, the near-far transition frequency is:

$F_{N\text{-}F} = 48/1 = 48$ MHz

Therefore, the entire spectrum of interest is in the near field.
From this point, the radiated H field is calculated, based on Eq. (2.19) or Fig. 2.7, for a 1 A-cm^2 magnetic moment. The following table shows the calculated values for a range of frequencies.

Frequency	*25 kHz*	*50 kHz*	*100 kHz*	*300 kHz*	*2 MHz*	*10 MHz*
I_p, in dBA	20	14	8	–2	–18	–30*
H_0, in dBμA/m (1A-cm^2) at 1 m	17	17	17	17	17	18
Area correction (dBcm^2)	40	40	40	40	40	40
H_{total} in dBμA/m	77	71	65	55	39	28
VDE 871 (translated to 1 m)	62	56	50	40	28	20
Δ dB	15	15	15	15	11	8

*Spectrum hump due to overshoot

Notice that this is a NB type of measurement, since the 10kHz receiver bandwidth is smaller than the 25kHz SMPS frequency. So, the actual read-out will be in rms value, i.e. $\sqrt{2}$ (3dB) less than the table values.

This power supply (or the equipment using it) will be off spec up to a few megahertz if no shielding or radiation control is effected. A size reduction of 10 times (20 dB) is necessary to be within specs, with a small (5 dB) margin. This means that the effective radiating area of the primary switching loop should be brought down to 10 cm^2. This is feasible but requires a drastic modification of the circuit layout.

For the MIL-STD-461, RE0 1 test, the distance being 7 cm. Eq. (2.19) gives:

$$H\mu A/m = 7.96 \text{ Iamp x } Acm^2 / D^3m = 23.10^3 \ \mu A/m$$
$$= 88 \text{ dB}\mu A/m, \text{ for } 1 \text{ A-cm}^2$$

The calculated values are shown in the following table:

Frequency	*25 kHz*	*50 kHz*
I_p, in dBA	20	14
H_0, in dBμA/m (1A-cm^2) at 7 cm	88	88
Area correction (dBcm^2)	40	40
H_{total} in dBμA/m	148	142
RE01, in dBμA/m	56	54
Δ dB	92	88

RE01 compliance is required only in specific cases of very low magnetic ambient requirements; e.g., anti-submarine warfare environments. For such applications, the power supply needs about 90 dB of H field reduction—a huge suppression requirement. By repackaging the circuit on the PC board, the radiating area can be brought down to a few square centimeters; i.e., about a 30 dB reduction. The rest has to be provided by a thick shield over the SMPS module, plus some additional shielding of the host machine.

8.2 EFFECT OF ACTUAL CURRENT WAVEFORMS

In general, SMPSs use square waves with steep slopes to minimize transistor power dissipation, thus increasing efficiency. This may not be an optimized choice overall: if maximum efficiency is the only driving parameter in the design, it will result in a proliferation of harmonics that will require additional EMI-suppression components. It is not exceptional to see SMPSs whose active size shrinks by "hot rod" design, but whose overall size and weight increase due to the many filtering components (especially magnetics) that must be added to make up for the additional noise.

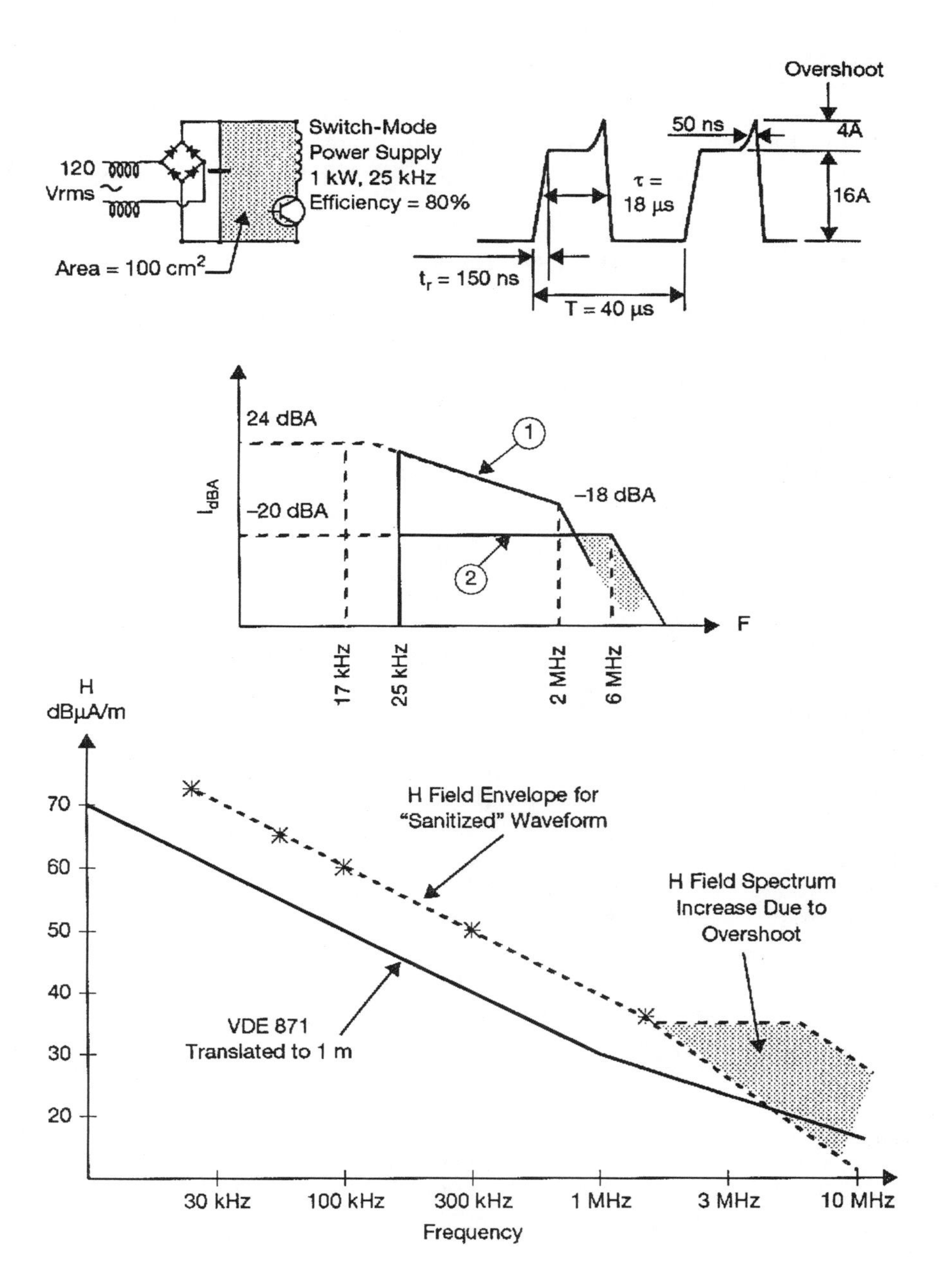

FIGURE 8.2 Power Supply Radiation of Example 8.1. Current spectrum I corresponds to straight trapezoidal waveform, without overshoot.

If some sacrifice in efficiency is acceptable, pulse shaping and corner rounding can strongly limit the EMI spectrum above the megahertz region. A good example of this is evident in the resonant converter, where the basic switched wave form approaches a sine wave, with a much more limited spectrum (see Fig. 8.3). Applications using a buck converter are also generating less magnetic field, because the primary current has an isoceles triangular shape, whose spectrum decreases by $1/F^2$.

8.3 PACKAGING AND CIRCUIT LAYOUT

Up to about 100 W of secondary power, SMPSs are generally small enough to be packaged on a PC board or a small, compact module. Above this range, the SMPS is generally a hybrid of printed and hard-wired assemblies, housed in an open or six-sided metal frame. The general principles discussed in Chapter 2 apply here. One of the driving ideas is to reduce, by all practical means, the magnetic moments (current x area). The following methodology is recommended:

1. On your schematic or wiring diagram, highlight in color all the connecting wires that carry changing (not necessarily alternating) currents (see Fig. 8.4).
2. When visualizing loop areas, consider them in the three dimensions.
3. Identify the high dI/dt paths and minimize their inductance.
4. Watch for "hidden" radiating loops such as snubbers (small areas, but high peak currents and wide spectrum).
5. On all identified loops, keep the area to minimum, and always try to "pair" a trace or wire with its return; this will decrease:
    a. emissions from high dI /dt circuitry
    b. the susceptibility of sensing circuitry

The best way to accomplish this is by using strip-line style PC board wiring.

6. For connecting transformers and bulky components to the PCB, use flat conductors. The best approach is to have transformer leads coming out as flat straps for direct PCB mounting (see Fig. 8.5).
7. When two components or circuits are carrying equal currents at the same time, orient them such that their magnetic fields are in 180° out of phase. This reduces (ideally, nullifies) their net magnetic moment (see Fig. 8.6).

8.3.1 Magnetic Leakages from Transformers and Chokes

Magnetic components are optimized for best efficiency and minimum heat, not for minimum EMI. The designer tries to achieve the best usage factor of his magnetic core by working close to (but not in) saturation. Leakage inductance is often a functional parameter and cannot be controlled just for the sake of EMC.

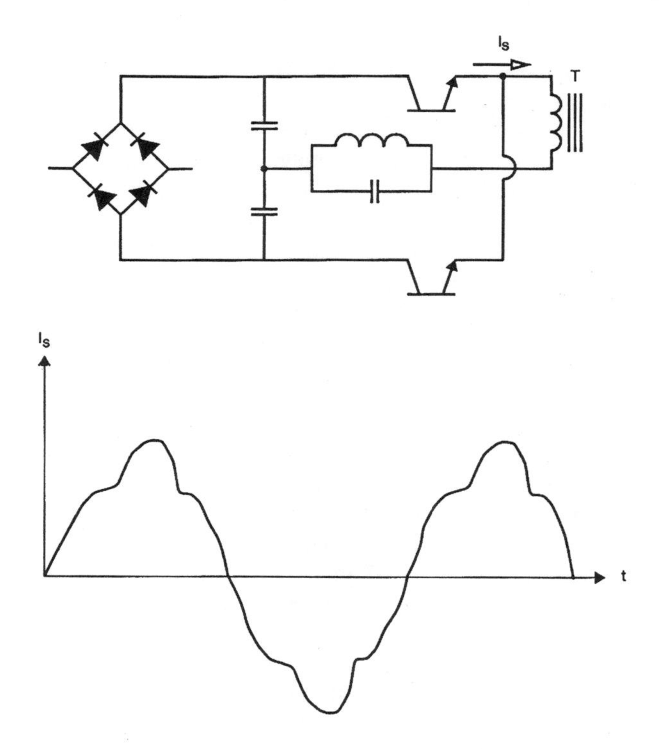

FIGURE 8.3 One Type of Resonant-Mode Converter, Causing Less High-Frequency Harmonics

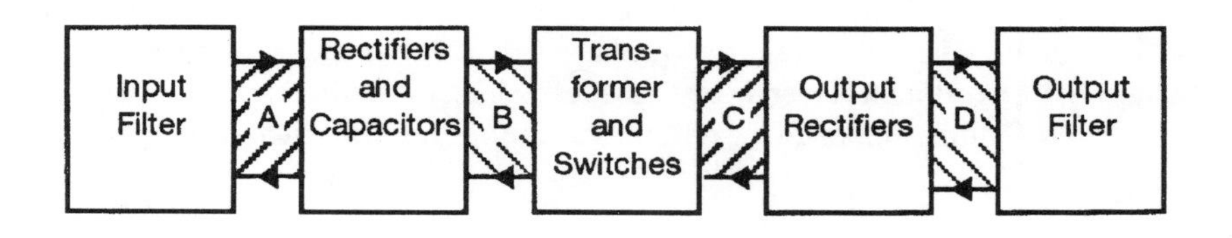

FIGURE 8.4 Example of SMPS Wiring Diagram Used to Identify and Control Magnetic Moments

However, trade-offs exist:

- Try to use magnetic materials that show a hysteresis cycle with a soft "knee", instead of a square cycle
- For low-voltage, high-current windings, use flat conductors instead of round ones.
- Prefer closed or semi-closed shapes for magnetics (see Fig. 8.7).
- If magnetic leakages are still too high, a transformer shield may be needed. A transformer whose full load leakage exceeds 1 mG at 10 cm distance (equivalent to 0.08 A/m or 98 dBμA/m) on the first harmonics in the 20 to 50 kHz range has no chance to meet the H field limits of MIL-STD-461 or VDE 871. A transformer whose leakage is simply 10 times larger (10 mG at 50 kHz and 10 cm distance) will start to cause internal EMI problems if there is nearby sensitive analog wiring. At this frequency, such a field induces 0.3 mV per cm^2 of exposed circuitry.

The simplest kind of magnetic shield (the "poor man's shield") is the shading ring. This is a closed copper band, centered mid-height on the bobbin. The optimum width has been found empirically to be about half the bobbin height. Its operating principle is that the induced current in the shorted ring creates a cancelling field against the leakage field. The copper must be thick enough and well soldered to withstand the Joule effect. In the predominant leakage directions, the field is reduced by a 2 to 3 factor. If more attenuation is needed, the ultimate solution is a closed box made of iron or ferrous material.

It is important to note that the inductors used for conducted EMI suppression may require a similar treatment around their E-style magnetic cores. Otherwise, the designer will have traded his conducted noise for a radiated emission problem!

8.3.2 The Power Supply PC Board

On power supply boards, it is a good practice to leave copper on all unused areas instead of etching it away. These copper lands will be used for the positive and negative dc voltage on the rectified primary side, and for the heavy current outputs. If the cost of a double-layer board is acceptable, the radiating loop sizes can be reduced by running opposite current traces on top of each other rather than side by side.

8.3.3 Secondary Loops

The secondary loop carries a pulsed current whose shape can be extremely distorted, especially with rugged designs like the one in Fig. 8.8. Such a current shape bears a high harmonic content. To reduce radiation from this loop, one must control its area, as mentioned previously. In addition, the current waveform should be spread and smoothed. Full-wave rectification already reduces the problem by decreasing I peak and increasing τ/T. But more is gained by adding an inductor between the rectifier and the capacitor. This causes the current spectrum to roll off at a rate of $1/F^2$ (-40 dB per decade) instead of 1/F.

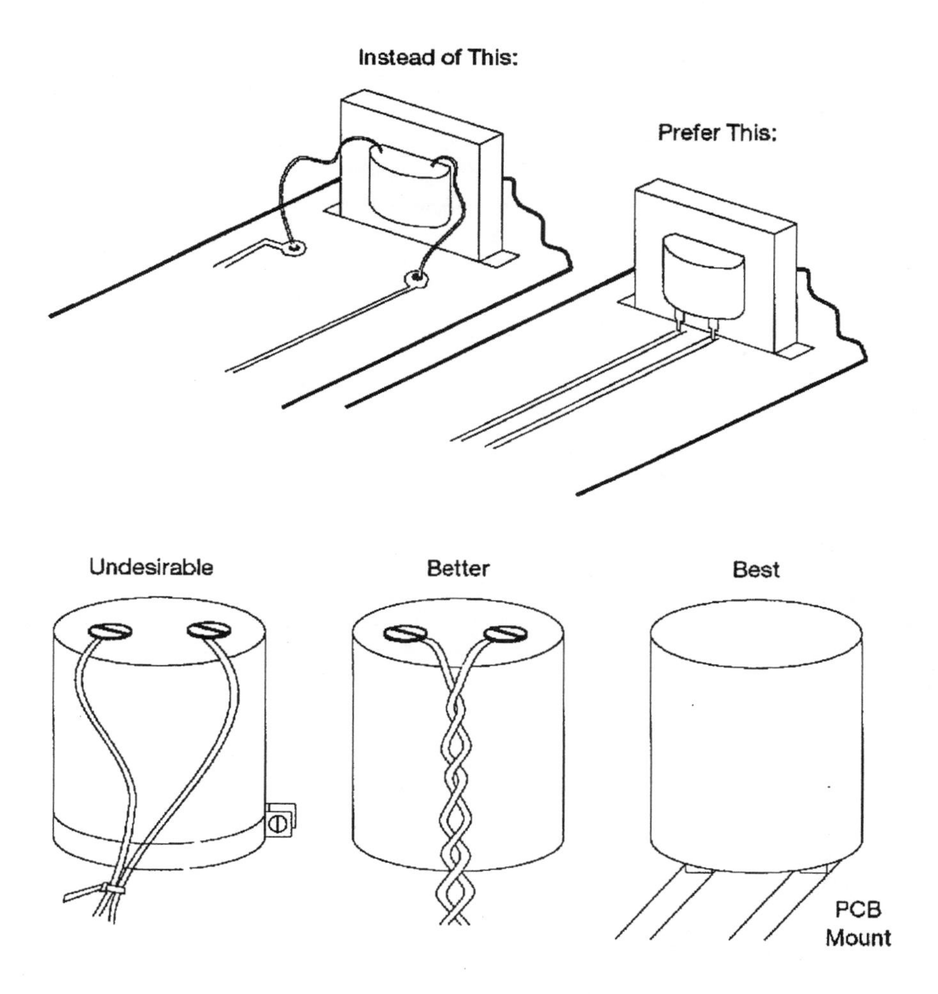

FIGURE 8.5 Minimizing Component Lead Loop Areas

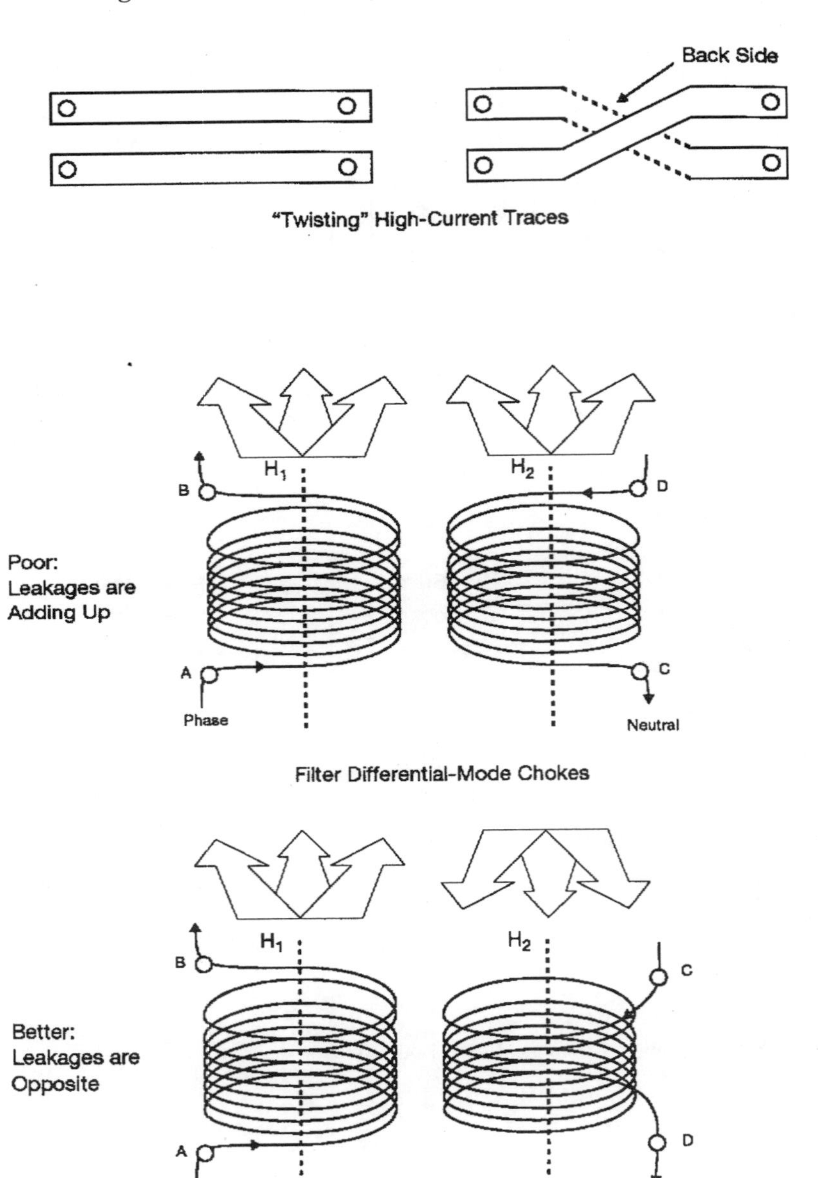

FIGURE 8.6 Neutralizing Identical Magnetic Moments. By 180° reversing the "neutral" filter choke, its leakage field now opposes that of the "phase" choke.

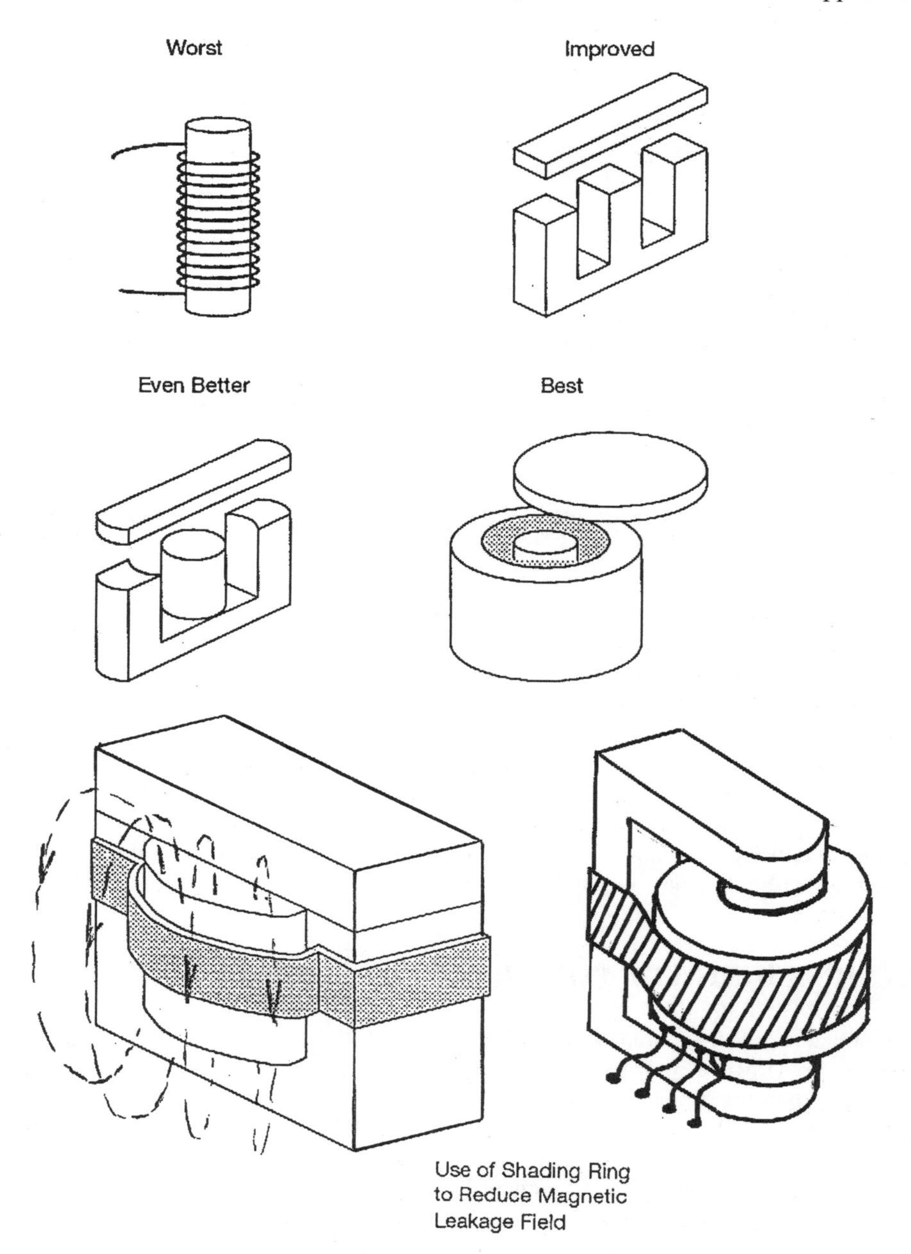

FIGURE 8.7 Reducing Transformer Leakage Fields

8.3.4 Electromechanical Packaging

All the guidelines described so far need to be translated into a compact electromechanical design that minimizes radiating loop surfaces in three dimensions. Figure 8.9, part a, shows a poor SMPS layout. In this illustration:

- The unfiltered ac input radiates in the entire host machine before being filtered by G, causing internal and external EMI.
- The primary diode bridge (labeled A) and electrolytic capacitor (labeled B) form a radiating loop (50/60 Hz pulsed, non-sinusoidal current).
- Capacitor B, transformer T and switching transistors S form a large and mostly detrimental radiating loop.
- Output filter section F will pick up switching harmonics, by radiation.

In comparison, Fig. 8.9, part b, shows a good layout:

- The ac input is cleaned up by filter G at the SMPS barrier, with no input-to-output recoupling
- B, T and S are closely packed such that the HF loop is minimal. The dc output diodes and ripple filter F are distant from the input switching loop. A large copper plane under B, T and S is used for the bulk dc (rectified ac) return. It also helps to decouple the heat sink (collector) noise from the emitter common before it goes to the chassis. Wider traces are used for transformer output.

Whenever possible, there is an advantage to keeping the input and output terminals on the same face. Provided that the primary section is correctly decoupled internally from the output, all terminals on the same side will reduce the HF current flow across the entire power supply chassis, where it would have more of an opportunity to radiate through slots and discontinuities. (This does not apply to integrally shielded SMPS.)

8.4 SHIELDING THE POWER SUPPLY MODULE

A more detailed discussion of shielding materials and practices will be found in Chapter 10. However, we will summarize here some shielding guidelines that are peculiar to power supplies.
Below about 100 W and 100 kHz switch frequency, a SMPS generally does not need to be enclosed in a shield, provided that the packaging precautions described earlier have been incorporated. Unless stringent internal EMC requirements exist (e.g., the proximity of analog circuits, magnetic heads amplifiers etc.), a PCB with open-frame power supply packaging is usually sufficient. Notice, though, that even an open-frame design can be a fairly useful Faraday shield to prevent CM currents from a "HF -hot" heat sink to circulate into the entire equipment chassis (see Fig. 8. 10). In part a, the heat sink is a voltage-driven antenna with respect to ground. Notice that "earthing" the heat sink would decrease the E field but increase I_{CM} back to the power mains, aggravating conducted EMI.

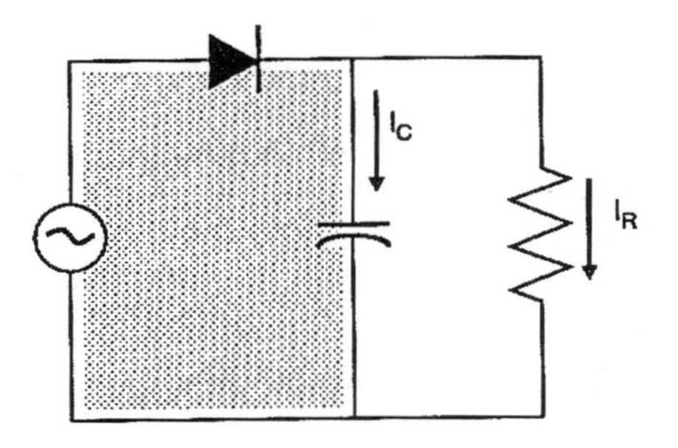

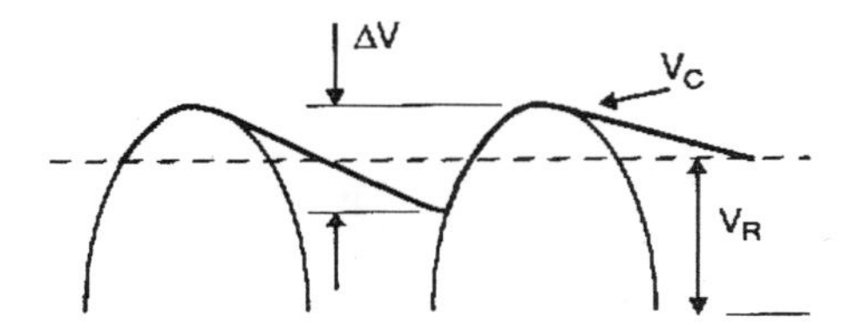

$$I_C = I_{R\,(average)} \times \frac{T}{\tau} = 8 \text{ to } 20 \times I_{R\,(typical)}$$

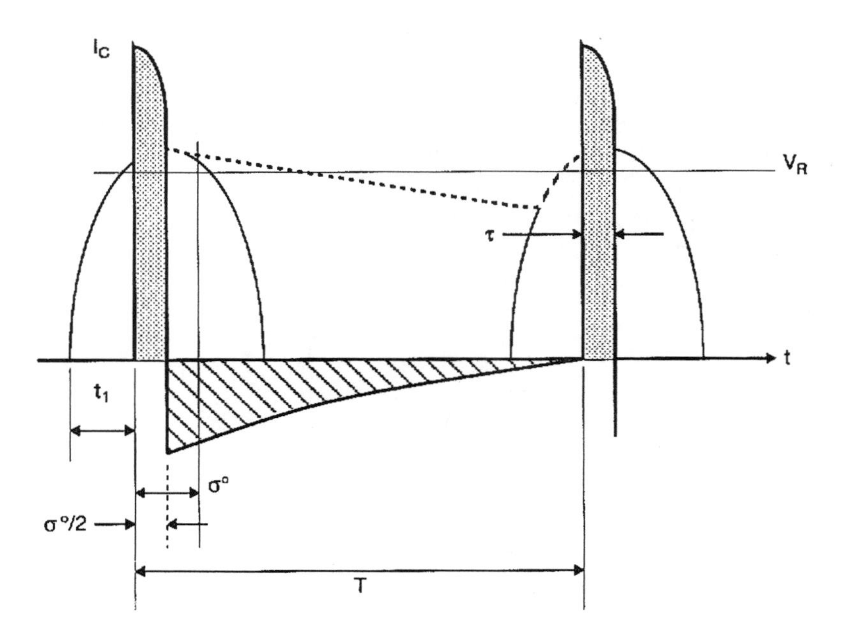

FIGURE 8.8 Secondary Loop Current into Filtering Capacitor

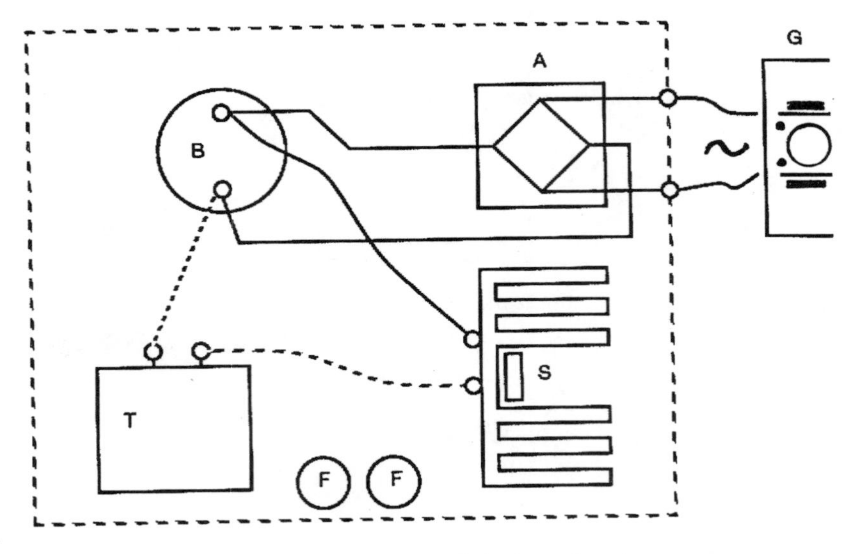

a. Example of a Poor Layout

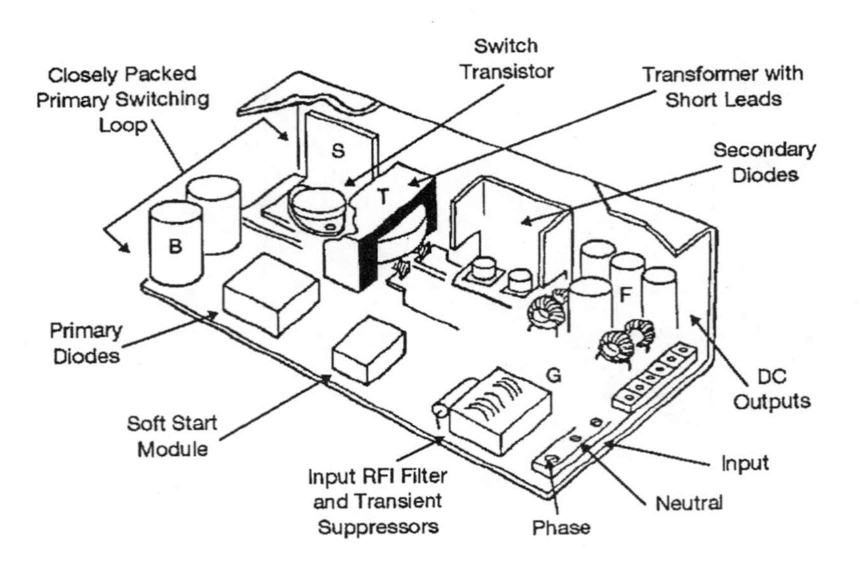

b. Example of a Good Layout

FIGURE 8.9 Poor and Good Switch-Mode Power Supply Layouts.

In Fig. 8.10, part b, the I_{CM} path is contained within the frame, even if the box is not 100 percent closed. With a continuous metal sheet, there is virtually no voltage drop along the B-to-G path, and the frame does not reradiate.
At higher power and/or frequencies, the SMPS radiation becomes a greater concern, and real shielding has to be considered. At the first harmonic of the switching frequency, electromagnetic fields with a predominantly magnetic term (i.e., low impedance fields, at a distance much shorter than $\lambda/2\pi$ from the source) are very hard to shield.
Against very low-frequency (< 10 kHz) magnetic fields, it is necessary to use thick iron (at least 1 mm thick) or permeable materials like Conetic®, having a μr larger than 1,000.
Above a few hundred kilohertz, any metal like copper, aluminium or iron, with at least 0.8 mm (30 mil) thickness, is intrinsically an excellent shield. At 100 kHz, for instance, a 0.8 mm thick aluminium plate provides 55 dB of attenuation against an H field source at 5 cm.
In the 10 to 100 kHz region, a good trade-off for performance and weight consists of:

- a layer of good conductive material (copper, zinc or tin) facing toward the H field source, providing reflection loss
- a barrier of ferrous material providing good absorption loss

This combination is realized by copper-clad, tinned or zinc-coated iron.

Next in importance to the choice of the shielding material is the realization of the shielded box itself. If the SMPS stands entirely on a PCB, a ground plane on the external face of the board can act as one face of the shield, and the metal housing needs to be only a five-face one. Avoid large perforations, especially near the high dV/dt or dI/dt sources. Eliminate long slots and seams. Cooling apertures should be arrays of round holes instead of long slots.
Do not mount magnetic components near openings. Ideally, to preserve the normal field attenuation through an opening, the radiating source should be at a distance, D, greater than 3 times the largest opening dimension, *l*.
To assemble shield edges, prefer a continuous weld to spot welding. For screw mounted covers, keep the screw spacing small, consider the space between screws as a leaky slot, and respect the $D/l > 3$ criteria, as explained above.

8.5 EFFECT OF THE POWER SUPPLY FILTER ON RADIATED EMI

Whether the power supply filter is specific to a power module only (i.e., inside the machine) or acts also as a mains filter for the whole unit, there is a definite interaction between filter performance and the radiating profile of the equipment. There is a tendency to envision EMI as a two-sided coin, conducted and radiated, presuming that the two can be handled separately. Therefore, a power supply filter that has successfully helped in not exceeding conducted limits is labeled "good" and exempt from any further suspicion. In fact, mounting deficiencies or disregard for parasitic effects in PCB layout (see Fig. 8.11) may cause the filter attenuation to drop in the range above 20 to 30 MHz. Beyond these frequencies, conducted

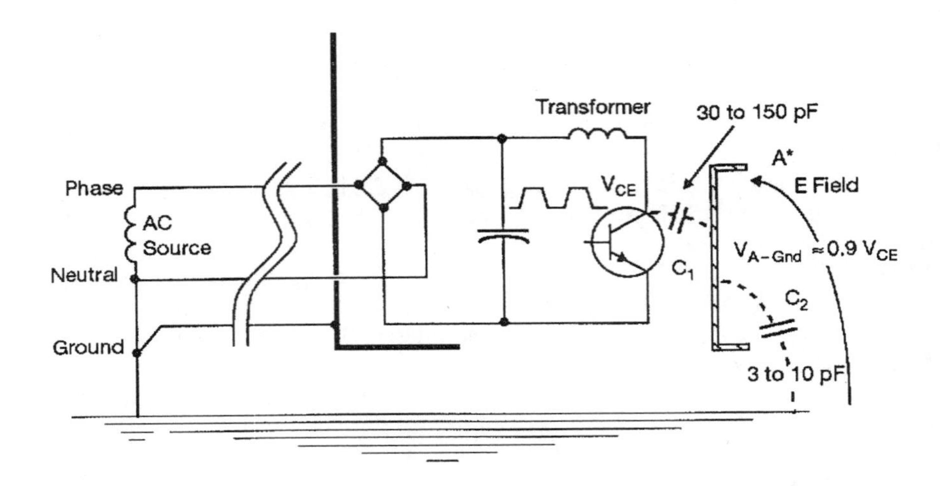

a. Open-Frame Power Supply
*Electrically "hot" heat sink becomes a voltage-driven antenna.

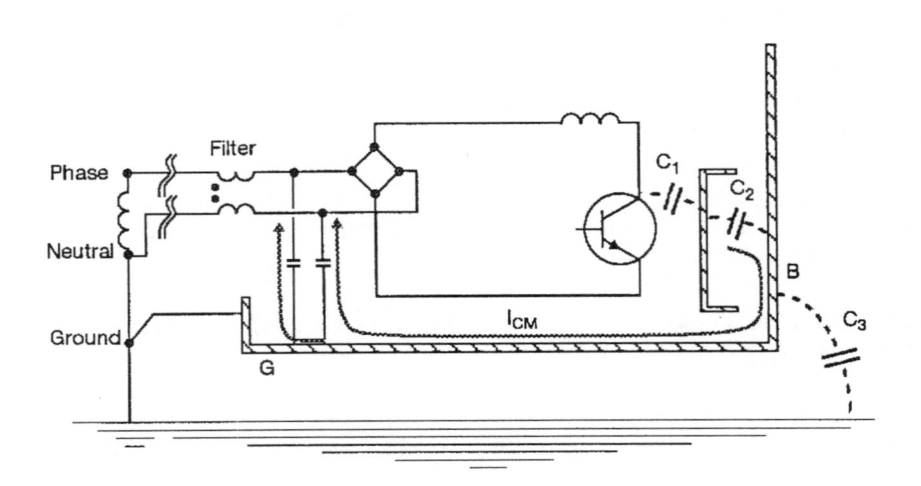

b. Closed-Frame or Folded-Frame Power Supply

FIGURE 8.10 Properly Designed Open-Frame Power Supply Act s as a Faraday Shield

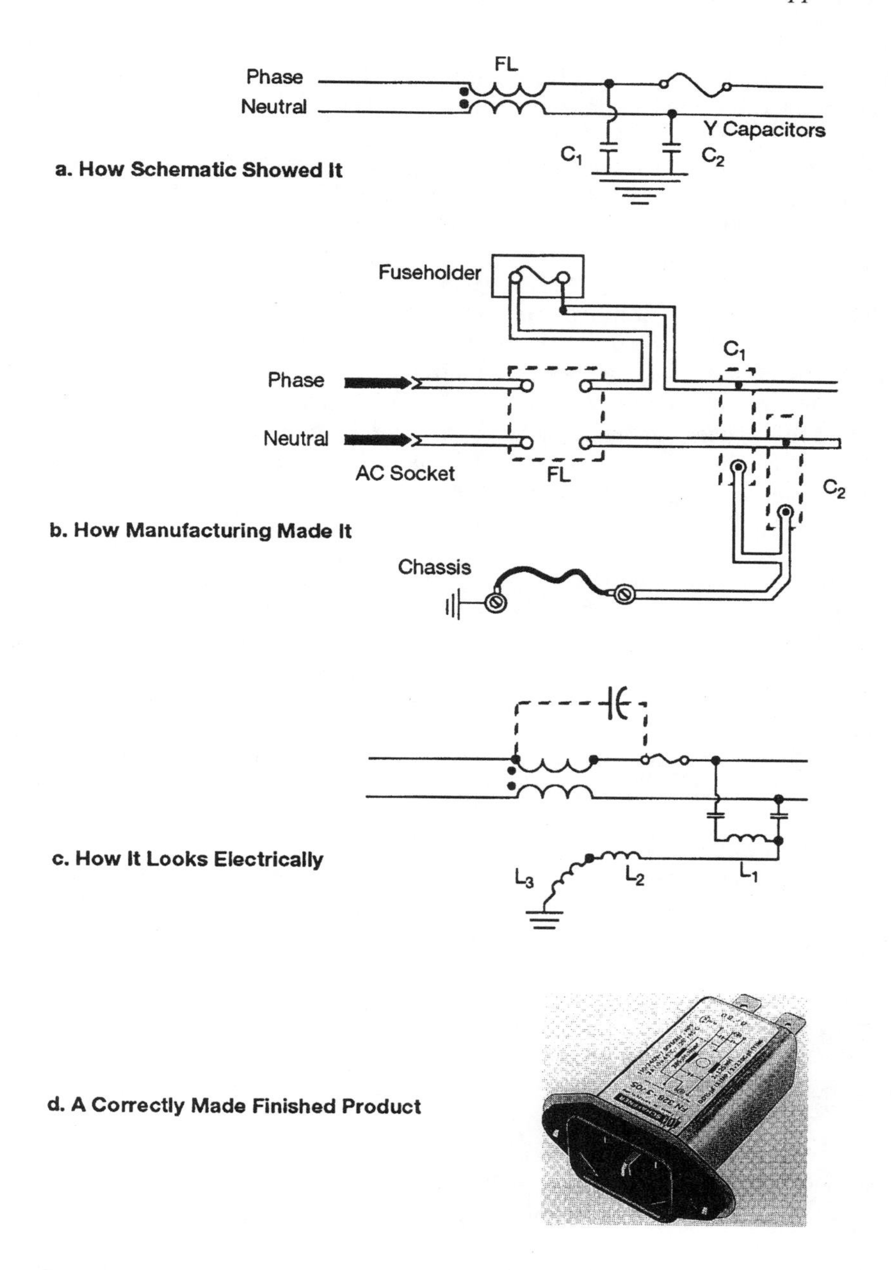

FIGURE 8.11 Filter Mounting Problems. The feed-through mounting (bottom) avoids recoupling. Photo courtesy of Schaffner, Inc.

specifications generally stop. Therefore, filter leakages could go undetected by the conducted test but still cause the equipment to radiate strongly by its power cord.

To avoid this, filter performance and mounting techniques should be scrutinized up to the high frequency range of clock harmonics (not just SMPS harmonics), typically up to 10 to 30 times the highest clock rate, or at least 100 MHz, whichever comes first. (see Fig. 8.12).
Quite typically, an unfiltered SMPS generates conducted EMI 40 to 60 dB above FCC/CISPR or MIL-Std limits. To attenuate* this noise before it radiates into the entire equipment and beyond, it is advisable to mount the filter components as closely as possible to the power supply input terminals, and separate noise-conducting leads from regulated dc and sense leads. The best way to achieve this is to use filters (either purchased or home made) that are packaged in a feed-through style. In this manner, any fortuitous recoupling of output and input conductors is avoided. The following example will show how compliance with conducted limits does not necessarily guarantee radiation compliance.

Example 8.2
A tabletop commercial equipment, with plastic housing, satisfies the conducted FCC class A limit at around 25 to 30 MHz with a 4 dB margin. Knowing that the conducted narrowband spectrum stays at approximately the same amplitude from 25 to 35 MHz, does the equipment meet the class A radiated limit with its unshielded 1,5m power cord, at about 0.80 m above ground? Repeat this exercise for MIL-Std 461, CE03 versus RE02 (cable at 5 cm above ground). The test data from these examples are illustrated in Fig. 8.13.

a) FCC class A Solution
For the FCC tabletop test setup, we have an indefinable loop, so we will use the worst-case assumption of an open-ended cable and apply Eq. (2.27). At 35 MHz, the conditions are far field (test distance = 10 m) with an electrically short cable ($l < \lambda/4$). We first determine the CM current, from the 50 Ω LISN impedance:

$$I\ \mathrm{dB\mu A} = (70\ \mathrm{dB\mu V} - 4\ \mathrm{dB}) - 34\ \mathrm{dB} = 32\ \mathrm{dB\mu A}\ \text{or}\ 40\mu\mathrm{A}$$

$$E\mu\mathrm{V/m} = \frac{0.63 \times 40\ \mu\mathrm{A} \times 1{,}5\mathrm{m} \times 35\ \mathrm{MHz}}{10\mathrm{m}} = 130\mu\mathrm{V/m}\ \text{or}\ 42\ \mathrm{dB\mu V/m}$$

Accounting for 5 dB ground reflection in the FCC test method, this comes to 47 dBμV/m. The limit is exceeded by 7 dB.

---

* Although commonly used, including by this author, the terms "attenuate" and "suppress noise" are generally misnomers. Except for the small amount of energy that is dissipated into heat, the filter in fact reroutes the EMI currents away from the equipment and power mains loops, and forces them to remain confined within the power supply compartment.

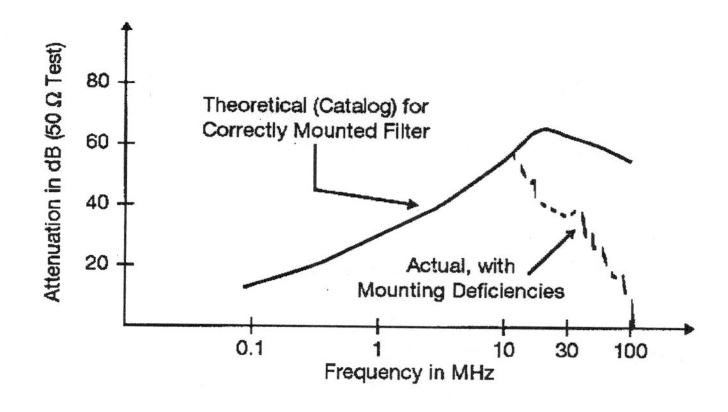

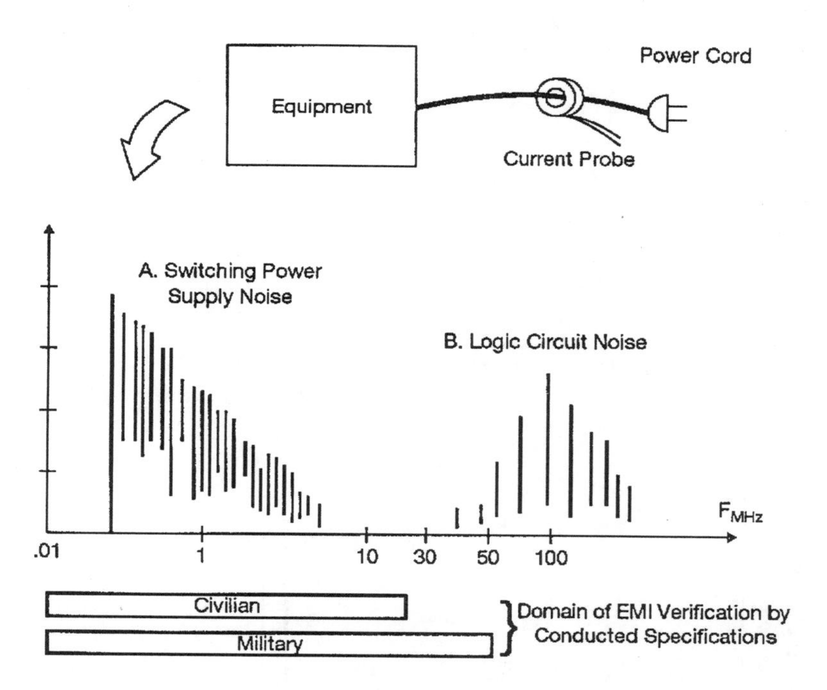

FIGURE 8.12 Power Line Filter Performances versus EMI Spectrum on Power Cord. If filter does not have good attenuation above 30 to 50 MHz, the A portion of the spectrum is correctly suppressed or conducted specification compliance, but the B portion contributes to radiated EMI.

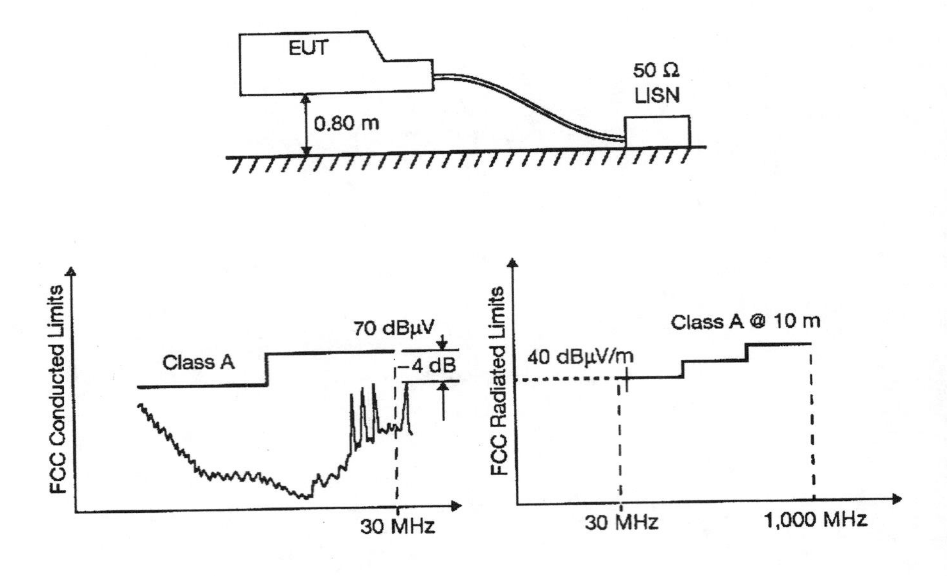

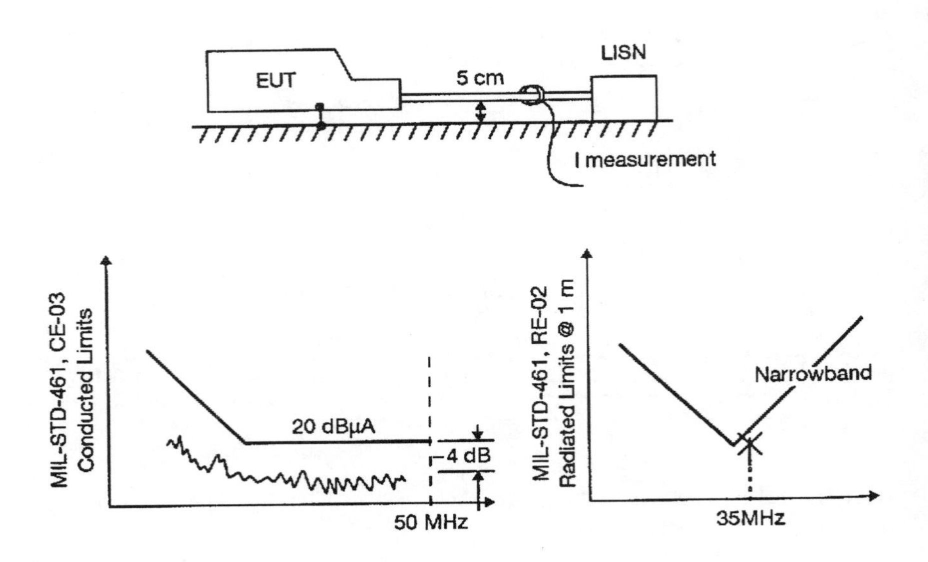

FIGURE 8.13 Test Data of Example 8.2, Conducted vs. Radiated Compliance

b) MIL-STD Solution
For MIL-STD-41, CE03, if the product is within specifications by a margin of 4 dB. this means a (20 - 4) = 16 dBμA, or 6 μA current exists between 25 and 35 MHz However, at a 1 m test distance, we are in the near field (D < 48/F). Also, no more than l = D can be entered for cable length (see Chapter 2, Section 2.5.3):

$$E_{\mu V/m} = \frac{0.63 \times I_{\mu A} \times A_{m^2} \times F}{D^2}$$

$$= \frac{0.63 \times 6\ \mu A \times (1\,m \times 0.05) \times 35}{1}$$

$$= 6.6\ \mu V/m \text{ or } 16.4\ dB\mu V/m$$

The limit is 22 dBμV/m. Therefore, we are within specified limits. Incidentally, we retrieve our criteria for CM currents as set forth in Chapter 2, Section 2.5.2, Examples 2.5 and 2.6.

Chapter 9

Reducing Radiated EMI via Internal Cabling and Packaging

Once the major building blocks have been designed for minimum EMI radiation, it must be ensured that the internal equipment wiring between subassemblies does not radiate excessive noise. Such wirings will be found to radiate as a function of the intentional signals they carry, plus the spurious signals gathered and insufficiently filtered at each subassembly.
As general rules:

1. Try to route internal cables close to internal frame members, metal compartments and conductive equipment covers, but not across large slots, louvers and seams.
2. Avoid crosstalk between high-frequency carriers that may be filtered or shielded when they exit the equipment, and other (innocuous) wiring that is not. A minimum separation of 2.5 cm between the culprit and victim pairs, and the height of neither one exceeding about 0.5 cm above the chassis, will guarantee a maximum crosstalk of $\approx$ 3 percent (-30 dB) in the worst possible scenario (high-impedance victim circuits, parallel length > $\lambda/4$ of the culprit signal).
3. Avoid creating large loops between hot wires carrying high dV/dt (more than a few volts per nanosecond) or high dI/dt (a few hundreds of milliamps per nanosecond) and their normal, or fortuitous, return conductors. When attempting to track down and reduce them, always picture these loops in three dimensions to figure out the loop contour.

9.1 CARD-TO-CARD AND BACKPLANE INTERCONNECTS

PCB to-PCB internal connections are usually made with ribbon cables or flexprint. To accommodate the largest possible number of conductors, designers tend to use only a single return wire and assign all other positions to signal wires. This is a poor practice, because the signal wires at the far edge from the return wire make a wide DM loop that:

- radiates efficiently
- exhibits strong crosstalk with the next conductors (see Fig. 9.1)
- is susceptible to ambient interference

For a typical ribbon cable 1 or 2 inches in width, such a loop is an efficient radiator (see Chapter 2).

The alternating 0V-signal-0V arrangement should be used systematically for rise times shorter than about 12 ns and clock frequencies or bit rates above 1 or 2 MHz. (This calculation is based on satisfying FCC/CISPR Class B limits with a 1.50 m flat cable, and the worst possible wire spacing of 5cm.) Therefore all signals with faster rates and rise times should use:

- one ground for each signal wire, or at least one ground wire running along each high frequency signal, at 1.27 or 2.54 mm spacing (see criterion above)
- a ribbon cable with ground plane
- a flex print with ground plane

Some vendors offer twisted ribbon cables. Twisted ribbons are efficient in reducing crosstalk and near-field radiated problems (emissions or susceptibility). Benefit with regard to radiated EMI specifications is generally limited, however , compared to the major improvement achieved by simply reducing wire spacing down to 1.2 or 2.5 mm. This is true for two reasons:

- the reduction in DM radiation by simply reducing wire spacing is so large that CM radiation generally takes over, masking any twisting improvement. (Twisting has no effect on CM current radiation.)

- the vendor often leaves an approximately 4 to 10 cm long untwisted segment every half meter or so to allow for easy mounting of self-stripping connectors. This seriously limits the twisting-derived attenuation.

Another risk of crosstalk with flat cables occurs with stacking. In this case, high-speed flat cables that do not exit the equipment can very efficiently contaminate I/O cables. Alternate grounds won't help very much; a better solution is to interpose a shield, use shielded flat cables, or insert a spacer of a few millimetres thickness to increase separation.

Flex print interconnects with copper planes on two sides give an even better shielding, especially if the two planes are connected regularly every few centimeters by through holes. This approaches the performance of conventional shielded cables.

Occasionally, the designer may use short segments of flat cables to connect one card that lies above a larger one (see Fig. 9.2), or two daughter cards on their front ends, in addition to their normal backplane interconnection. In this case, a potential; radiating loop is formed: because the ground references of the two boards are now connected by two paths, each signal current going from one card to the other can return not only by the expected path, but also by the alternate one. In addition, ground noise current will flow between the two top connectors. Radiation from such loops is difficult to prevent, so they should be avoided whenever possible. (Otherwise, the flat jumper cable should have a ground plane.

In all cases, all the unused wires should be connected to ground at both ends. A floating wire just invites more radiation and crosstalk. In extreme cases of internal

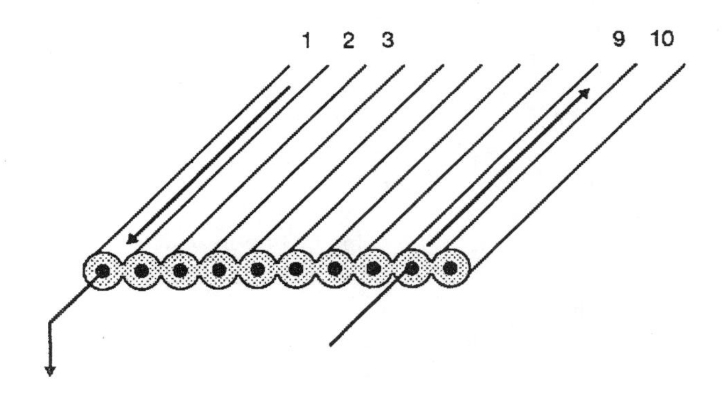

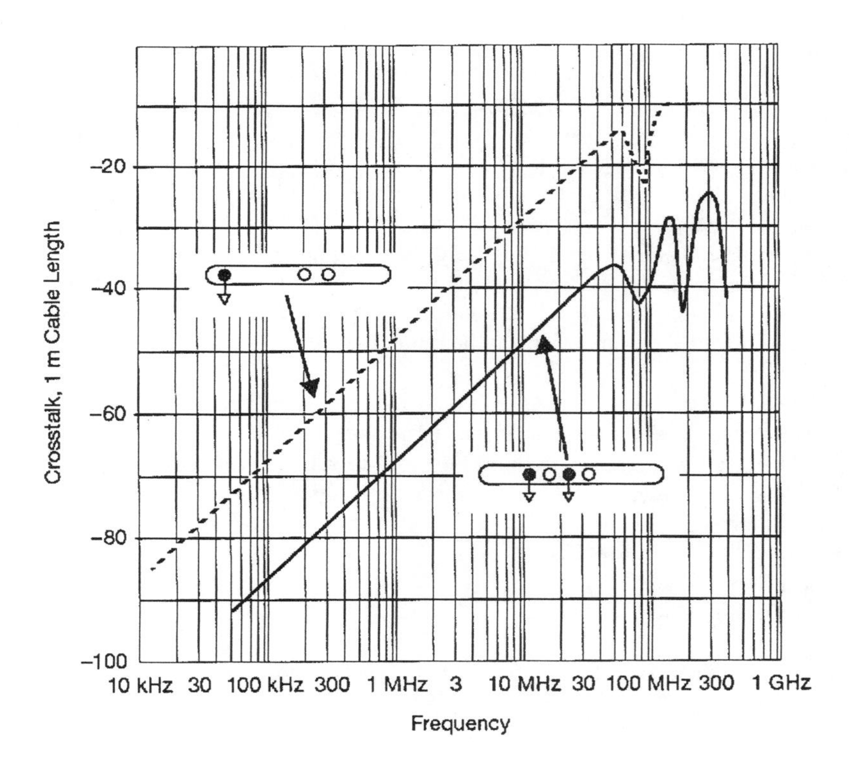

FIGURE 9.1 Radiation and Crosstalk with Flat Cables. Wire # 9 or # 10 forms a large radiating loop with the 0V return. The crosstalk curve on top is for two wires, 2.5 mm apart with a 0V return at a 25 mm distance.

flat cable radiation, ferrite suppressors can be installed over the cables. Such ferrites work by creating a high mutual inductance and resistive losses. Above a few megahertz, they increase the CM loop impedance, artificially balancing the pairs in the flat cable.

9.2 INTERNAL COAXIAL AND SHIELDED CABLES

A coaxial cable is essentially a two-conductor line in which one conductor happens to be concentric to the other. Therefore, there is no question that the shield to be connected both ends to the ground reference. In the year 2000, more than 70 years after invention of coaxial cable, the author in his role of EMC consultant often encounters coaxial shields that have been kept floated "to avoid making ground loops." Missing the shield connection, the HF current from the center conductor will have to worm its way back to the source, generally using every possible return route through chassis and bonding wires. In this process, the radiation is multiplied by several orders of magnitude, and the shield becomes useless.

Interfacing coaxial cables with PCBs should always be accomplished via a coaxial connector before the signal can be transferred into the strip or microstrip configurations of the PCB. Multi contact edge connectors used for PCB and mother boards are available with all or some of their contacts being of the coaxial type. If space or cost dictate, cheaper substitutes like coaxial ferrules (see Fig. 9.3) can be used, but radiation and crosstalk will be generated in these areas. A rather regrettable practice is shown in Fig. 9.3, part c, where a large portion of each shield is stripped away, then daisy-chained to a single Ground pin. As demonstrated in the following example, this type of termination can be quite detrimental.

Example 9.1
A coaxial cable carrying a video signal terminates on a PCB as shown in Fig. 9.3, part c. The stripped center wire and the shield-to-ground connection form a 4 cm x 2. 5 cm loop. The video signal has the following characteristics:

- peak voltage: 10 V
- fundamental: 6 V at 25 MHz
- harmonic #3: 2 V at 75 MHz
- load resistance: 50 Ω

What is the radiated field at 3 m distance at 25 MHz and 75 MHz?
From Eq. (2.22) or Fig. 2.6, for far-field conditions:

$$E_{\mu V/m} = \frac{1.3 \times V \times 10 \ \text{cm}^2 \times F^2}{3 \ \text{m} \times 50 \ \Omega}$$

$$= 300 \ \mu\text{V/m, or } 50 \ \text{dB}\mu\text{V/m at } 25 \ \text{MHz}$$

$$= 1{,}000 \ \mu\text{V/m, or } 60 \ \text{dB}\mu\text{V/m at } 75 \ \text{MHz}$$

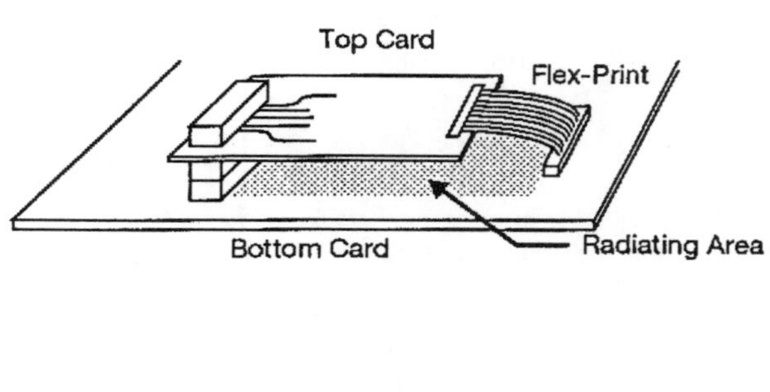

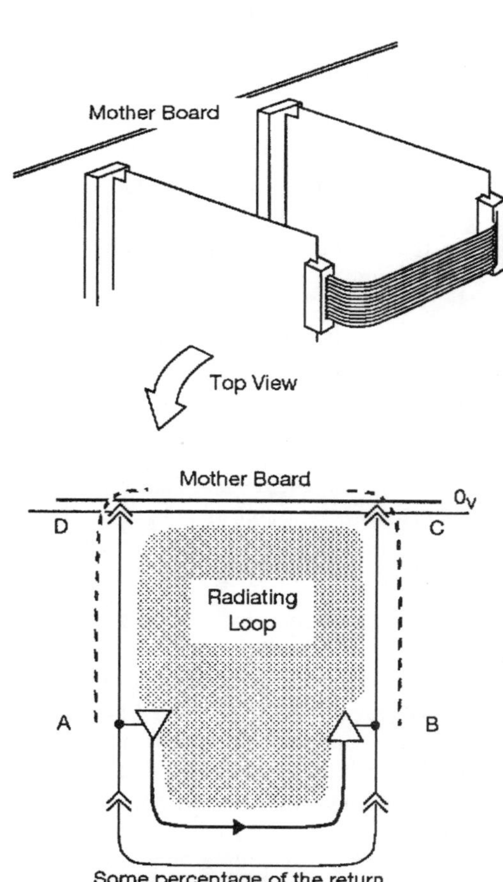

FIGURE 9.2 Some Hidden Radiating Loops with Card-to-Card Cabling

The second value exceeds FCC/CISPR Class B requirement by about 20 dB. (There is no requirement at 25 MHz for commercial equipment.) The MIL-STD-461 RE02 limit, translated at 3 m, is exceeded by at least 40 dB.To make it a negligible contributor to the radiated field (taking a 6 dB margin), the loop surface must be brought down to:

- ≤ 0.5 cm^2 for FCC Class B
- ≤ 0.05 cm^2 for MIL,STD-461, RE02

This virtually eliminates any such mounting.

For internal shielded pairs or shielded harnesses, the shield is no longer an active return, so there is no functional reason to connect it to the signal ground (0V). Because its role is to prevent some cables from radiating or picking up internal radiation, the best connection for such protective shields is at the chassis, at both ends. Each end of a cable shield should be at the same potential as the chassis or box that it enters or exits.

The following rules apply to internal cable shields other than coax:

1. If the chassis (or a sheet metal component box, conductive plastic enclosure, etc., that is bonded to the chassis) is available near the cable ends, connect both ends of shield to this part, via the metallic connector housing/receptacle, or use a short, wide strap. (Do not use "pigtail" wire.)
2. If rule 1 cannot be met, stay away from nonsense: do not run long wire leads to a remote chassis ground. A copper land connected to the chassis should have been provided on mother board edge (see Chapter 7), which is an acceptable place for a short, low-inductance shield connection.
3. If neither rule 1 nor rule 2 can be met, connect the shield ends to the ground reference plane or copper land of the corresponding PCBs or subassemblies. Do not use a thin trace for this. A shield is basically a "noise collector". Driving current from a noise collector into a copper plane is of no consequence, because Z_{plane} means milliohms up to 300 MHz. Instead, driving the same current into a ground trace can contaminate the whole signal reference.

There are few exceptions to rule 3 with regard to grounding shields at both ends "for lack of any better" method:

- If the cable carries low-level analog signals, tying the two ends of the shield to different ground references may inject objectionable noise into the enclosed wires, destroying high CM rejection obtained via coupling transformers, optoisolators or differential amplifiers. In this case, the input of the analog amplifier should be decoupled against HF. The cable shield will be grounded : a) on the amplifier side only, if the sensor can be floated, or b) on the sensor side only, if the sensor, magnetic head or other reference is already grounded. If the amplifier has a floated "guard" shield, the cable screen will be connected to it as well.
- If the cable carries mixed analog/digital signals, with an analog ground reference on one end and a digital ground on the other, it is presumed that the designer has

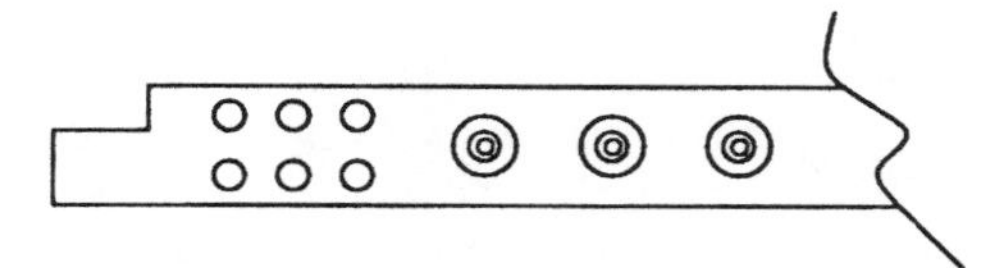

a. Best: Combination Edge Connector, with Coaxial and Regular Contacts

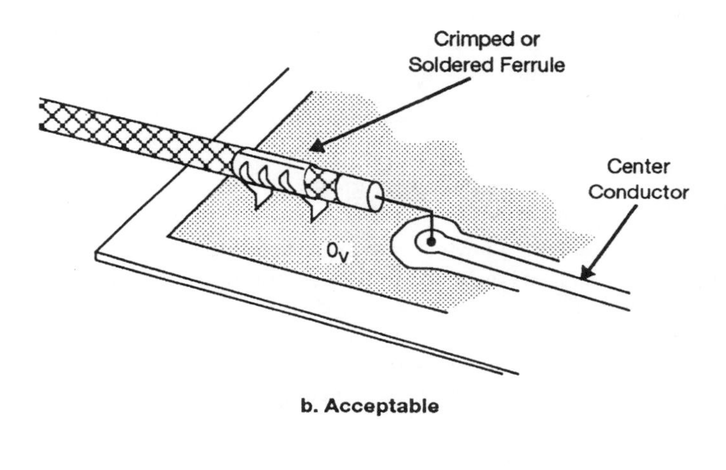

b. Acceptable

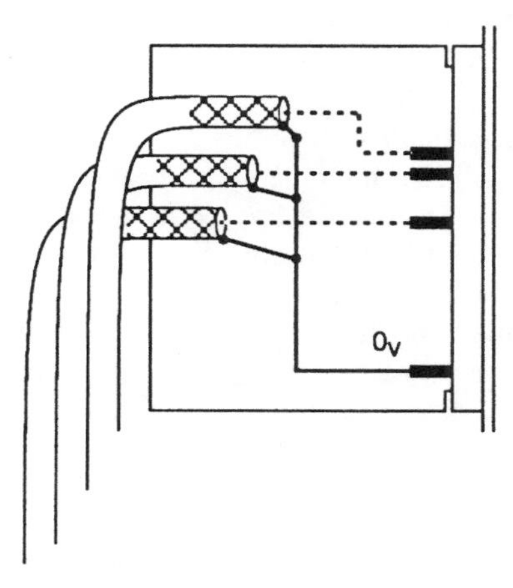

c. Poor

FIGURE 9.3 Connecting Coaxial Cables to PC Boards

provided some kind of isolation interface device or translator. Tying the two ground references together via the shield may create an objectionable loop. In this case, a shield aimed at protecting sensitive wire from radiation should be tied to the analog ground only. Conversely, a shield meant to prevent digital wires from inducing capacitive crosstalk into other wiring should be grounded to the digital ground only.

It can be said that the two above exceptions are aimed more at internal EMC than to outside radiation concerns. Shields connected in such ways are primarily electrostatic shields, with no effect in reducing radiated emissions above 30 MHz, let alone CM loop reduction. However, in this book we are trying to provide emission reduction guidelines that are not counter productive to noise immunity considerations.

9.3 SOME HIDDEN RADIATING ANTENNAS

In the packaging of some equipment, there exist radiating loops that are not easy for a non-specialist to catch. They usually involve the interconnect cabling between subassemblies, where the signal wiring and the ground returns are cabled in many directions. Looking simply at the schematic reveals nothing wrong, and it takes a closer look at the point-to-point wiring and layout to visualize the parasitic "antennas". A few examples of such radiating cable loops are shown in Figs. 9.4 and 9.5.

In Fig. 9.4, part a (top), we see the central processing unit (CPU) board being connected to a control card by a flat cable A-B. The same control card also gets its regulated supply from a pair of signal-ground Vcc and 0V wires on separate connectors C-D.

Although there is one signal-ground wire in the ribbon cable A-B, some percentage of the logic signal currents will return by the alternate path C-D, causing the loop A-B-C-D to radiate. The author has seen a few cases where the situation was even worse: the A-B ribbon cable had no ground wire at all, because the designer feared that two ground wires "would create a ground loop".

In the same figure, the display /keyboard card is interconnected to the control card and the CPU card by different flat cables. Here again, a small percentage of the digital pulsed currents in the link F-B will return by the ground conductor of the cable F-E, and vice versa, causing the entire loop A-B-F-E to radiate.

Finally, in the same figure, a flat cable goes to the I/O port H, reserved for an optional second printer that is not installed. No loop exists here, but we have an unterminated line that can still receive some signals from the CPU card. The open end will cause reflection and voltage standing waves, with a peak amplitude twice that of the normal signal. This is a radiating monopole.

Figure 9.4, part b, shows solutions to this problem, which must be considered early enough in the design stage. The A-B and C-D cables are run very close to reduce loop area. The C-D cable carries enough current to also supply power to the display/keyboard card through the control card. Connector E has been relocated so that the CPU card will interface with the display card through the control card PCB.

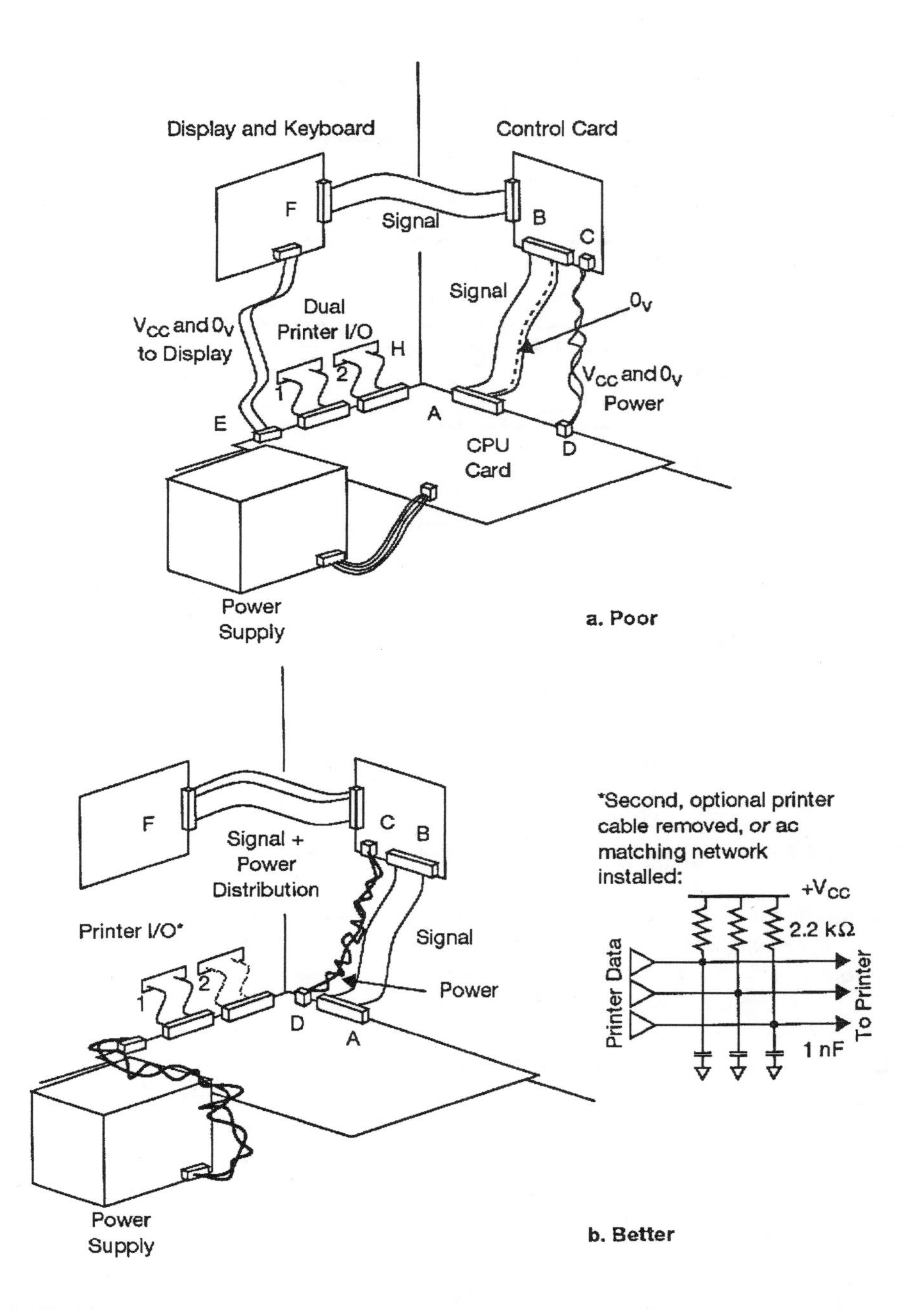

FIGURE 9.4 Some Large Radiating Loops in Packaging. Notice that most loops are not obvious at first glance

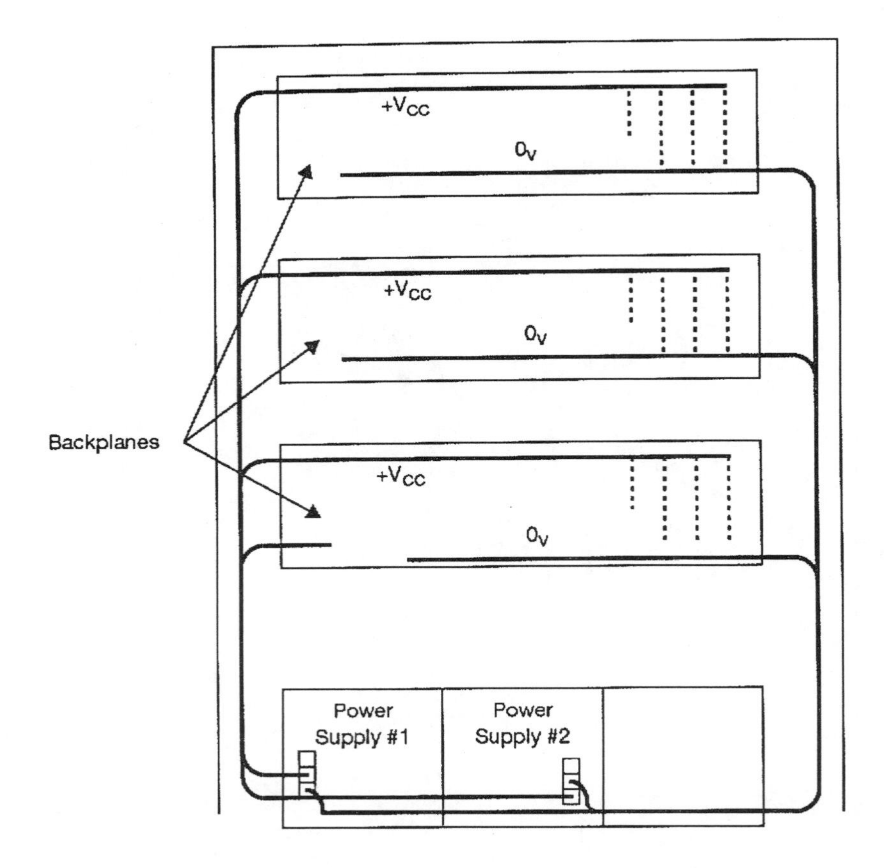

FIGURE 9.5 Other Large Loops on a Cabinet's Rear Wiring

This adds extra printed traces on the latter, but eliminates another cable loop. Another fix, quicker but less efficient, would have been to make the original cable E-F longer and reroute it, closely following the A-B-F path.
Finally, the cable to the second I/O port can be removed, to be installed only when the option is ordered, or the I/O lines may be terminated near the cable end by an RC network that achieves an approximate ac matching.

Figure 9.5 shows a huge loop, although not always considered to be one (because people do not expect dc currents to radiate). All the HF currents superimposed to the dc supply wires are flowing in loops that can be as large as the entire frame perimeter. These loops will radiate efficiently, especially if the cabinet is not entirely metallic, with proper seams leakage control. Such large dc distributions should be fed from one side of frame, preferably with flat bus bars, stacking the +Vcc and 0V bars to form a low impedance stripline. Then, they should immediately be connected to the backplanes' Vcc and 0V planes.

9.4 INTERNAL CONNECTION OF 0V REFERENCE AND CORRESPONDING GROUND LOOPS

The two subjects of 0V-to-chassis connection and ground loops cannot be dissociated. We have already shown (in Chapter 2, Section 2.5) that floating a 0V reference from the chassis can be a useful solution for opening ground loops below a few megahertz but becomes useless, and sometimes actually worse than grounding, above this frequency range. Grounding practices are as likely to create havoc as to solve EMI problems if they are driven purely by traditions or empiricism.

Considering these facts, it is mandatory that an overall, comprehensive grounding scheme for the system be determined to optimize between (a) LF immunity of low-level electronics (slow analog, audio), which would require a floating reference, and (b) HF immunity and emissions of digital, RF and video circuits. A given equipment is just one element of a system, and it must adhere to the best grounding scheme devised for this system.

When the equipment incorporates only the type (a), low-level circuits, keeping 0V floated from chassis and installing efficient HF decoupling on all analog inputs and outputs is the best solution. In any event, this type of circuit is never a cause of radiated RFI emissions. When the equipment incorporates only type (b), HF circuits, the best results are obtained by tying the 0V reference to the chassis, preferably near the I/O cable entry points.

Conflicts arise when, as is often the case, an equipment incorporates both (a) and (b) circuit types, plus a certain number of internal EMI sources (e.g., relays, motors, cooling fans, transformers, discharge tubes, lasers). As long as wiring lengths are less than $\lambda/20$, i.e., $l_m \leqslant 15/F_{MHz}$*, single-point or star grounding is often most achievable and compatible with other constraints.

The rationale for this magic $\lambda/20$ (variations such as $\lambda/10$ or $\lambda/50$ are sometimes mentioned) is as follows: the value $l = \lambda/20$ is the electrical length where any piece of round wire will represent approximately 100 Ω of inductive (series) impedance and 1,000 Ω of capacitive (shunt) impedance.

Therefore, this is the frontier up to which any conductor is still a wire (even though no longer a short circuit) and is still far from parasitic resonance.

Internally to an equipment, the star grounding eliminates the potential 0V-to frame loop that can either radiate or receive. Star grounding also prevents the signal return current from one building block from flowing through another building block's ground wire, which would create CM impedance problems. Therefore, each subassembly has its 0V floated from its compartment, and the compartment is hardwired, or preferably bonded, to the host equipment frame. There is no need, and it is generally counterproductive to waste copper and work time by making a star grounding of the subassemblies' housings within the mainframe. Such practices, in general, occur due to a poor understanding of the single point ground concept: signal and 0V ground loops to chassis are potential problems.

---

* "F" here means the highest frequency of EMI concern, which is not necessarily the highest signal repetition rate but the equivalent frequency calculated by $1/\pi t_r$

Chassis-to-chassis, or green-wire to green-wire, loops pose no threat. Equipment that appears to display lessened noise problems when the housings of its internal boxes are star grounded is really just revealing that the bonding of all its mechanical parts is poor or nonexistent.

When a good electrical bonding of all internal parts is restored, by scraping off the paint and tin plating all metallic mating surfaces under screws and bolts, and when the several frames or racks in a bay are made equipotential by several bolts or wide straps, noise problems are reduced more efficiently and across a higher frequency domain than with hardwired "stars".The only exception to this would be subassemblies containing very high voltages (above tens of kilovolts) or very high LF currents (above a 1kA level, at 50/60 or 400 Hz). Such large items need to have their chassis star-grounded with heavy-gage cables or flat braids because in the case of HV arcing or by mere induction, too much current would flow in the mainframe.

When the highest frequency of concern reaches tens of megahertz, the single point concept becomes more or less a utopian concept because:

1. Grounding conductors become too impedant.
2. Floated circuits become randomly "grounded" through their stray capacitance to the chassis.

For instance, at 100 MHz, corresponding to a rise time of 3.2 ns, our λ/20 rule would dictate that the branches of the star must not exceed 15 cm, which is generally impractical. Since this situation occurs more and more frequently as logic speeds increase, the following is recommended:

1. .If for some reason there is an absolute requirement to keep all 0V references floated (this still happens frequently, as some procureme specifications are based on the state of the art of the 1960s):

a. Decouple each logic 0V reference to the frame ground at the mother board level using a capacitor rated at a few nanofarads and installed with minimum lead inductance (leadless components are best). This still preserves low-frequency isolation (more than 10 kΩ at 10 kHz) but achieves HF grounding at a controlled location.

b. If the above is insufficient, complement it with a CM ferrite block slipped over the power supply and internal wiring to add series insertion loss above a few megahertz.

2. If floating is not mandatory, connect the logic 0V to chassis with a low-impedance strap or, preferably, with several screws.

It does not matter if the 0V-to-chassis connection is permanent or only a high-frequency bypass; the name of the game is to short out, to low-impedance chassis, all spurious currents that otherwise would tend to use external cables to return to their sources. This is designed to work from the inside out, to control emissions but the principle works just as well for the reciprocal, i.e., to prevent outside EMI currents from getting into the PCB. This brings up an additional question involved in the grounding of the 0V reference: *the issue is not simply whether to make the connection of the 0V to chassis, but also where to make the connection.*

Focusing on emissions, the dilemma is depicted in Fig. 9.6. In part a) of the figure, the 0V is connected to chassis at the regulated dc supply terminal. This seems to satisfy the general idea of a star connection. However, the far end of the PCB (labeled A) tends to be "hot" with respect to the chassis due to the HF impedance of ground traces and perforated ground planes. If cable ports exist in this area, I/O cables become common-mode driven antennas. Decoupling capacitors on the I/O port can help in cleaning up the external cable, but their return current, flowing back to the power supply, have to run across the chassis via path C-B. If the frame and covers have seams and slots, these will radiate (see Chapter 10).

In Fig. 9.6, part b), the reverse approach is used. The "A" side of the PCB is no longer "hot" near the I/O ports, but all SMPS switching noise (mostly CM) flows through the PCB to return to the chassis. As long as the SMPS switching spectrum does not extend above 10 to 30 MHz, this creates little outside radiation. But it does create a very noisy 0V distribution, leading to internal EMI, or at least reducing the allowable noise margin.

Figure 9.6, part c), shows an optimum trade-off: the power supply, its power cord entry and all I/O cables have been grouped on the same face of the equipment frame. The A side of the PCB in the I/O area sees minimal CM voltage to chassis as it is the grounded end, with a maximum current but virtually no voltage. This is why we have repeatedly insisted that this connection be as short and direct as possible.

9.5 I/O CONNECTOR AREAS

Having stated that all I/O connectors and cables entries should preferably be grouped on the same side of the equipment, this face will be the "hot plate" where many RF currents associated with I/O decoupling will flow. But it is easier to make one face particularly RF tight than to treat all six faces of the cubicle equally.

For radiation control (and immunity as well) all internal leads (wires, printed traces. etc.) arriving at the equipment skin should be decoupled as closely as possible to the point at which they cross the barrier.

When there is a significant* length of internal cabling from the PCB to the connector area, this decoupling has to be made at the connector, because if it is done just at the PCB/mother board, it is likely that the cleaned-up segment of cable will pick-up HF noise after it has been filtered and then reradiate it outside. A good way to visualize this, although extreme, is in the way shielded EMC test rooms are built.

---

*What is "significant" depends, of course, on the frequency of concern. Let's say that, for the VHF range (30 to 300 MHz), which is a prime concern in radiated EMI, a cable length in excess of 10 cm becomes a significant coupling length.

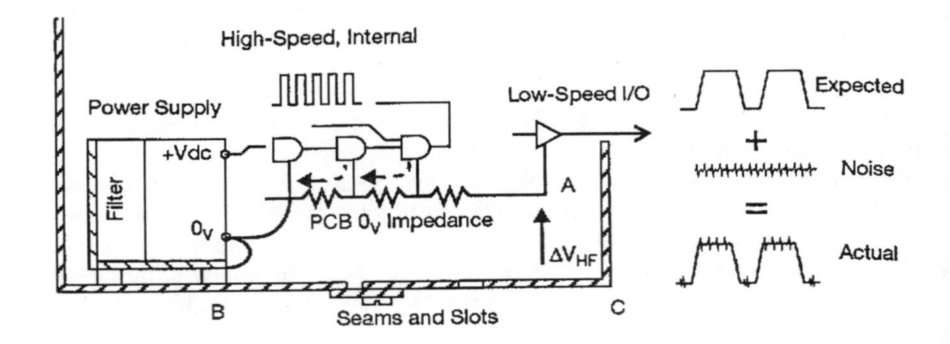

a. CM Pollution of Low-Speed I/O Lines by the Hot Side of PCB 0_V

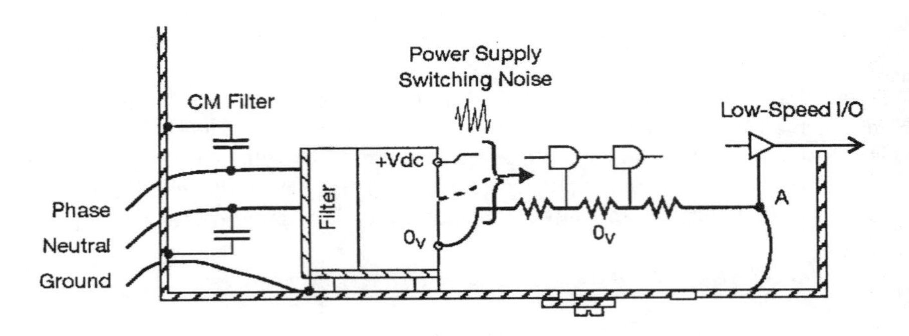

b. CM Pollution of PCB by SMPS Noise, 0_V Grounded at Far End

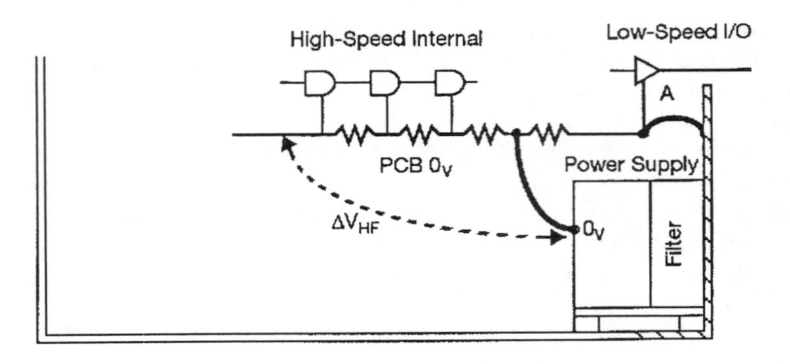

c. Best Trade-off between a and b

FIGURE 9.6 Common-Mode Pollution of PCB and I/O by Internal Sources

Absolutely no cable can enter or leave unless it is perfectly shielded or filtered at the very point of penetration. Decoupling at the I/O port level can be done using either of two approaches:

1. Apply discrete filtering to each individual conductor, especially where there is a permanent connection rather than pluggable contacts. This can be done with purchased signal filters (see Figs. 9.7 and 9.8) for high performance, or homemade filtering with discrete capacitors (as in Fig. 9.9) if a maximum attenuation of 20 to 40 dB, up to about 150 MHz, is sufficient.

2. Use filtered connectors whereby each contact is filtered by miniature ferrites and multilayer capacitor arrays. The price of such connectors was prohibitive two decades ago, when the applications were mostly military, and the quantities were low. As of 1999, a 15-position sub-D connector, with metallic shell and filtering on all contacts, costs no more than $15 in quantities. This compares favorably with the parts cost and labor required to prepare, solder and check the mounting of 15 discrete "pi" filters or ceramic capacitors and ferrite beads (see Fig. 9.10).

Very often, on smaller equipment, no hard-wired interface exists between PCBs and I/O connectors, and the receptacles of those components are directly mounted on the card edges. In some cases, the piece of internal cable is so short that there is no room (or justification) to install filter components anywhere other than on the PC board. In this case, all traces terminating on I/O connectors should be decoupled as described in the previous paragraphs to avoid spurious RF leaks, especially CM, on the I/O cabling.

Printed circuit board permits some economical and efficient mounting of filter components. If one-pole filtering (20 dB/decade) is deemed sufficient, simple ceramic capacitors are enough. SMT components allow for economical, and non inductive mounting. The preferred method is to have these capacitors connected to a "chassis ground" copper land on the PCB edge. If such commodity has not been provided, the capacitors can be connected to the ground plane, which itself should be connected to chassis, nearest to the I /O ports (see Fig. 9. 12).

The feed-through coaxial filter of Fig. 9.7 realizes an ideal capacitor, eliminating completely the parasitic inductance. But it does not lend itself easily to mass production PCB mounting. Several manufacturers have developed an SMT-equivalent of the feed-through, called the 3-terminal capacitor. In SMT size #1206, it has a parasitic inductance of only 0.1nH, and allows the signal line to jump over the perpendicular ground bus (Fig. 12.A).

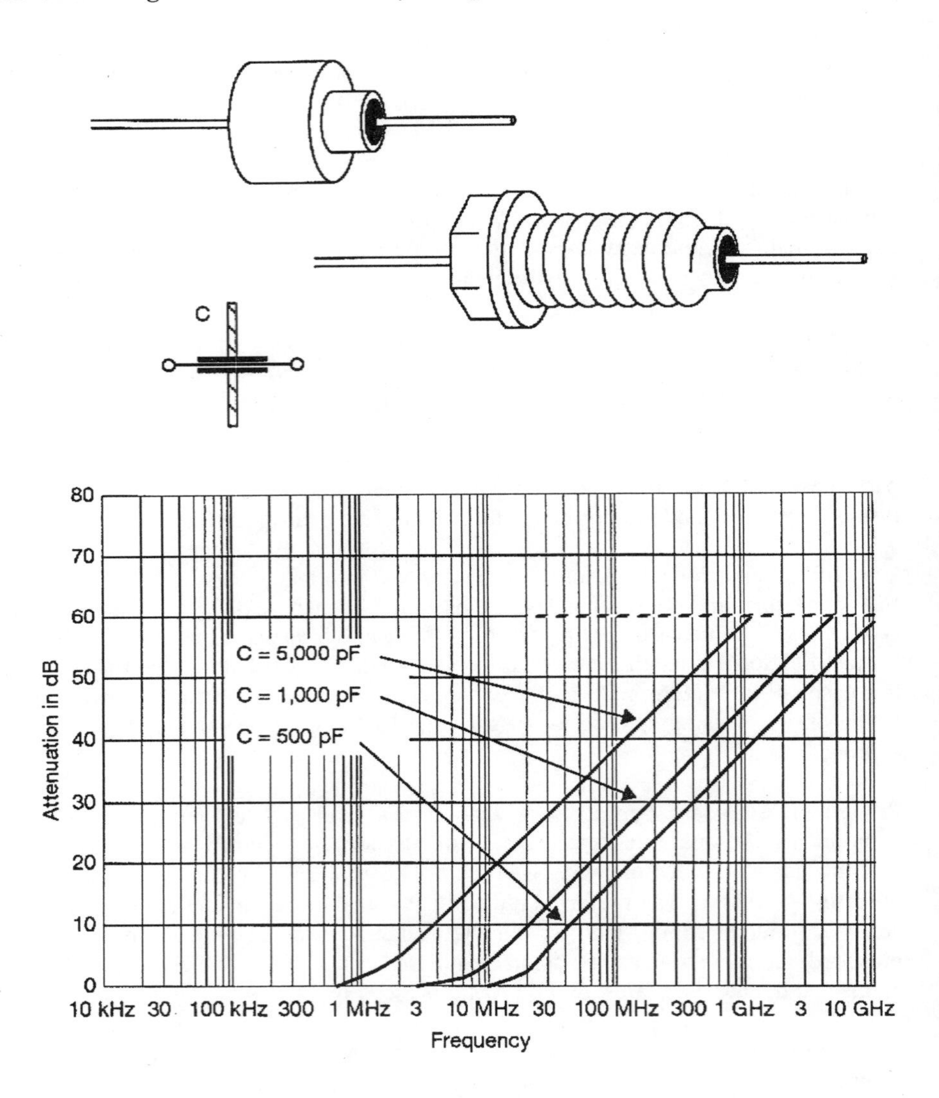

Note: Graph shows typical attenuation in a 50 Ω/50 Ω configuration. For other impedances, correct the curve data using the following expression:

$$20 \log \left[\frac{R_L R_g}{25(R_L + R_g)} \right]$$

FIGURE 9.7 Feed-Through Capacitive Filters

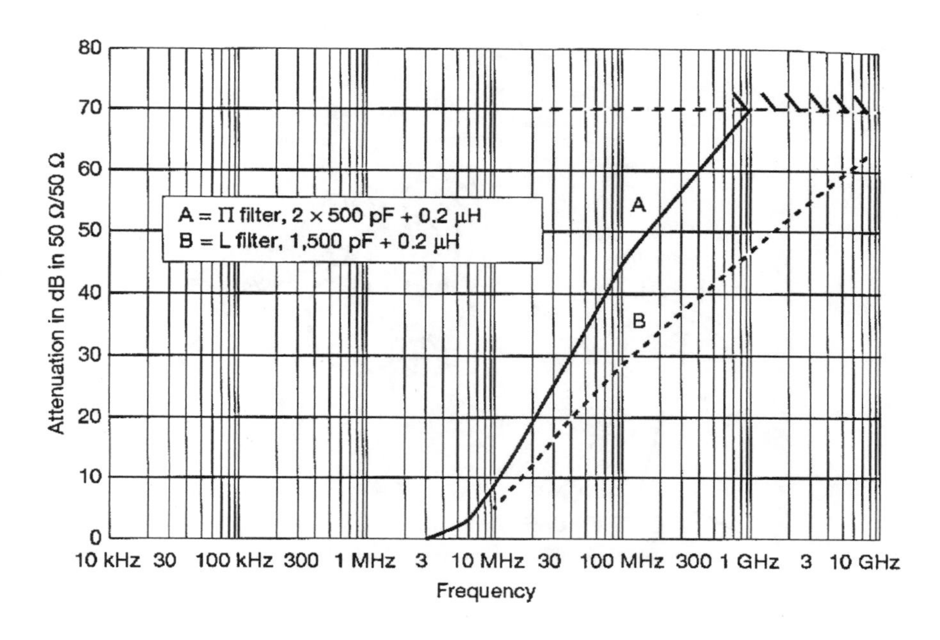

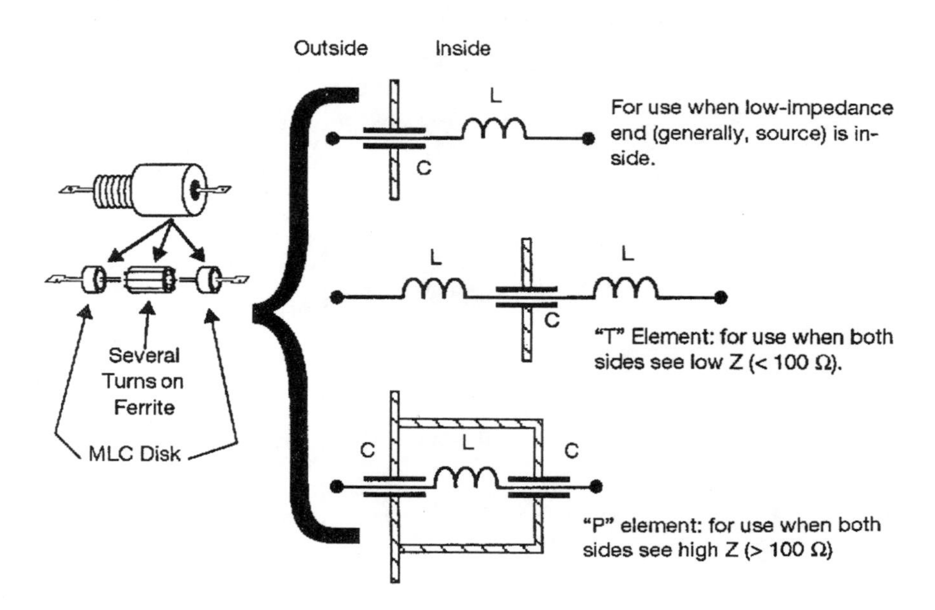

FIGURE 9.8 Feed-Through Filters, Two- and Three-Stage for Low-Current Signal Leads

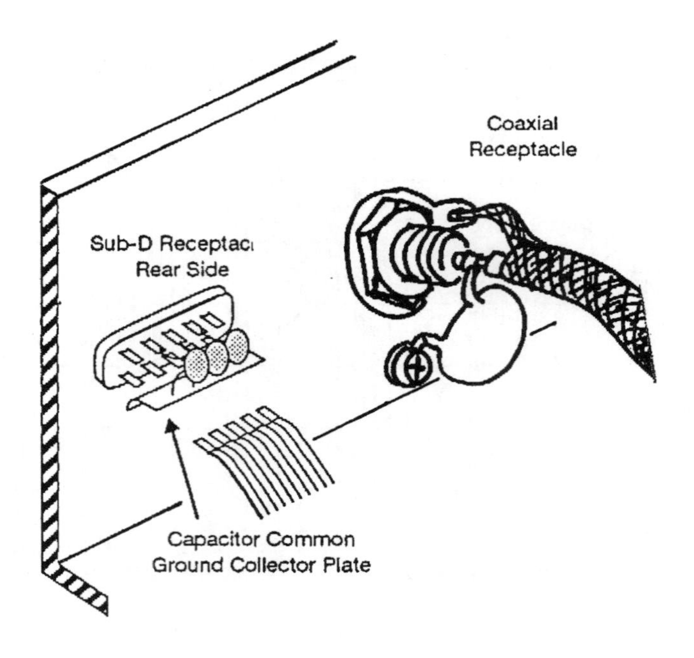

"Home-Made" Decoupling for I/O Ports

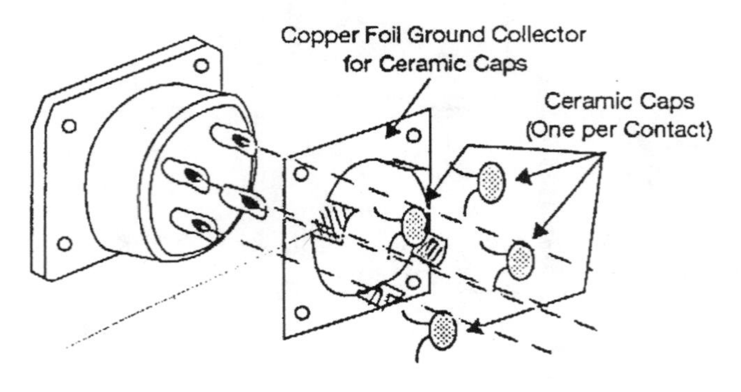

MIL Circular Connector, "Home-Made" Capacitive Decoupling

FIGURE 9.9 "Homemade" Decoupling of I/O Connectors Using Discrete Capacitors

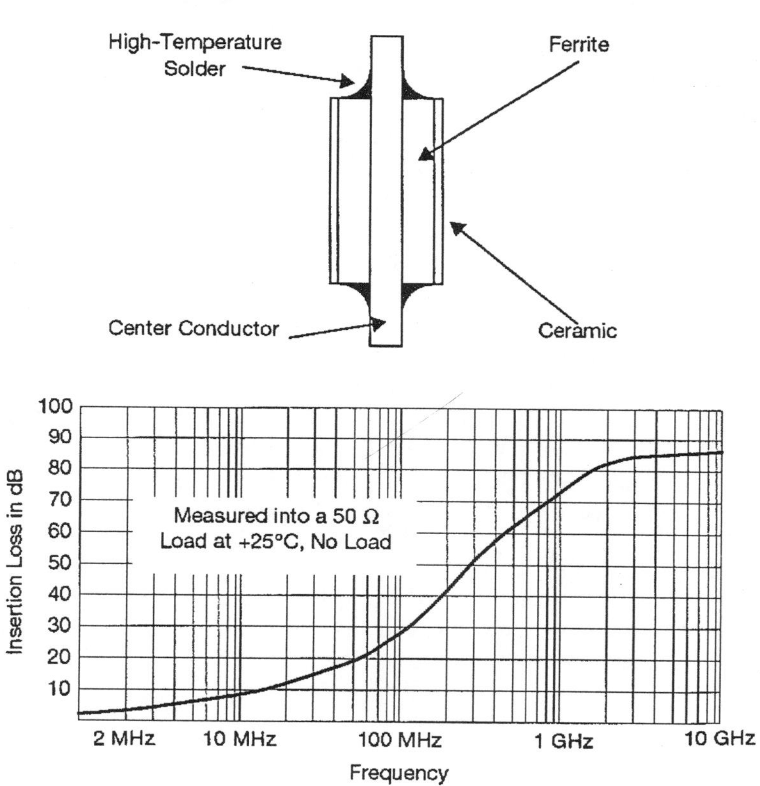

FIGURE 9.10 Example of a Commercial Filtered Connector (Source: AMP / Spectrum Control, Inc.). Attenuation for Pi filter contacts, 1,200 pF.

Trying to clean-up PCB traces of some undesired HF contents, one should be careful if relying solely on capacitive filtering. A single capacitor across an I/O trace will shunt the HF current back to its source, returning through the PCB ground network. This will reduce the HF contents escaping to the driven line, but may increase the PCB common mode ground noise, in the same proportion. The whole attempt may be a frustrating experience since a noisier PCB ground will drive more CM current into the external cabling.
Fig.9.11 shows two classical, treacherous pitfalls of this sort. In part A), a 200pF bypass capacitor has been installed at the I/O interface, to smooth the edges from the original driver output. But ground segment B-G, far from being a perfect plane is a heavily slotted one, or a set of ground traces, with a total inductance of 10nH, i.e. 10Ω @ 150 MHz.
Therefore, capacitor C plays as an exciting source to segment B-G, with point B raising at a fraction of total output $V_{A\text{-}G}$. Ground noise $V_{B\text{-}G}$ becomes a CM driving voltage for the I/O cable, and eventually *for the entire system*, since it can now feed other cables, including the power cord. The problem goes from bad to worse if several parallel outputs are filtered, because return currents now accumulate along the same ground impedance B-G.
In part B), the capacitor has been choosen to prevent the I/O trace, then the external pair, from being polluted by the X-talk from a nearby clock trace. Let us assume $V_{A\text{-}B}$ = 100mV @ 150 MHz, such as the X-talk coupling impedance to the 100Ω load behaves as a 1mA current source. Without any filtering, since the external link is balanced to 5%, $V_{A\text{-}B}$ is injecting a CM current equal to:

$$I_{cm} = 5\% \, x(100mV / 100\Omega) = 0.05mA$$

50μA is 16 times more than our 3μA criteria for not exceeding FCC class B (see Sect. 4.3). When the 200pF by-pass is added, it represents 5Ω @ 150MHz, thus reducing $V_{A\text{-}B}$ by 20 times, i.e. about what we needed. But the entire 1mA is now sinking through the ground segment B-G. With the same 10nH inductance representing 10Ω @ 150MHz, $\Delta V_{B\text{-}G}$ = 10Ω x 1mA = 10mV. Taking an average CM characteristic impedance of 300Ω for the external cable (see Sect. 2.5.1) this is causing a 33μA external current, still ten times more than our criteria!

In both cases, by using brute force capacitive bypass, we have traded one mode of external cable excitation for another, with no or little benefit. Of course, increasing the capacitor value would bring nothing, eventually exacerbating the problem. So, as attractive and unexpensive as they are on PCBs, capacitive-only filters should be used only when:

- a good, low impedance ground plane exist between the HF source and the I/O interface area.

or, at least,

- if the PCB ground is tied to a metal chassis plate very close (within centimeters) to the I/O filtering point.

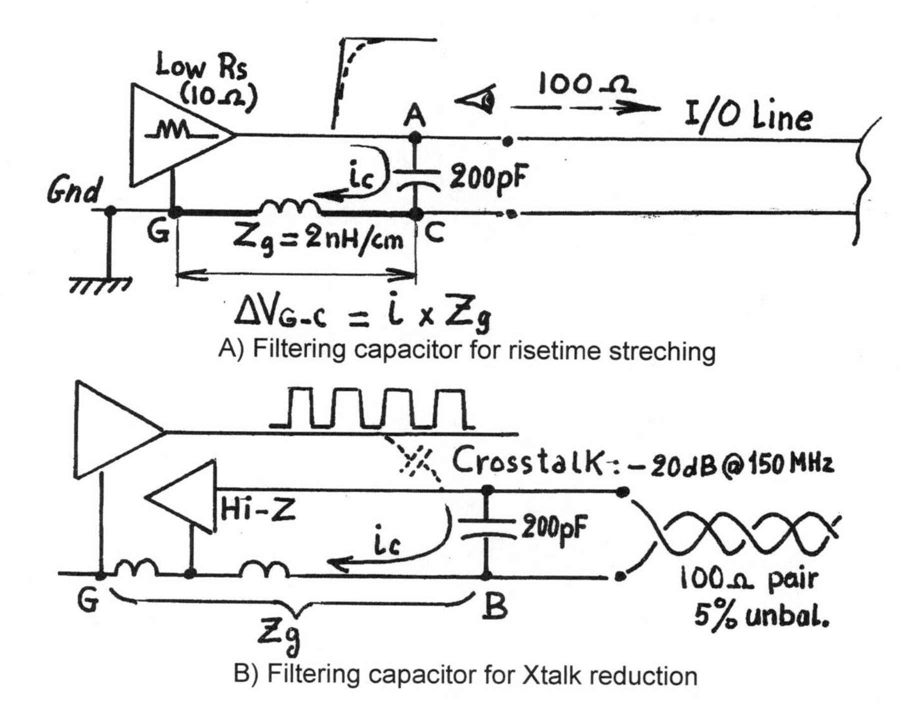

FIGURE 9.11 Some pitfalls with capacitive-only filters. Ground impedance noise$V_{C\text{-}G}$ is acting as a CM drive for the external cable. In A), problem is even worse if the filtered ouput is a parallel bus with 4,8 lines etc ... adding as many times i_c. $V_{G\text{-}C} = \Sigma\ i_c \times Z_g$

If the HF source is a low impedance device, adding a series R or L component will prevent the capacitor acting as "current sucker". Seasoned EMC practitionners often use the "100Ω - 100pF" rule of thumb whereby a 100pF capacitor, looking toward the high impedance side, is teamed with a 100Ω resistor or ferrite bead looking toward the lower impedance side. A simple resistor is an economical choice with ICs drawing only one mA. When a PCB trace is carrying more than 10mA, the ferrite is preferred, as explained next.

For more attenuation, or a steeper attenuation slope, decoupling capacitors can be complemented by discrete or SMT ferrites to form L, T or Pi filters. With LC filters, the ferrite should always be looking toward the low-impedance side of the circuit (generally the source), and the capacitor toward the high-impedance (generally the load).
Ferrite beads encapsulated with ceramic capacitor, forming a three-lead "T" filter are readily available from several vendors (see Fig. 9.13).

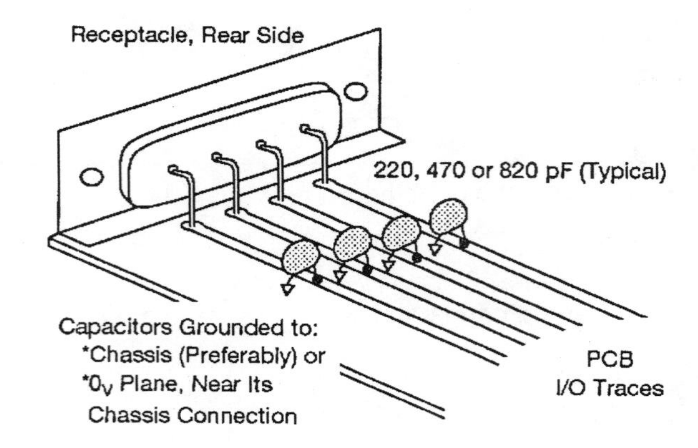

Filtering with Discrete Capacitors

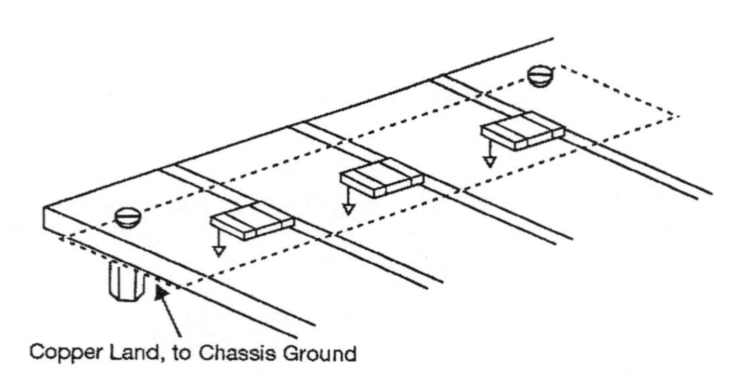

Filtering with Surface-Mount Capacitors

FIGURE 9.12 Simple Methods of Capacitive I/O Filtering at the PCB Edge (continued next page)

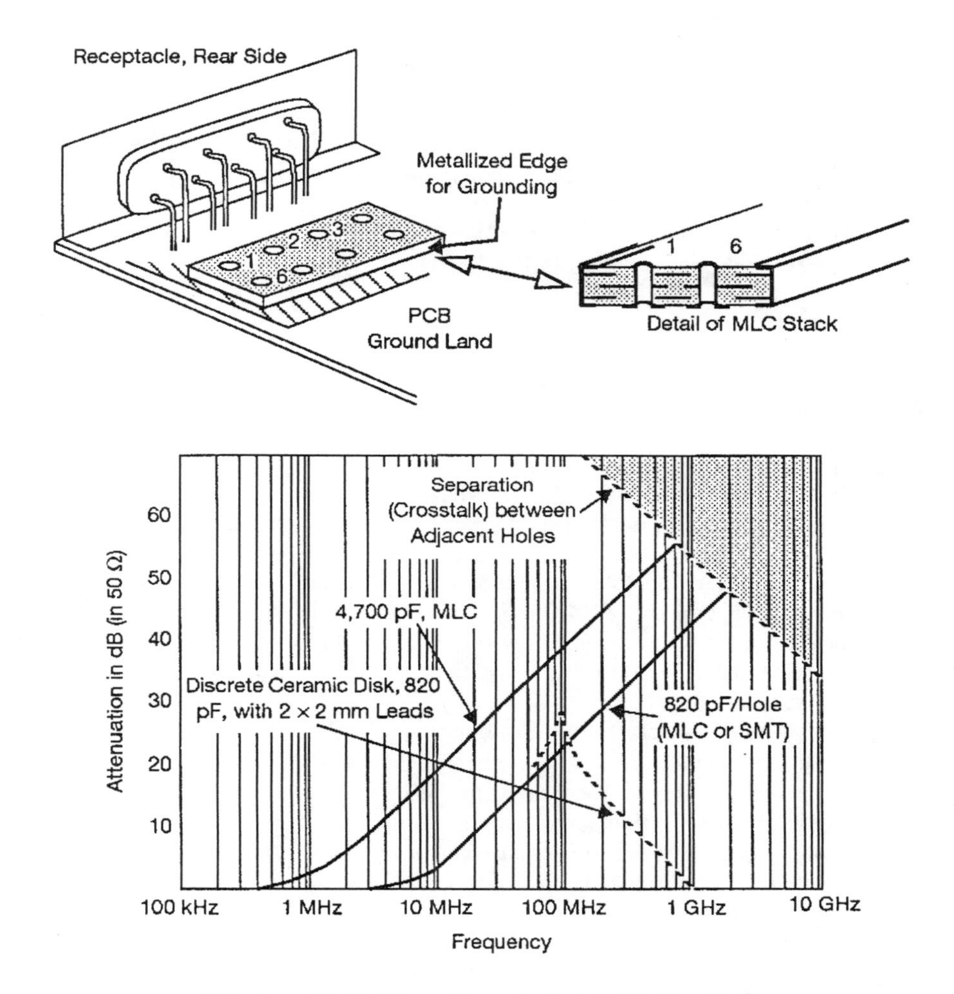

Filtering with Multilayer Ceramic Array—The Top Right Curve Indicates the Amount of Pin-to-Pin Crosstalk

FIGURE 9.12 (continued)

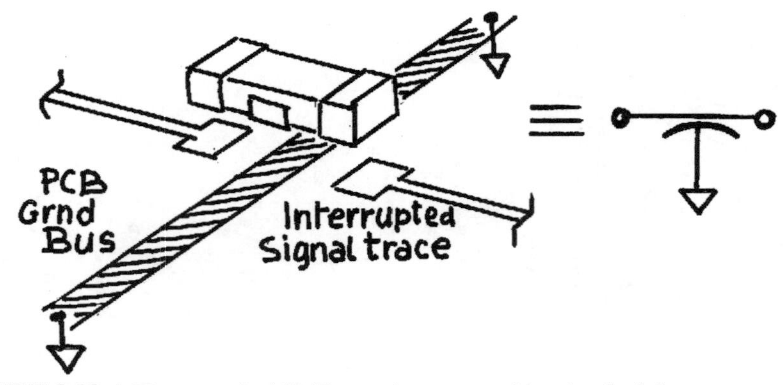

FIGURE 9.12.A Three terminal SMT capacitor, approaching the feed-thru concept

Careful mounting is necessary to avoid:
• performance degradation above 30-100 MHz
• serious line mismatch if output signal speeds require transmission line treatment

Another way of blocking CM currents consists in using longitudinal CM chokes (sometimes improperly called "baluns"). These are available in styles ranging from large toroids down to miniature multiline DIP CM transformers. Since these components belong to the ferrite family, they will be covered in Chapter 11, with cable EMI suppression.
Filtered connectors also exist in PCB-mount style. In this case, one must make sure that the metallic connector housing, which is the capacitor common, is positively grounded to the chassis (or, by default, to the PCB ground plane). The author has seen occasions where a filter connector housing was simply riveted to an epoxy glass board, with no contact to ground whatsoever. Not only was the filter useless (as discovered by the technician), but the filter capacitors' floating common created crosstalk between the different lines.

Capacitor values, or more generally the filter characteristics, should be calculated at least crudely and not picked up by chance, as is too often the practice. The following steps are recommended:

1. Identify the total resistance (or impedance) across the line to be filtered by taking R_{source} and R_{load} in parallel. If the line is electrically long, replace R_{load} by Z_0, the characteristic impedance of the line. For example: for R_s (driver) = 120 Ω and R_L (receiver input) = 1,500 Ω, take R_t = 110 Ω.
2. Identify the highest frequency, Fmax, to be processed.
3. On a balanced pair, the maximum tolerable value of each line-to-chassis capacitor, for no visible signal distortion, is given by the condition:

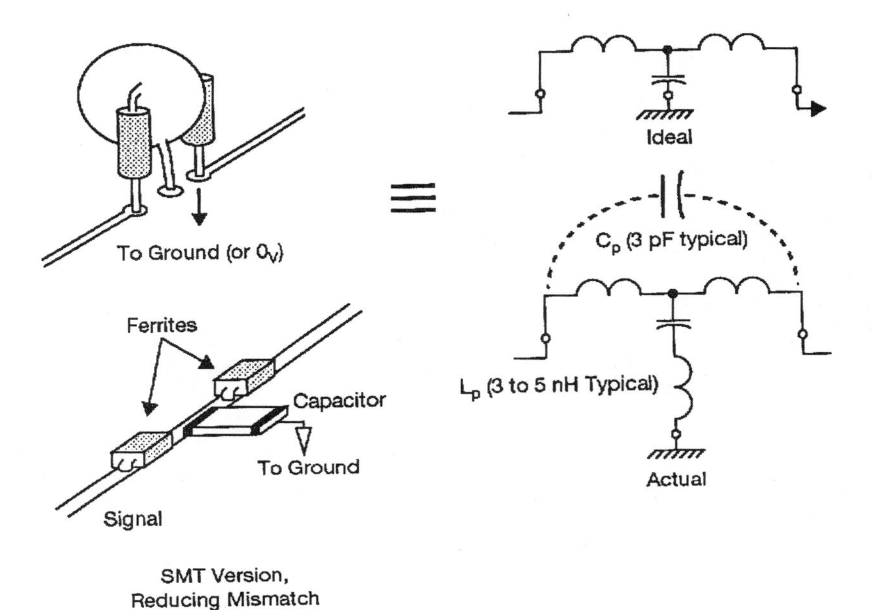

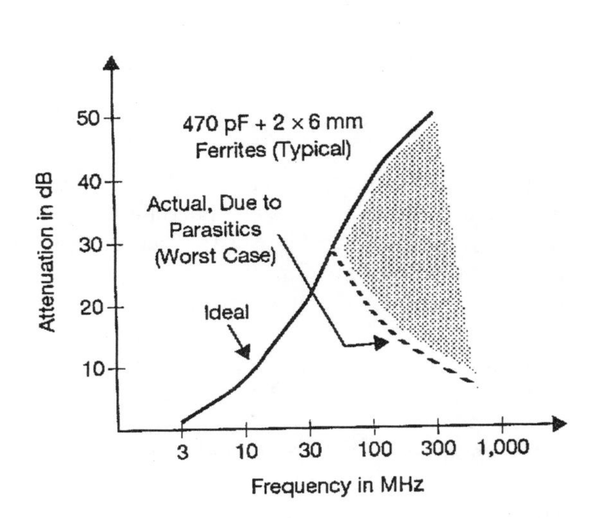

FIGURE 9.13 "T" Filters for PCB Mounting

$$\frac{1}{2\pi\, F_{max} \times \frac{C}{2}} \geq 3 \times R_t \qquad (9.1)$$

Which can translate simply as:

$$C_{max} = \frac{100}{F_{max} \times R_t} \qquad (9.2)$$

for, C in nanofarads and F in megahertz

For instance, if the highest bandwidth to process is 3 MHz, for the 110 Ω calculated previously applying (9.2) :

$$C_{max} = 100/(110 \text{ x } 3) = 0.3\text{nF}$$

4. If, instead of frequency, the rise (or fall) time of the useful signal is known:

$$C_{max} = 0.3\ t_r/R_t \qquad (9.3)$$

with,

C in nanofarads, for t_r in nanoseconds
C in picofarads, for t_r in picoseconds

For instance, if the typical rise time is 100 ns in our 110 Ω example applying (9.3),

$$C_{max} = 0.3 \text{ x } 100 /110 = 0.3 \text{ nF}$$

5. If a slight degradation of the rise time (stretching, or corner-rounding) is tolerable, up to 3 times the calculated C_{max}, corresponding to the 3 dB point in the attenuation curve, can be used (see Fig. 9.14). Therefore, in most practical cases, the value calculated by Eq. (9.2) or Eq. (9.3) can be easily rounded up to the next standard value. In our example, 330 or 470 pF are probably adequate.

6. If the EMI frequency is too close to the cutoff frequency, a one-pole filter may not be sufficient. Use a two- or three-pole (T, Pi) filter.

7. Check the service voltage and surge withstanding voltage against the application.

8. Select a capacitor series with tight tolerances. Large tolerances (such as -0/+50 percent) would not hurt for ordinary decoupling but could introduce line unbalance in differential links. Prefer NPO ceramic grade, for precision and stability.

9. Install the capacitors with the shortest leads possible, trimmed close to the capacitor body, or use SMT components.

10. Check that the frequency domain to be attenuated is not significantly within the parasitic (inductive) region of the capacitor.

11. Make sure that all I/O lines have been decoupled in the same zone. One single line (even a dormant one) left unfiltered can couple capacitively or magnetically with the others (see Fig. 9.15).

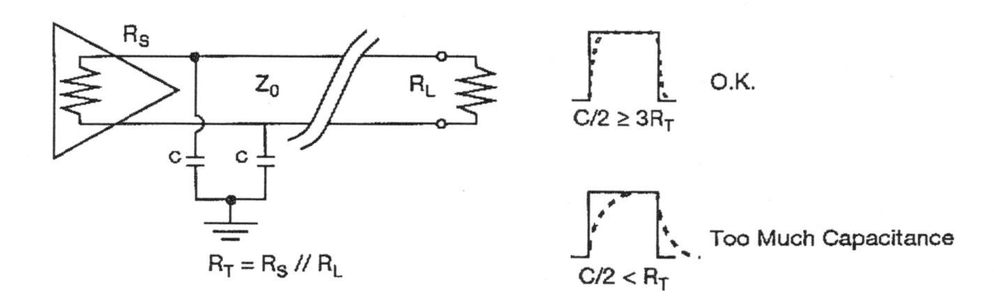

	Slow Interface (Typically 10–100 kB/s)	*Fast Interface (Typically 2 MB/s)*	*Slow CMOS*	*TTL*
Rise Time (t_r)	0.5–1 µs	50 ns	100 ns	10 ns
Bandwidth	320 kHz	6.0 MHz	3.2 MHz	32 MHz
R_T (typical)	120 Ω	100 Ω	300 Ω	100–150 Ω
C_{max}	2,400 pF	150 pF	100 pF	30 pF
C_{max} (if a slight t_r degradation is tolerable)	7,000 pF	450 pF	300 pF	100 pF

FIGURE 9.14 Maximum Permissible Values for Common-Mode Decoupling capacitors

Example 9.2: Filter Selection
An I/O signal on a balanced pair uses rise times of 20 ns. The source internal resistance (high-to-low) is 100 Ω, driving an electrically long external line with 120 Ω characteristic impedance.
Conducted tests of the prototype (or EMI analysis) have shown that excessive current will exist on an I/O cable in the 50 to 70 MHz region, with an excess of 20 dB above the desired objective. Determine the optimal filter.

Solution:
The equivalent bandwidth for 20 ns is:

$$F = 1 / \pi t_r = 16 \text{ MHz}$$

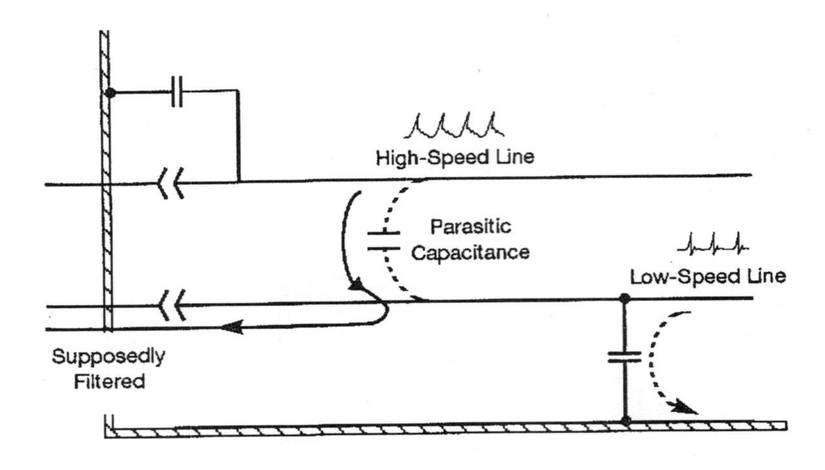

FIGURE 9.15 Back-Door Reentry of Noise when Decoupling Components are not Grouped in the Same Area

The total impedance for 120 Ω in parallel with 100 Ω is 54 Ω. Attempting a single capacitive filtering, we have, for each capacitor:

$$C_{max} = 0.3 \times 20\text{ns} / 54\ \Omega = 0.110\ \text{nF}$$

At 50 MHz EMI frequency, 110 pF represents 28 Ω of impedance; i.e., only about half R_t. So the attenuation will be roughly 6 dB, compared to the 20 dB objective.
A multiple-element filter is needed. Choosing its cutoff frequency about 20 percent above the useful bandwidth of 16 MHz gives:

$$F_{co} = 1.2 \times 16 = 19\text{MHz}$$

$$F_{EMI} / F_{co} \approx 50 / 19 = 2.6$$

A three-stage, 60 dB/decade “Tee” filter will provide, at 50 MHz, an attenuation of:

$$AdB = 60 \log 2.6 = 25\ \text{dB}$$

9.6 A FEW OTHER RADIATED EMI SOURCES

So far, we have been concentrating on SMPS, RF oscillators and digital circuits as major EMI sources. A few other devices can cause steady intrasystem or intersystem radiated interference. The following is a less than exhaustive list of such sources:

- CRTs
- electroluminescent displays
- fluorescent tubes (used inside certain categories of equipment)
- electric motors (dc and ac)
- printer head solenoids
- laser emitters
- ultrasonic generators
- plasma and controlled arc generators
- X-ray generators (in addition to their ionizing radiation)

It is difficult to provide characteristic values of the radiated levels for these components, as they depend strongly on the type and manufacturer of these OEM devices.
When equipment incorporates such devices, it is recommended that the designer obtain from the vendor a radiated field profile of the product—the vendor normally will have performed some appropriate test. If no such data is available, a radiated EMI test should be carried as soon as possible by an independent lab or the prime user himself. This will serve to identify if (a) the device will cause the host machine to exceed its relevant specification level and (b) the device could cause internal EMI. In either circumstance, a proper shielding must be provided.
As an example of such OEM emissions, the following table shows the radiated field levels at 10 cm from a small black-and-white CRT—Fields are those due to the CRT Circuit itself, and not to associated control logic.

F	*16 kHz*	*30 kHz*	*100 kHz*	*300 kHz*
$H_{(A/m)}$	0.1	0.07	0.02	$0.4.10^{-3}$
$E_{(V/m)}$	11	6	1.2	0.05

Chapter 10

Box Shielding

All the former efforts being accounted, a conductive box may constitute the ultimate barrier against radiated emissions. However, more than often, electronic cabinets or equipment housings were traditionally designed per one, or a combination of the following approaches (which, in other circles, would be quoted as receipes for failure):

1. Make the enclosure similar to earlier versions that are known, or presumed to be EMI-free. Then, to confirm expectations, test the box when a prototype is available.
2. Starting from the ground up, design and construct a box per mechanical, aesthetic, cost and accessibility requirements and test it as above.
3. Do as above, but perform only the mandatory emission tests. Do not test for susceptibility unless a specific purchasing specification calls for it.

Such a strategy - or lack of strategy - allows the final test to govern the outcome of a design and results in one or more of the following:

1. Time and money are wasted during the hit-or-miss process.
2. Components or techniques which are not optimized become integral parts of the product.
3. EMC overdesign occurs, with its plethora of cost, weight and maintainability additions.
4. EMC underdesign occurs because tests sometimes give a less than complete simulation of all possible EMI situations.

This being said (and regretted), the designer who prefers an analytical approach faces the following questions

1. How much attenuation (if at all) should the enclosure provide?

2. How can one design an enclosure to meet the attenuation requirements before any prototype exists?
3. If item 1 is not known (as is usually the case) how can it be quantified?

In light of the above, a deterministic approach to the EMC design of the equipment enclosure is needed. The two being related, we will address a combined emission and susceptibility strategy, then concentrate on emissions for the application part. This strategy is derived from Ref. 27, which offers a complete compilation of present knowledge on this subject.

10.1 HOW TO DETERMINE BOX ATTENUATION REQUIREMENTS

Using the flow diagram of Fig 10.1, the designer first asks whether the required shielding effectiveness (SE) is already known across a defined spectrum. While SE requirements generally are not known, there are cases where procurement specifications or test data from a similar equipment dictate the amount of shielding needed. If the SE requirement is known, the routine is bypassed, except (eventually) for adding an appropriate safety margin (exit at the bottom of the figure).

Since the needed SE is usually unknown, the flow diagram covers three cases:

1. Shielding for susceptibility hardening

a. Determine the ambient threats (e.g., LF magnetic field, electric field), frequencies and amplitudes. This is based on the product's intended application and location, and provided by applicable immunity specifications. For a new application, if no adequate specifications exist, a site survey is required.
b. Compute, or evaluate on a breadboard prototype, the interference situation via the coupling of fields to internal cables and PCBs. This includes the in-band and out-of-band response of victim circuits.
c. The desired SE is the difference in decibels between the imposed threat and the "bare-bones" susceptibility of the unshielded equipment.

2. Shielding for emission control

a. Compute (see Chapter 2), or measure on development prototypes, the radiated emission levels for each major subassembly to be housed in the box, excluding I/O cables. (Their radiation needs to be addressed and resolved separately from box shielding.) For each frequency interval of at least one decade (half-decade intervals are strongly recommended) record the highest calculated or measured field level up to approximately 10 x F_2. F_2 represents the highest significant frequency of the voltage or current spectrum (for instance, $1/\pi tr$ for pulsed signals).
b. If several amplitudes are in the same range, compute their combined effects. Once the radiated field envelope is drawn across the spectrum for the unshielded electronics, it is compared to the applicable civilian or military specification.

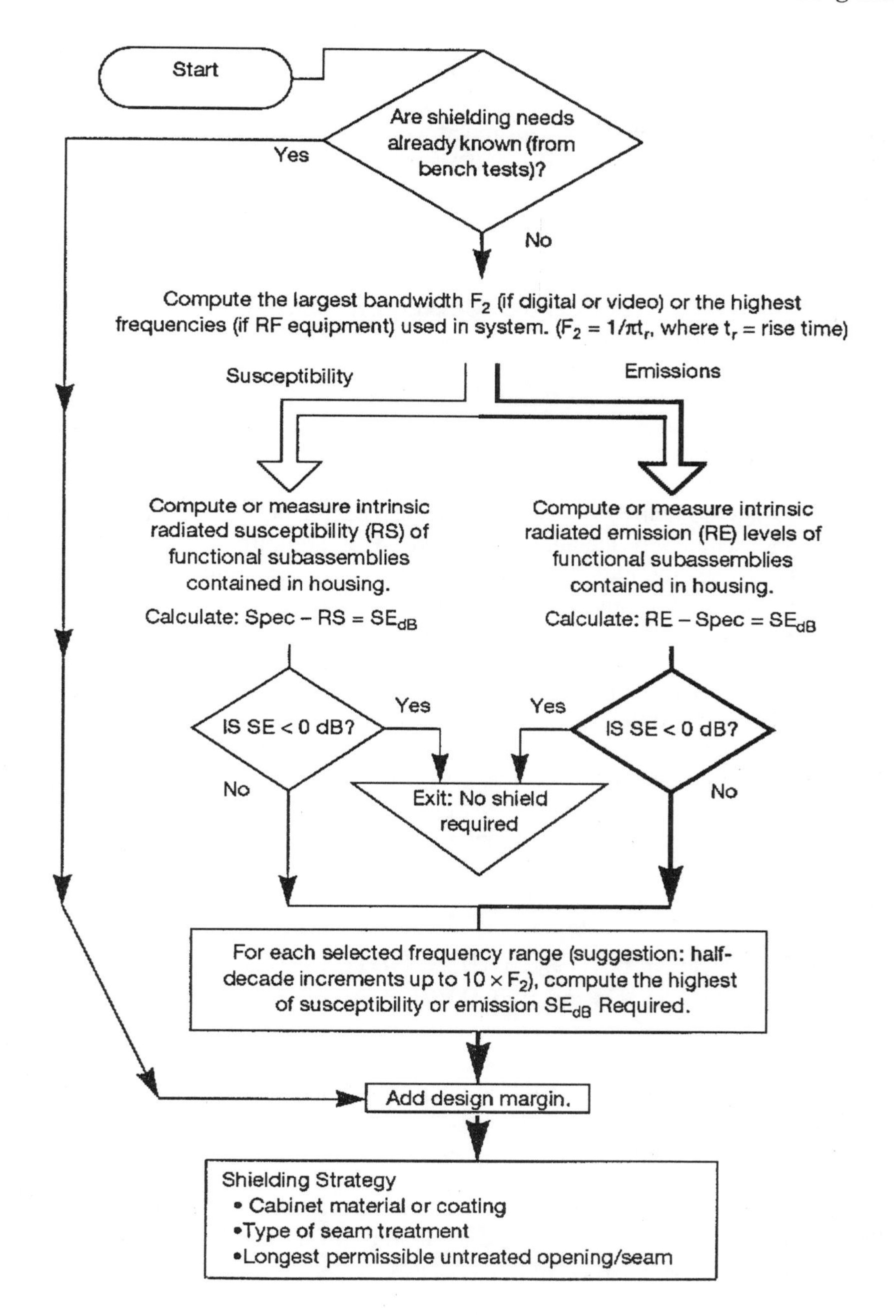

FIGURE 10.1 Flow Diagram for Shielding Design. Emphasis is on radiated emission shielding. in the right-hand branch. (From Ref. 15)

3. Optimized shielding for susceptibility and emission control.

After items 1 and 2 have been determined, compare in each frequency interval the susceptibility SE_1 and emission SE_2, and retain the toughest of the two requirements. "Toughest" does not necessarily mean the highest figure in dB. For instance, 20 dB of SE against a near-field magnetic source may be harder to achieve than 60 dB against an E field or plane wave at the same frequency.

The SE requirements having been established, what remains is to select or verify:

- the cabinet material
- the way apertures and seams will be treated
- the surface treatment/finish if specific corrosion and longevity requirements exist.

Over, the years, clock frequencies have increased constantly. For emission control, the present 300 MHz or higher spectrum span obliges the designer to consider possible leakage from any slot that exceeds a few centimeters height or width.
Decades ago, empirical methods often led to a "steamroller" approach where equipment housings resembled a vault. Such techniques, although effective, increase manufacturing and hardware costs and complicate maintenance and accessibility. In addition, aesthetic and weight considerations prohibit the use of certain shielding materials. Typically, then, the designer will look for shielding techniques that are economical and remain unaltered after intensive use across equipment lifespan. At the risk of overgeneralizing, emission SE requirements are usually less demanding than immunity SE requirements, particularly for military environments. Even with a mediocre design of the PCB and internal packaging (which means the shield will have to make up for internal deficiencies), SEs in the 10 to 40 dB range for civilian applications, and in the 30 to 60 dB range for MIL-461, typically will be required at the worst offending frequencies.

10.2 SOME SHIELDING BASICS: SHIELDING EFFECTIVENESS OF MATERIALS

Although a comprehensive coverage of shielding theory is far beyond the scope of this book a few guidelines are provided on how and why shields work and examples are given of when they do not. The reader who wants to know more about the principles and applications of shields is invited to refer to the much more complete book of Ref. 27.

Shielding effectiveness (SE) is defined as the ratio of the impinging radiated power to the residual radiated power (the part that gets through).

For E fields:

$$SE\ (dB) = 20 \log E_{in} / E_{out}$$

For H fields:

$$SE\ (dB) = 20\ \log H_{in} / H_{out}$$

If shields were perfect, E_{out}, H_{out}, and, therefore, Pout would be zero. In practice, a shield is merely an attenuator that performs on the basis of two principles: absorption and reflection (see Fig. 10.2).

Absorption increases with:
- thickness
- conductivity
- permeability
- frequency

Reflection increases with:
- surface conductivity
- wave impedance

To evaluate *absorption,* or penetration losses, one needs to know how many skin depths the metal barrier represents at the frequency of concern, knowing that the field intensity will decrease by 8.7 dB (or will lose 63 percent of its amplitude) each time it has to go through one skin depth.

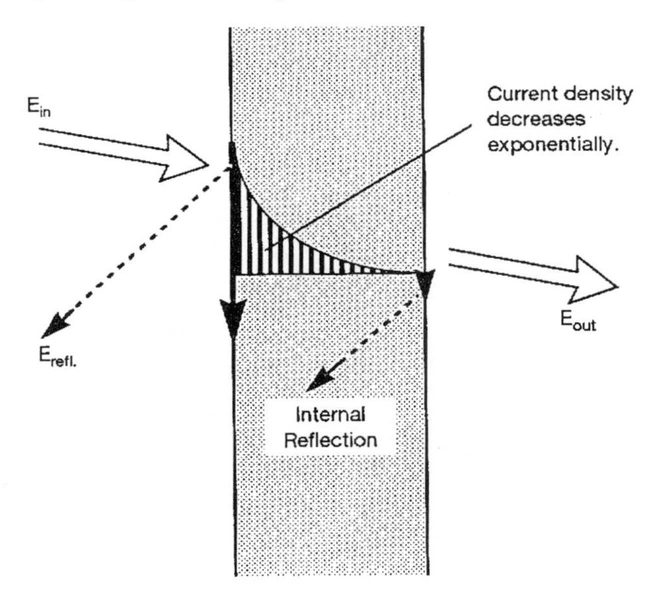

FIGURE 10.2 Basic Shielding Mechanisms

Entering all the electrical constants, we come to a simple expression for absorption loss:

$$A_{dB} = 131\ t\sqrt{F\mu_r\sigma_r} \qquad (10.1)$$

where,
t = thickness of conductive barrier in mm
F = frequency in MHz
μ_r = permeability relative to copper = 1 for non-magnetic materials (Fig.10.3)
σ_r = conductivity (the inverse of resistivity) relative to copper

For example, a 0.03 mm (1.2 mil) aluminium layer will offer an absorption loss at 100 MHz of:

$$AdB = 131 \times 0.03\sqrt{100 \times 1 \times 0.6} = 30.4 \text{ dB}$$

This is equivalent to a field strength reduction factor of $(10)^{30.4/10}$ = 33 times.

Looking at Eq. (10.1) leads to a few remarks:

1. *For nonmagnetic materials* (μ_r = 1), the penetration losses increase with conductivity, σ_r . Since no metal offers better conductivity than copper (except for silver, with σ_r = 1.05), any nonmagnetic metal will show less absorption than copper. Zinc, for instance, for which σ_r= 0.3, will exhibit, for a thickness of 0.03 mm (1 mil), an absorption loss at 100 MHz of:

$$AdB = 131 \times 0.03.\sqrt{100 \times 1 \times 0.3} = 20 \text{ dB}$$

2. For magnetic materials (μ_r > 1), the penetration losses increase with μ_r. On the other hand, their conductivity is less than copper's. Since μ_r for steel or iron is in the range of 300 to 1000, while σ_r is about 0.17, a definite advantage exists for magnetic materials. However, above a few hundred kilohertz (ferrites excepted), μr generally collapses to equal 1, while σ_r is still mediocre.

To evaluate *reflection*, one must know if the shield is in near- or far-field conditions. If far-field, the reflection loss is given, neglecting internal multiple reflections, by:

$$R_{dB} = 20\ \log\frac{(K+1)^2}{4K}, \text{ where } K = \frac{120\pi}{Z_b} \qquad (10.2)$$

Which, for K > 3, simplifies as:

$$R_{dB} = 20\ \log\frac{120\pi}{4Z_b} \qquad (10.2a)$$

Near-field conditions, where the shield is closer than λ/6 to the source, are the most critical. For pure electric fields, because their wave impedance is high, it is relatively easy to get good reflection properties because the field-to-shield mismatch is large. For nearby magnetic fields, the wave impedance is low, and it is more difficult to get good reflection.

For *near-field* conditions, the reflection losses are equal to:

For E fields,

$$R_{dB_{(E)}} = 20 \log \left(\underbrace{\frac{120\pi}{4Z_b}}_{\substack{\text{far-field} \\ \text{reflection} \\ \text{term}}} \times \underbrace{\frac{\lambda}{2\pi D}}_{\substack{\text{near-to-} \\ \text{far-field} \\ \text{correction}}} \right) \quad (10.3)$$

For H fields,

$$R_{dB_{(H)}} = 20 \log \left(\underbrace{\frac{120\pi}{4Z_b}}_{\substack{\text{far-field} \\ \text{reflection} \\ \text{term}}} \times \underbrace{\frac{2\pi D}{\lambda}}_{\substack{\text{near-to-} \\ \text{far-field} \\ \text{correction}}} \right) \quad (10.4)$$

$$= 20 \log \frac{2\ DF}{Z_b} \quad (10.4a)$$

where,

Z_b = shield barrier impedance in ohms/square
F = frequency in megahertz
D = distance from radiating source in meters

(How does one know if, at distance << λ, the field is more electric or magnetic in nature? By looking at the radiating source, one might gather an idea of the predominant mode: circuits switching large currents (such as power supplies, solenoid drivers and heavy current logics) generate magnetic fields. Conversely, voltage-driven high-impedance or open-ended lines create electric fields.

Figures 10.3 and 10.4 show shielding properties of some materials.

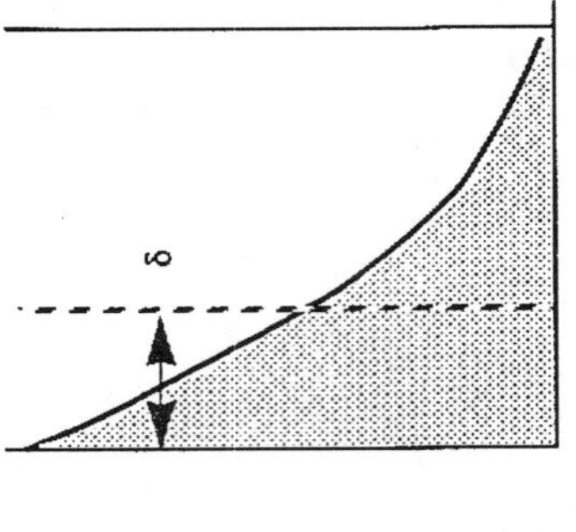

	σ_r	μ_r*	Skin Depth, S δ 60 Hz	10 kHz	1 MHz	100 MHz
Copper	1	1	8.5 mm	0.66 mm	0.066 mm	6.6 μm
Aluminum	0.6	1	11 mm	0.85 mm	0.085 mm	8.5 μm
Steel (ordinary cold-rolled)	0.16	200	1.4 mm	0.14 mm	0.1 mm	16 μm
Steel (low-carbon)	0.1	1,000	0.85 mm	0.06 mm	0.04 mm	20 μm
Conetic	0.032	20,000	0.33 mm	0.058 mm	0.058 mm	38 μm
μmetal	0.028	20,000 to 60,000	0.34 mm	0.058 mm	0.058 mm	40 μm

Absorption Losses (dB)															
	Copper			*Aluminum*			*Zinc*			*Steel**			*Nickel** $\sigma_r = 0.2$		*Copper Paint*† $\sigma_r = 0.04$
Thickness (mm)	*0.01*	*0.1*	*1*	*0.01*	*0.1*	*1*	*0.01*	*0.1*	*1*	*0.01*	*0.1*	*1*	*0.01*	*1*	*0.05 mm (2 mil)*
30 MHz	7	70	700	5.2	52	520	4	40	400	3	28	200	3	31	7
100 MHz	13	130	>1,000	9.5	95	950	7	72	720	5	50	500	6	58	13
300 MHz	22	220	>1,000	17	170	>1,000	12	125	>1,000	9	88	880	10	98	22

*Although steel and nickel are magnetic materials, their relative permeability μ_r collapses to ≈ 1 above a few hundred hilohertz. This has been considered in calculating δ and losses.

†Nonhomogeneous metal

FIGURE 10.3 Skin depths and Absorption Losses for various Materials (reminder : Absorption does not depends on the E,H or plane wave nature of the field)

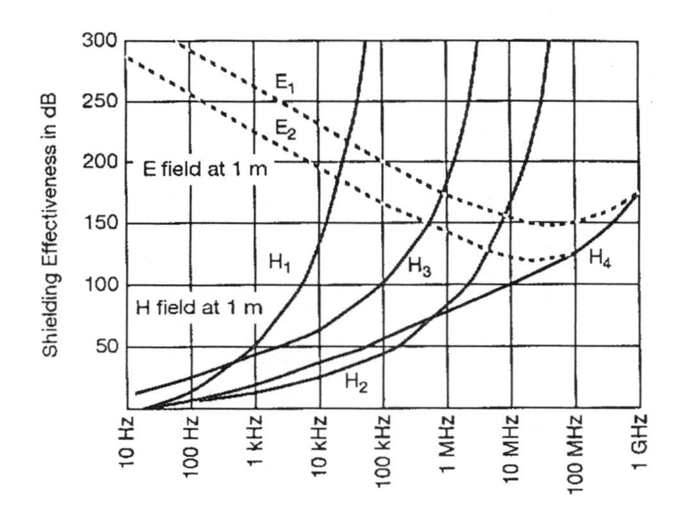

FIGURE 10.4 Total Shielding Effectiveness (Absorption + Reflection) of a few Typical Materials, against E or H field Sources at 1 m distance. H_1 = steel, 0.8mm; H_2 = steel, 0.025 mm; H_3 = copper, 0.8 mm or aluminium, 1 mm; H_4 = copper, 0.025 mm; E_2 = steel, 0.025 mm; E_1 = copper or aluminium, 0.025 mm.

10.3 SHIELDING EFFECTIVENESS OF CONDUCTIVE PLASTICS

Plastic housings provide no shielding whatsoever. Therefore, unless the PCBs and internal wiring have been hardened sufficiently, the plastic must be made conductive. Several metallizing processes exist, as summarized in Fig. 10.5, along with their average 1990 costs. Since, as discussed in Section 10.2, thin coatings exhibit rather low absorption loss, their only chance to work is by reflection. Based on reflection loss only, Fig. 10.6 shows the shielding effectiveness of thin coatings versus their distance/wavelength relation with the source. (A more detailed explanation can be found in Ref. 15). If shielding effectiveness in the range of 30 to 40 dB is desired, especially against low-impedance sources, a conductive process giving 1 Ω/sq or less must be selected.
The case of emission shielding with metallized plastic is more critical. Since the sources are inside the box, the conductive surface very often will be in the near-field conditions. At 10 cm from a source, one has to wait until 500 MHz is reached to be in a far-field situation. Because many of the radiating sources (e.g., current loops, capacitors discharges, bus drivers) exhibit less than 377 Ω impedances, the reflection performances of metallized plastics will be less than their far-field figures.

	Surface Resistance in Ω/sq	SE_{dB}, Plane Wave	Cost range ** (1993) $/m2 (0.1 – 100)	Comments
Electroplating	Depends on Metal			Good adhesion to complex forms Easy bonding of metal pieces
Electroless Plating, Nickel, Copper (inside and outside)	1 0.3	40 50		Good adhesion to complex forms Easy bonding of metal pieces
Metal Spray (zinc arc)	0.2	54		
Vacuum Deposition	Depends on Metal			
*Carbon Coating (0.5 mil = 12.7 μm)	10 to 100	a few dB to 20 dB		Easy to apply Mediocre SE Adequate for ESD Drainage
*Graphite Coating (2 mil = 51 μm)				
*Copper Coating (2 mil = 51 μm), Overcoated with Conductive Graphite (0.2 mil = 5 μm)	0.2	54		Good cost/performance Resistant to oxidation if topcoat or passivation
*Nickel Coating	1	40		Easy to apply Resistant to oxidation
*Silver Coating (0.5 mil = 12.7 μm)	0.01	80		Easy to apply with conventional spray equipment
Filament-Filled Plastic	N.A.	40 for 1 Ω-cm Vol. Res.		Mediocre refl. losses in near field Good absorption > 100 MHz No additional process

*Water-based solutions available. †Includes application cost. For paints/spray, manual operation is assumed. For $/ft^2, divide figures by 10.

FIGURE 10.5 Summary of Conductive Plastic Processes

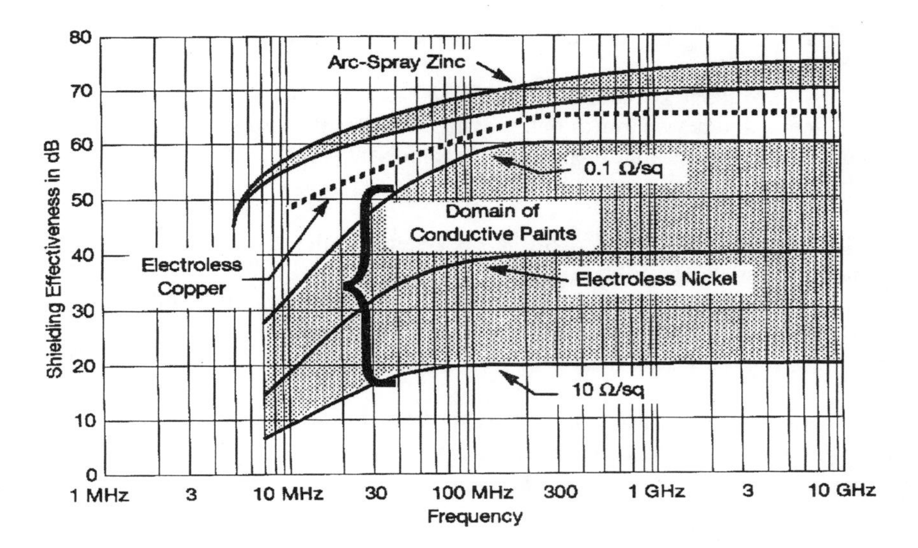

Note: Performances are given against an H field source.
For magnetic sources at closer distances, subtract

$$20 \log \frac{30 \text{ cm}}{D_{cm}}$$

FIGURE 10.6 Shielding Effectiveness of Conductive Plastic Processes (per standard 0.30 m distance test. From Ref. 27).

In addition, and in contrast to the susceptibility case, the reflected wave does not vanish in the exterior but comes back after a two-way trip inside the box. If this re-reflected wave comes in phase (which will occur as soon as the two-way trip reaches $\lambda/2$; i.e., box length reaches $\lambda/4$) we have a resonant cavity, and the effective SE will be less than calculated.

This is why, ironically, a high conductivity and a thin layer, giving excellent reflectivity and no absorption, is not the best barrier against emissions. Less conductivity but a thicker coat can give better results provided that:

$$t_{new} / t_{old} > \sigma_{r\,(old)} / \sigma_{r\,(new)}$$

In this expression, t_{new} is the thickness of the thicker, less conductive barrier, and t_{old} is the thickness of the thinner barrier. In this respect, conductive plastics with volume conductivity provided by conductive particles are more efficient than thin conductive paints or films.

10.4 FIELD ATTENUATION THROUGH APERTURES

Housings, unfortunately, cannot be made like continuous metal cubes. They have slots, seams and other apertures that inevitably leak. Like in a chain, a shield is only as good as its weakest link; therefore, it is important to know the weak points in the shields to establish some realistic objectives.

• At low frequencies, what counts is the nature of the metal used, its thickness, conductivity and permeability.
• At high frequencies, where any metal would provide hundreds of dB of shielding, this is never seen because seams and discontinuities completely spoil the metal barrier (see Fig. 10.7).

From Babinet's theory, a slot in a shield can be compared to a slot antenna which, except for a 90° rotation, behaves like a dipole (see Fig. 10.8). When the slot length reaches λ/2 (no matter how small the height), this fortuitous antenna is a perfectly tuned dipole; i.e., it reradiates outside all the energy that excites the slot inside. It may even exhibit a slight gain (about 3 dB). Below this resonance, the slot leaks less and less as frequency decreases.

A simplified expression for aperture attenuation below λ/2 resonance is (from Ref. 27):

$$A_{dB} \approx 100 - 20\ \log \ell_{mm} - 20\ \log F_{MHz} + 20\ \log\left(1 + 2.3\ \log \frac{\ell}{h}\right) + 30\ \frac{d}{\ell}$$

$$\approx 0\ \text{dB for } \ell \geq \lambda/2 \qquad (10.5)$$

This is the worst-case far-field attenuation, for the worst possible polarization (i.e., in general, actual attenuation will be better).

The first three terms represent the reflection loss of a square aperture, due to the mismatch of the incident wave impedance (120π Ω for far-field conditions) with the slot impedance.

The equivalent circuit for a slot is an inductance (see Fig. 10.8), until it resonates with edge-to-edge capacitance.The forth term is the "fatness factor" of the slot, to take into account the influence of h. Notice that h plays a secondary role by the logarithm of l/h. A slot 100 times thinner will only radiate 5 times less than the equivalent square aperture, and not 100 times less. Some typical values of this factor are:

0 dB for l = h (square aperture)
10dB h/ l =0.1
15 dB h/ l = 0.01
18 dB h/ l = 0.001

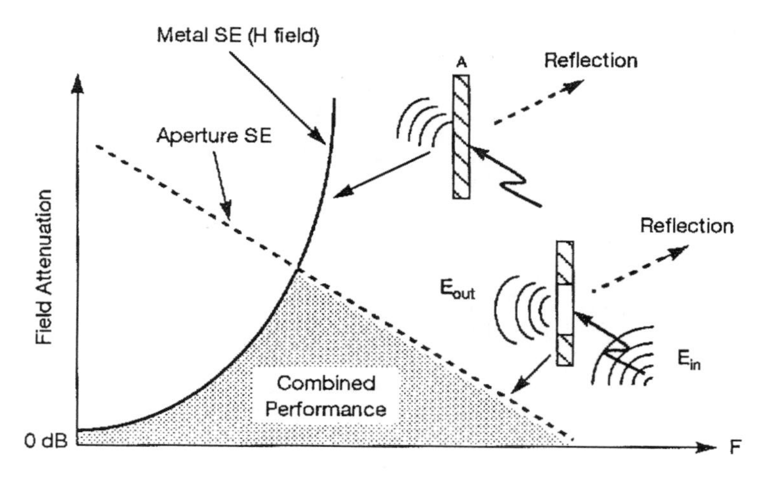

FIGURE 10.7 Attenuation of Field by an Aperture vs. Attenuation by a Metal Barrier

The last term is the guided wave attenuation term (as it would happen in a real waveguide below its operating frequency). This term is directly given in decibels. It has only some influence if d is a significant fraction of l . For ordinary sheet-metal enclosures, this term is negligible. For small holes, or artificially "lengthened" holes (see Fig. 10.9), a significant attenuation is added.

Example 10.1

Calculate, for 1 to 1,000 MHz, the far-field attenuation of the openings in Fig. 10.10:

1. display aperture, 60 x 20 mm
2. Cooling fan aperture, 100 mm x 100 mm
3. cover seams l = 300 mm, h = 0.3 mm
4. cooling slots l = 100 mm, h = 3 mm, thickness of metal ribs, t = 3 mm
5. open slots on front panel edges: l/2 = 15 mm, h = 5 mm (remaining open slot of length l/2 leaks as much as a closed slot of length 2 x l).

The metal thickness on all faces is 2 mm. It will be assumed that in this whole frequency domain, the metal SE is much higher than the slot SE.

1. *display*

SE = 100 - 20 log 60 mm - 20 log FMHz + 20 log(1 + 2.3 log 60/20)
= 64 - 20log F + 6
= 70 dB - 20 log F
= 0 dB above $\lambda/2$ resonance, at F = 150 x 10^3/l = 2,500 MHz

2. *cooling fan*

SE = 100 - 20 log 100 mm - 20 log F (no fatness term)
= 60 - 20 log F
= 0 dB above $\lambda/2$ resonance, at 1,500 MHz

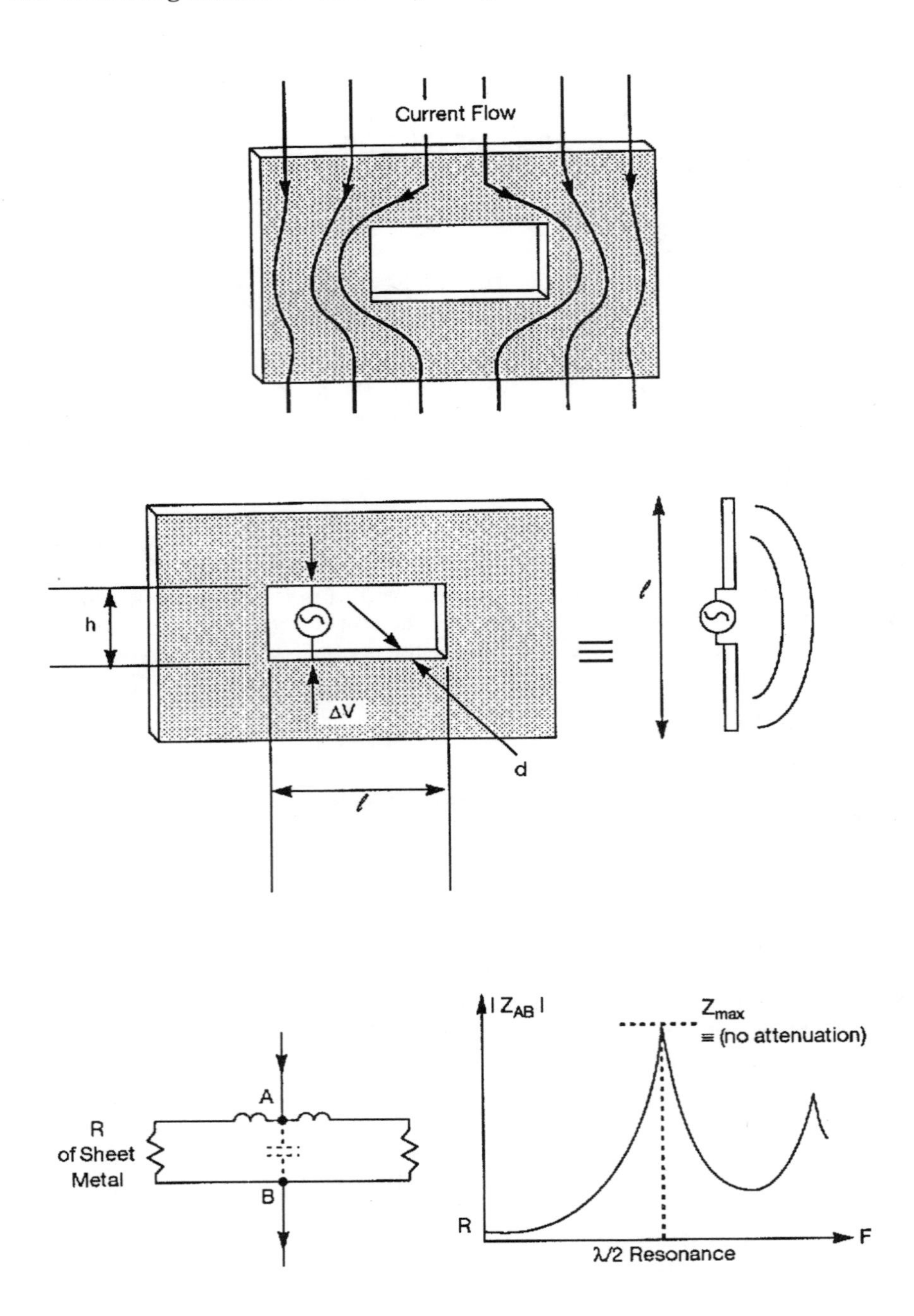

FIGURE 10.8 Effect of Discontinuity in a Shield. At low frequency, the slot is approximately shoe: reflection is significant.

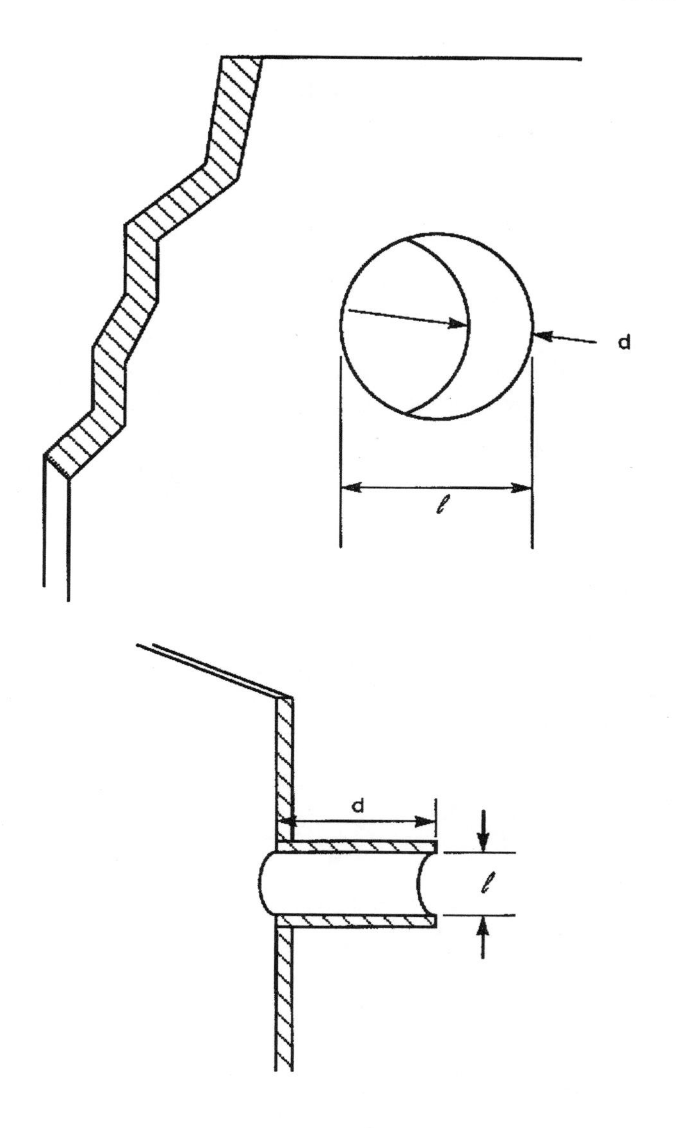

Absorption Loss, AdB ≈ 30 (d/*l*)

For $l < \lambda/2$	d = 0.3 *l*	8 dB
	d = *l*	30 dB
	d = 3 *l*	90 dB

FIGURE 10.9 Additional Attenuation Offered by Lengthened Holes (Waveguide Below Resonance)

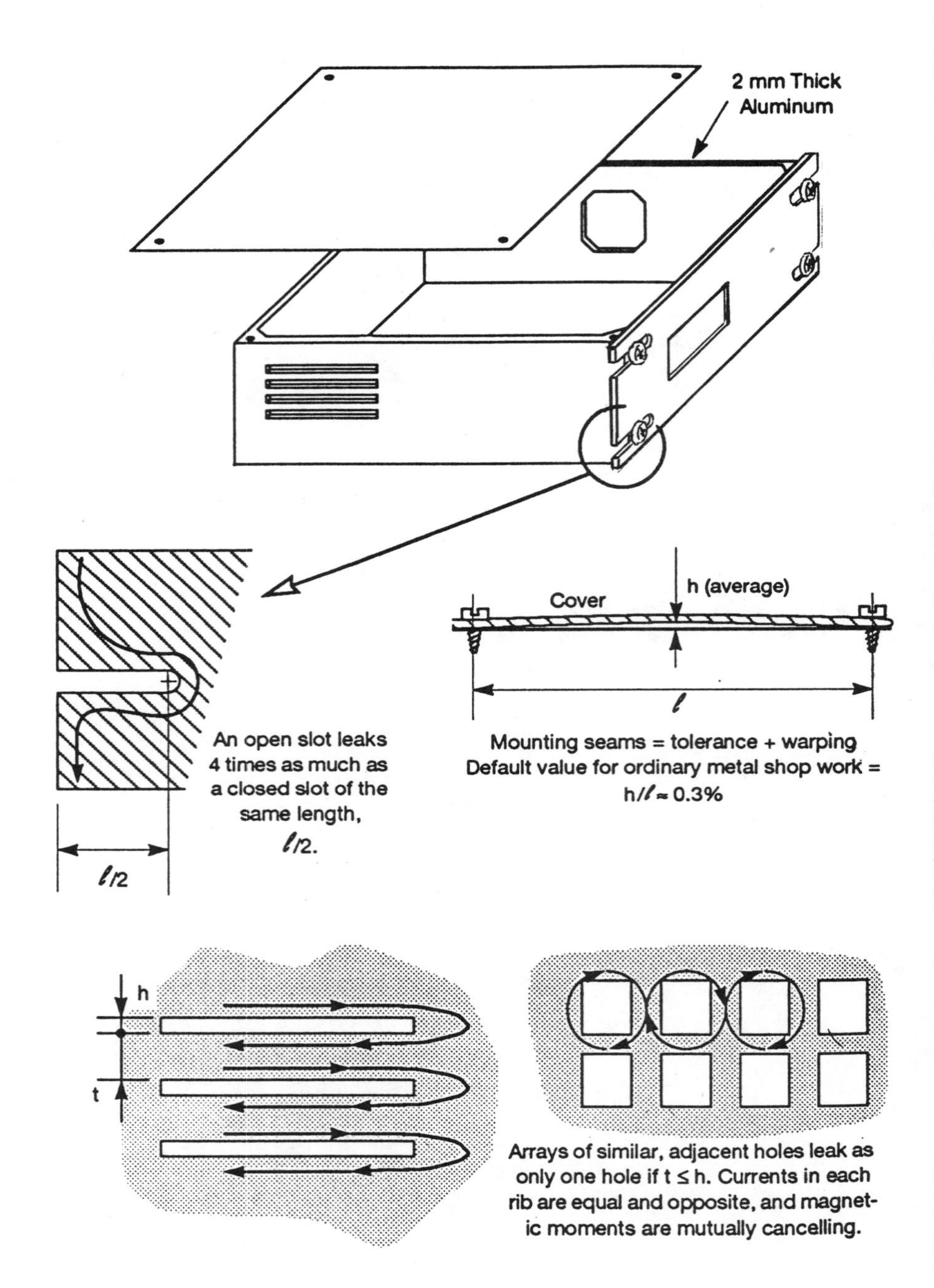

FIGURE 10.10 Box from Example 10.1, with Typical Leakages

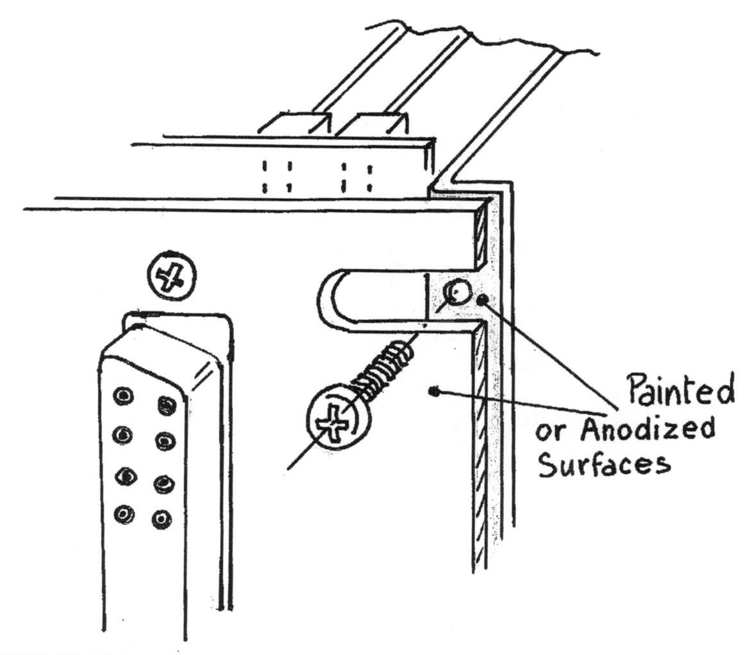

FIG.10.10A Detail of open-sided slot leakage. Such slot could leak if the mounting hardware (screw & flanges) does not restore a good electrical closure of the end gap.

3. *cover seams*

SE = 100 - 20 log 300 mm - 20 log F + 20 log(1 + 2.3 log 300/0.3)
= 50 - 20 log F + 18 dB (fatness term)
= 68 - 20 log F
= 0 dB above $\lambda/2$ resonance, at 500 MHz

4. *cooling slots*

SE = 100 - 20 log 100 mm - 20 log F + 20 log(1 + 2.3 log 100/3)
= 60 - 20 log F + 13
= 73 - 20 log F
= 0 dB above resonance, at 1,500 MHz
(only one slot is considered because t is not larger than h. Ref. to lower part of Fig. 10.10)

5. *open-sided slot*

(This will leak as a 4 x 15 mm slot, if open end does not make good contact withbox edge))

SE = 100 - 20 log 60 - 20 log F + 20 log(1 + 2.3 log 60/5)
= 64 - 20 log F + 8 dB
= 72 - 20 log F
= 0 dB above 2,500 MHz

Notice that this simple edge slot leaks practically as much as the display opening if it is excited by an internal, nearby source.

These five expressions will give us the attenuation of all these apertures at any frequency.

10.5 ALTERATIONS OF THE IDEAL "HOLE-IN-A-WALL" MODEL

The above calculations for both metal SE and aperture SE are assuming a rather academic situation where:

• the metal wall has quasi infinite dimensions, or at least very large versus the source-to-shield distance such as the current density in the plain shield (before any leakage exist) would be uniform,

• the reflected wave does not encounter any opposite wall, causing multiple reflections.

Reality is different:

a) Electronic cabinets have finite dimensions, causing current concentrations at the edges.

b) Radiating sources can be rather close from the box walls openings, such as a wide portion of the concerned frequency span is in a near field condition.

c) the box will behave as a cavity excited by internal sources, if one or several of the circuit frequencies meet the natural resonance frequencies.

Item a) is of no great consequence, except that some apertures might be in regions of the box walls where there is practically no current, such as the aperture will not be excited. Calculated SE will be pessimistic, making our predictions conservative. Items b) and c) are addressed next.

10.5.1 Effect of source proximity on aperture leakage

As previously mentioned, aperture SE in the near field departs significantly from its plane wave expression of Eq. (10.5). Since the wave impedance = 377 Ω, the reflection term will be affected. It will be higher (greater SE) for a predominantly electric field, and lower for a predominantly magnetic field.
In Chapter 2 we discussed that the wave impedance with actual radiators cannot possibly be higher than the "ideal" electric dipole, nor lower than the "ideal" magnetic loop, and in most cases is bound by the radiating circuit impedance Z_c. Setting aside the case of pure E field, which is academic for virtually all radiated EMI problems, it can be demonstrated that when the radiating source within a box is in near-field conditions (i.e., distance $D_m < 48/F_{MHz}$), two conditions may apply:

1. If $Z_c > (7.9/D_m F_{MHz})$, Eq (10.5) is modified and becomes:

$$SE_{(near)} = 48 + 20 \log Z_c - 20 \log \ell_{mm} F_{MHz}$$

$$+ 20 \log(1 + 2.3 \log \ell/h) + 30\, d/\ell \qquad (10.6)$$

2. If $Z_c < (7.9/D_m F_{MHz})$, Eq (10.5) is modified and becomes the attenuation of a slot against an ideal H field loop:

$$SE_{near} = 20 \log \frac{\pi D}{\ell} + 20 \log \left(1 + 2.3 \log \frac{\ell}{h}\right) + 30 \frac{d}{\ell} \qquad (10.7)$$

Notice that this last expression becomes independent of frequency, as long as the near-field criterion and condition 2 both exist.
In most cases, condition 1 is true, which gives better SE than condition 2. Condition 2 can be considered as the worst conceivable lower boundary of aperture SE, against pure H-field sources.

As a recap of the previous sections, Fig. 10.11 shows skin attenuation versus slot attenuation for typical emission conditions, with D = 10 cm. It is clear that, very rapidly, as F increases, the SE of any box skin is bypassed by aperture leakages, which become the governing factors for overall box SE.

10.5.2 Effect of box natural resonances

For a rectangular metal box with dimensions *l*, w, h, the natural resonance frequencies of the TExxx modes are given by:

$$F_{res} = 150 \sqrt{[(m/l)^2 + (n/w)^2 + (p/h)^2]}$$

For *l*, w, h in meters.
The terms m, n, p are integer numbers which can take any value, but no more than one at a time equal to zero. For instance, with the box example of Fig 10.10, the first natural resonances exist at:

F_1 = 707 MHz
F_2 = 1,120 MHz
etc..,

At these specific frequencies, an empty metal box could exhibit resonances with a Q factor as large as 10 (20dB), which could result in a negative SE (an apparent "gain").
Hopefully, electronic equipment boxes are never empty but filled with PCBs, components and cables which are behaving as many lossy elements, so that the measured Q stays within 0 - 10dB, with typical values of 6dB. This doubling of the inside field results in an apparent 6dB drop of the theoretical SE, at every self-resonance frequency.

If the box is large in dimension, with relatively small apertures, the 6dB notch of the first resonances occur at rather low frequencies where SE is high enough to afford

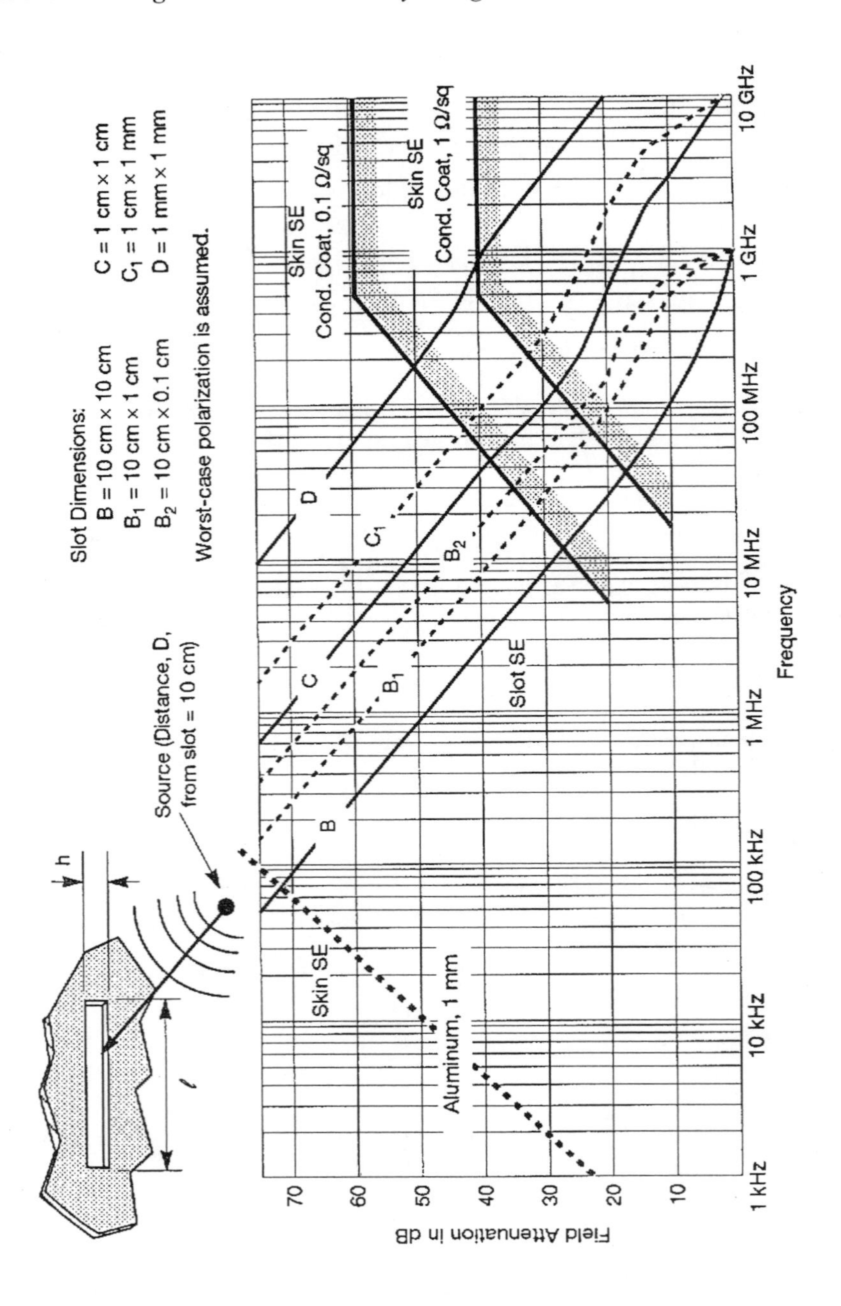

FIGURE 10.11 Skin Attenuations (aluminium and conductive paints) vs. apertures attenuations, facing a low impedance source at 10cm. A 100Ω source impedance is assumed. If source is closer, SE will decrease proportionally.

this loss. But when it comes to resonant frequencies which are no more than one octave below the cut-off frequency for the largest opening, the box SE will deteriorate to 0dB faster than expected. Some manufacturers of shielding hardware are offering lossy composites with surface resistivity > 10Ω/sq, which can be used like an anechoic coating inside the enclosure (Ref: Contex® Lossy felt by MILLIKEN Co.)

10.6 METHODS OF LEAKAGE REDUCTION AND APERTURE TREATMENT

A conductive housing already has the basic advantage of being a naturally efficient barrier. All the talent of the designer should be aimed at not spoiling this barrier with excessive leakages. Leakages (i.e., poor SE) result from:

1. seams at mating panels, covers, etc. (a frequent cause of SE spoiling)
2. cooling apertures
3. viewing apertures for displays, meters, etc.
4. component holes: fuses, switches, shafts
5. cable entries

10.6.1 Mating Panels and Cover Seams

The general, simple rule is:
All metal parts should be bonded together.
A floated item is a candidate for re-radiation.

For cover seams, slots, and so forth, how frequently they should be bonded is a question of the design objective. Figure 10.11 shows that a 10 cm leakage is worth about 20 dB of shielding in the neighborhood of 150 MHz. If the goal is closer to 30 or 40 dB, seams or slots should be broken down to 3 cm or 1 cm. For permanent or semi-permanent closures, this means frequent screws or welding points or an EMC conductive gasket. For covers, hatches and such, this means flexible contacts or gaskets.

The following is a sequential organization of these solutions. As efficiency increases, cost increases as well.

• If only minimal shielding effectiveness is needed (in the 0 to 20 dB range), the simplest technique is to have frequent bonding points and, for covers, short flexible straps made of flat braid or copper foil as shown in Fig. 10.12. This solution bonds only on the hinge side, but if no noisy cables or devices are located near the opposite side of the hinge, this can be sufficient. For this unbonded opposite side, a wise precaution is to use a grounded lock or fastener.
The $\lambda/20$ criterion shown in Fig. 10.12 means that, for a maximum emission frequency of 100 MHz, the distance between jumpers should stay within 15 cm for a 20 dB shielding objective, and up to 45 cm if a 10 dB reduction is sufficient.

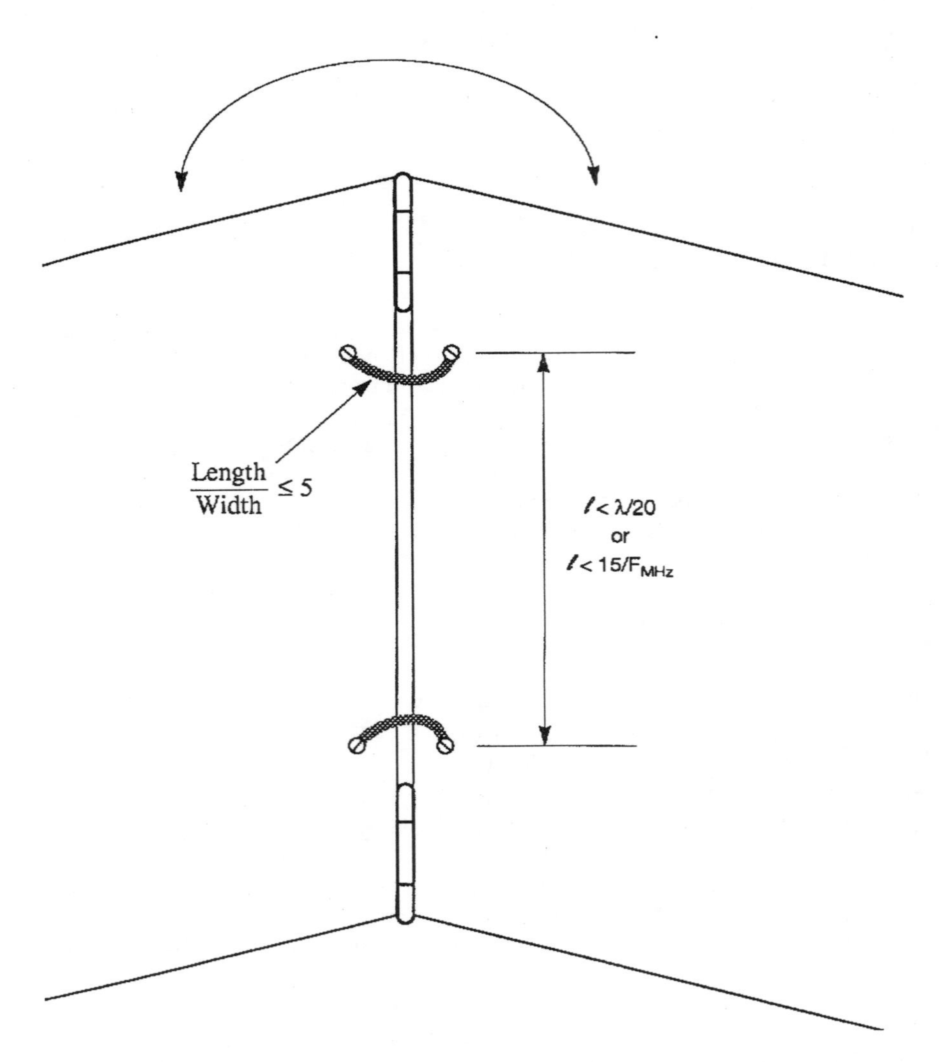

FIGURE 10.12 Leakage Reduction by Frequent Seam Bonding (For Moderate Shielding Needs)

• If bonding at only the hinged side leaves an excessive length of ungasketed seams, additional bonding points are necessary. In this case, the techniques shown in Fig. 10.13 can be used. Part a), shows an example of a soft spring, several of which would be scattered along the edges of the cover. For durable performance, the spring contact riveting must be corrosion free, which may render that solution more difficult to apply than it would seem.

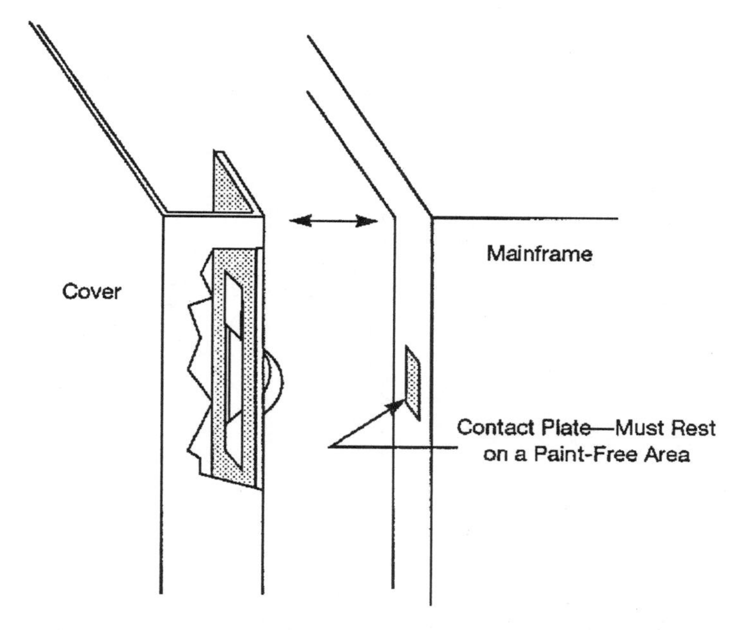

a. Captive beryllium-copper springs are located along cover edges. When closed, they mate with abutting frame edge. Contact plates can be nickel or tin plated or made from adhesive conductive tape.

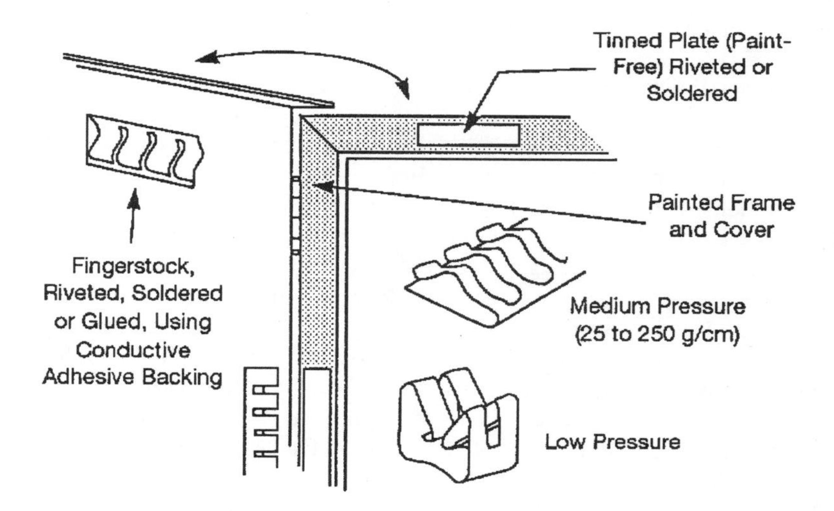

b. Partial bonding by knife-edge or regular fingerstock

FIGURE 10.13 Maintaining Shield Integrity by Regular, Flexible Bonding Points

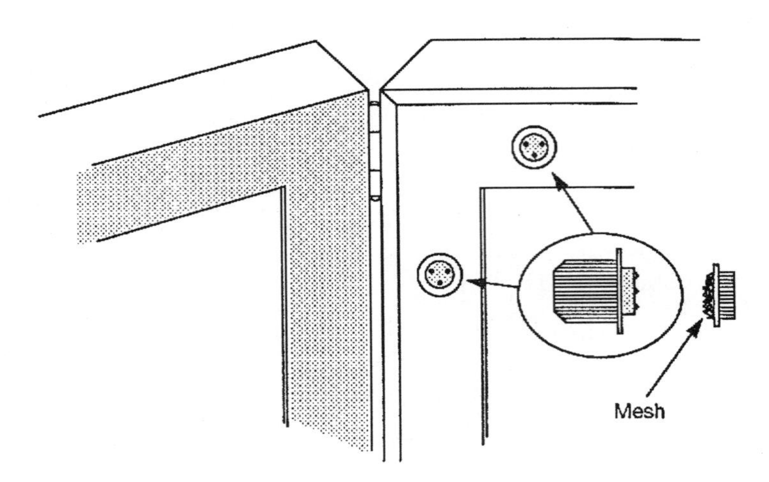

FIGURE 10.14 Press-Fit Grounding Buttons

A variation of this, shown in Fig 10. 13, part b), is to use sections of spring contacts called fingerstocks. Several types of fingerstocks are available,such as low-pressure, knife edge and medium-pressure styles. They require an adequate control of pressure through tight manufacturing tolerances but they are extremely dependable.

a. captive beryllium-copper springs are located along cover edges. When closed, they mate with abutting frame edge. Contact plates can be nickel or tin plated or made from adhesive conductive tape.

b. Partial bonding by knife-edge or regular fingerstock

A third technique, shown in Fig. 10.14, is an interesting alternative that takes minimal surface preparation. The grounding "buttons", which are fairly compliant to gap variations due to their loading, are mounted simply by press fit or a threaded stud. These are available with spring for near 100% compliance.
If a higher grade of shielding is required (20 to 60 dB), a continuous conductive bonding of seams is necessary, since an SE of 40 dB at 300 MHz ($\lambda/2$ = 50 cm) would require screws or rivets every 0.5 cm! These continuous conductive joints are available in several forms and stiffnesses (see Fig. 10.15). The hollow rubber gasket is less expensive to use because its wide elasticity compensates for large joint unevenness and warpage. The price for this is a lesser contact pressure, hence higher resistivity. Therefore, it is best used as a solution for the lower side of the SE range. Here again, a good quality mating surface can be made by applying conductive tape (see Fig. 10.16) over the metal surface before painting. Then a piece of masking tape is pressed over the conductive foil, and the metal surfaces can be painted, after which the masking tape is carefully peeled off. Metal braid or mesh-type gaskets provide higher shielding, close to or beyond the upper side of the required SE_{dB} range.

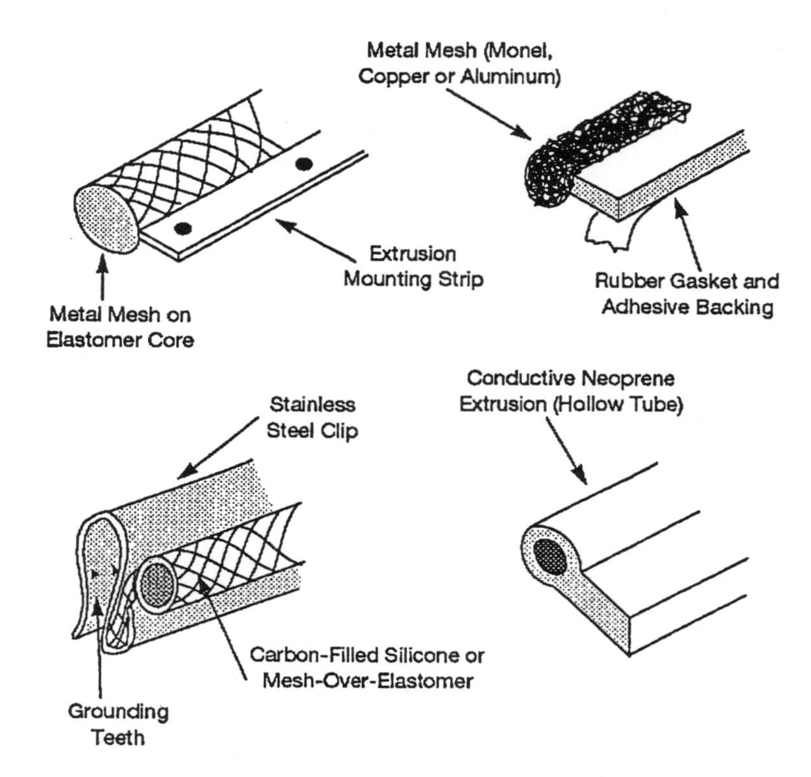

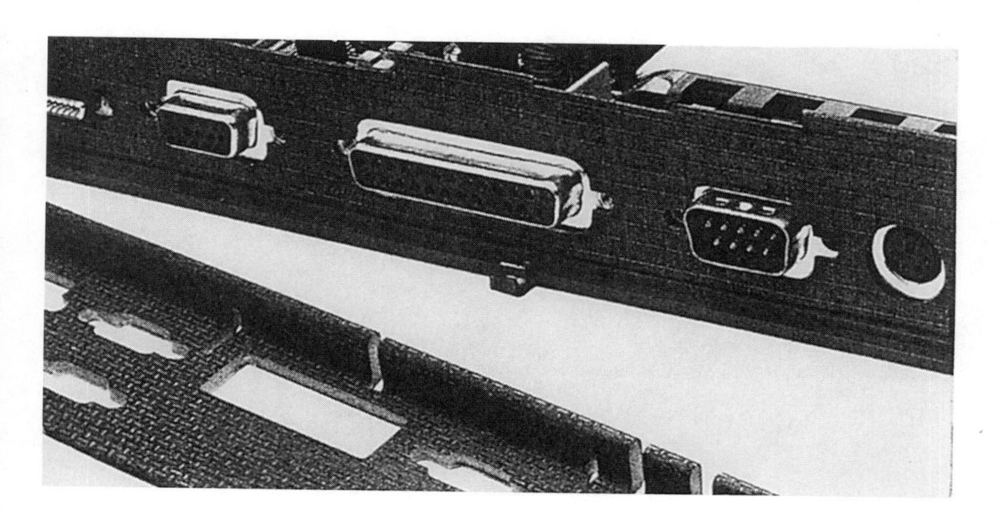

FIGURE 10.15 Compressible RF Gaskets and Mounting Styles. Bottom : Conductive textile gaskets with Foam Core (From Schlegel Co.)

Finally, if an even higher hardening level is necessary, the ultimate solution is shown in Fig. 10.17. This solution is the most efficient because 100 percent of seam becomes a very good conductive joint. Beside its direct cost, it adds the need for a strong locking mechanism to ensure good, even pressure on all of the spring blades. This method is applicable to both rotating (hinged) or slide-mating surfaces.

Use of rubber, mesh or spring fingers types of gasket requires an adequate design of covers and box or frame edges to provide:

• a seating plane or groove for the gasket

• proper mechanical tolerances to avoid gasket overpressure at some points (lower tolerance gap), causing permanent gasket flattening, and underpressure at others (higher tolerance gap), hence insufficient contact.

For metallized plastic housings, the seam treatment needs only to be commensurate to the box skin SE, which is generally more modest (typically less than 50 dB below 100 - 150 MHz).
Mating edges can be designed to provide a decent electrical continuity if the conductive coating is sufficiently resistant to abrasion. This is done by using tongue-and-groove or other molded profiles for assembly (See Fig. 10.18). The relative compliance of plastic provides the necessary contact of the conductive surfaces.

FIGURE 10.16 Conductive, Adhesive Tapes Used to Create a Good Conductive Area for Localized Shielding, or Contact Area. Surface resistance can be as low as a few milliohms per square.

10.6.2 Shielding for Cooling Apertures

Several techniques can be used to restore shield integrity at convection or forced-air cooling openings (see Fig. 10.19):

1. Break large openings into several smaller ones. This has the advantage of costing virtually nothing if the holes are produced during stamping or molding of the box walls. It also can put the source at a relative greater distance, compared to the aperture size, eliminating some proximity effect. The improvement is:

$$\Delta dB = 20 \log (\text{old length} / \text{new length})$$

Or, simply, $\Delta dB = 20 \log N$, if N is the number of identical holes that are replacing one larger aperture.
This is done by replacing long slots with smaller (preferably round) apertures. If some depth can be added to the barrier such that $d \geqslant l$, the waveguide attenuation term in Eq. (10.5) becomes noticeable, improving SE.

2. Install a metal screen over the cooling hole. This screen has to be continuously welded or fitted with a conductive edge gasket having an intrinsic SE superior to the overall objective.

3. Install a honeycomb air vent if an SE superior to 60 dB is required above 500 MHz and up to several GHz, along with a low aerodynamic pressure drop.

CRTs, alphanumeric displays, meters and the like are often the largest openings in an equipment box, offering the lowest SE of all packages. On the other hand, typical high-frequency sources are seldom mounted right on or behind display panels.

Compared to the typical RF "hot plate" represented by a filter mounting panel, or I/O connectors and I/O cards area, experience shows that many equipments can tolerate rather large, unshielded apertures on their user's display panel, while a 10 times smaller slot in the cable entry zone would radiate significantly. In a sense, the intrinsic SE of any aperture being calculable, its radiation still depends on whether it is excited.
Since one does not know, a priori, how RF currents will be distributed on the box's inner skin, we will keep with the conservative assumption that viewing apertures are as prone to leak as any other ones. The shielding solutions are:

1. Finely knitted or woven wire mesh, on top of or sandwiched in the glass, plexiglass or other material. Densities of up to 12 wires/cm (knitted mesh) or up to 100 wires/cm (woven product) are obtainable. The performance can be derived from the curves of Figs. 10.20 and 10.21. The denser mesh offers more SE because the individual holes are smaller, but this is at the expense of transparency.

2. Transparent conductive film, where a thin film of gold or indium oxide is vacuum deposited on the transparent substrate. The film thickness has to be low (10^{-3} to 10^{-2} microns) to keep an 80 to 60 percent optical transparency, but the thinner the film,

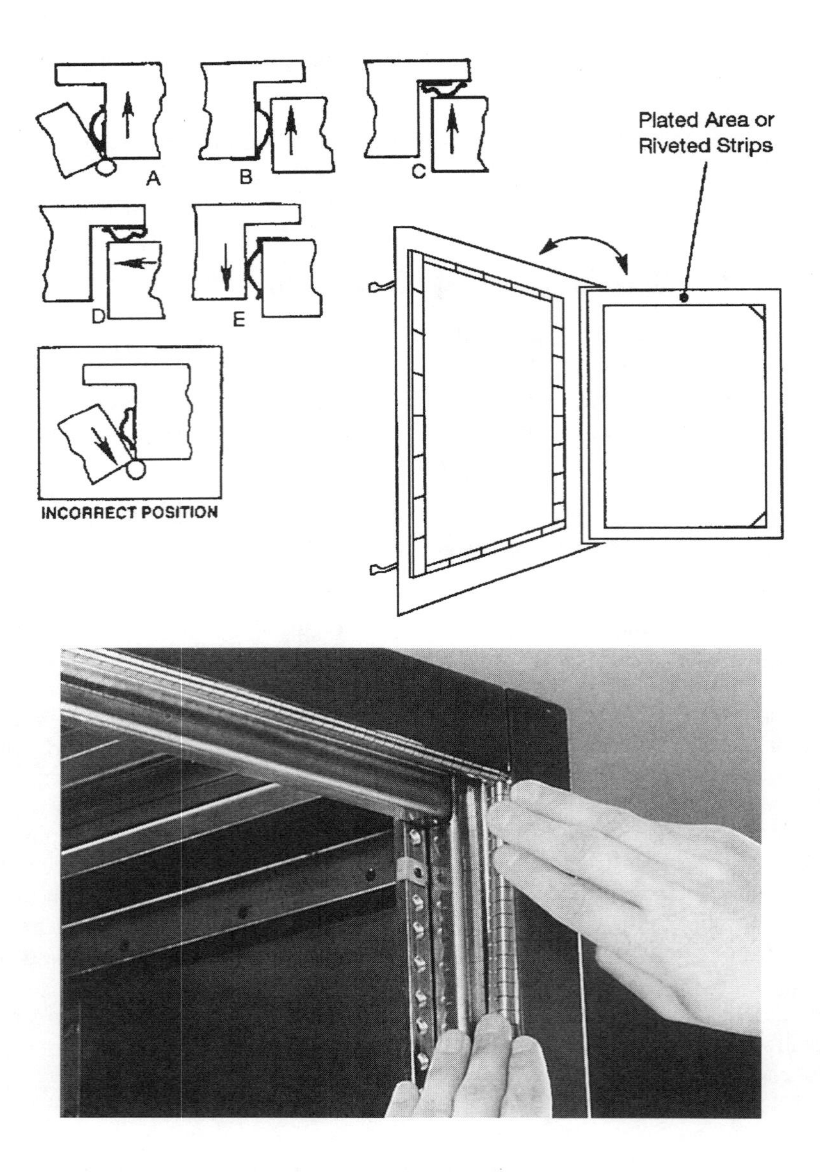

FIGURE 10.17 Fingerslocks with 100 Percent Perimeter Coverage (courtesy of Instrument Specialties Co.) (continued next page)

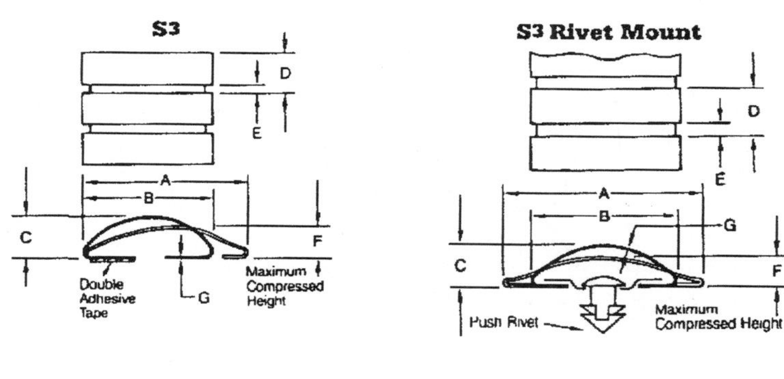

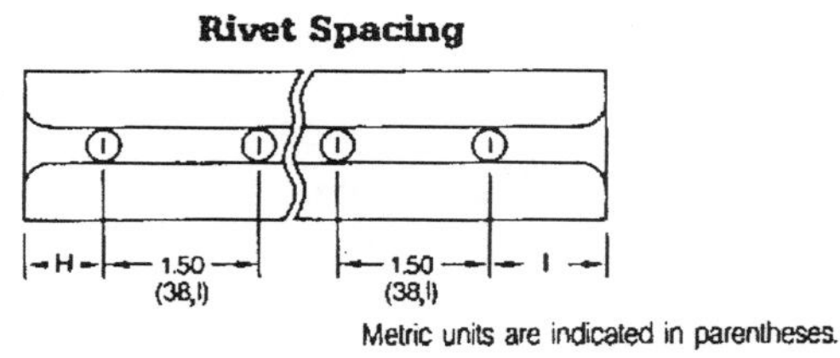

FIGURE 10.17 continued

the more the surface resistance. Typical transparent coatings have surface resistivities in the range of 50 to 5 Ω/sq, corresponding to far-field SE of 8 to 26 dB. Near E-field SE would be better.

3. Shielding the display from the rear side: the display is shielded behind the box panel by a doghouse, which is equipped with feedthrough capacitors for connecting wires (see Fig. 10.22).

In all three of the solutions enumerated above, an EMI gasket is needed at the shield-to-box joint. Often, one is already fitted by the shielded window vendor.

10.6.4 Shielding the Component Holes

Holes for potentiometers shafts, switches, lamps, fuseholders and the like generally are small. But their mere presence in the middle of metal pieces that have picked up CM current from inside the box will enhance the radiation phenomenon.

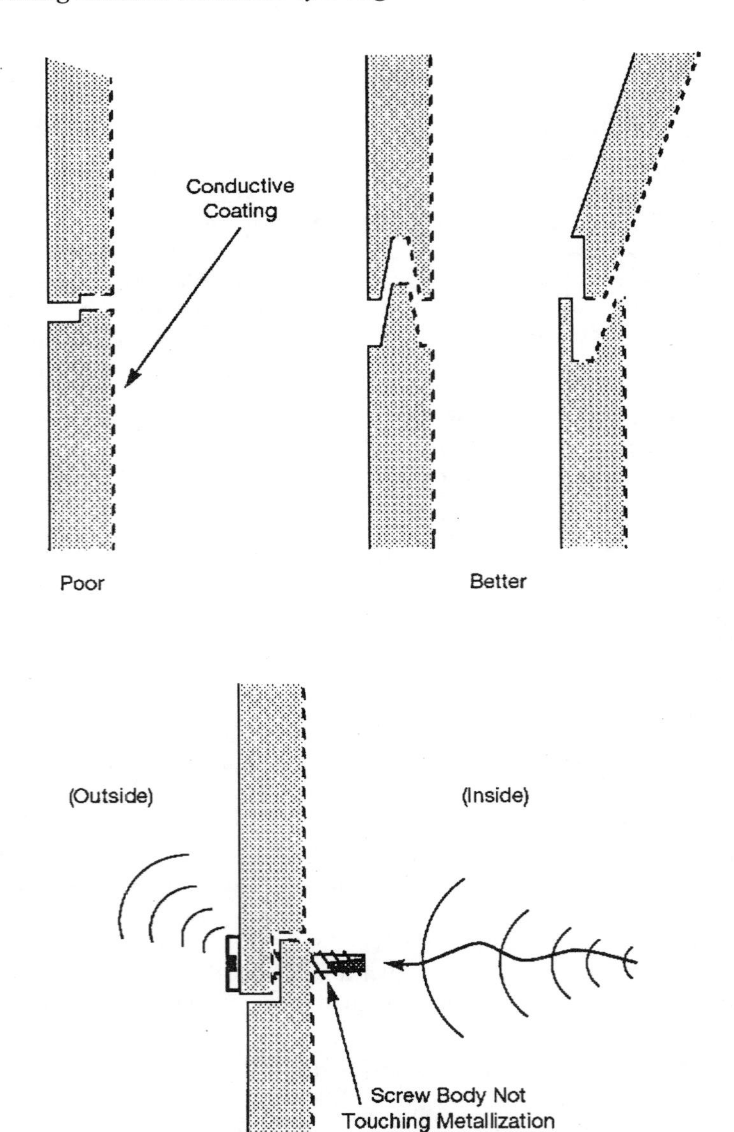

FIGURE 10.18 Metallized Plastic Box Design Against Radiation. Conductive coating should extend far enough into the tongue-and-groove shape to create a good, continuous contact (but not too far to avoid ESD problems). Long, protruding screws must be avoided because they can become reradiating antennae.

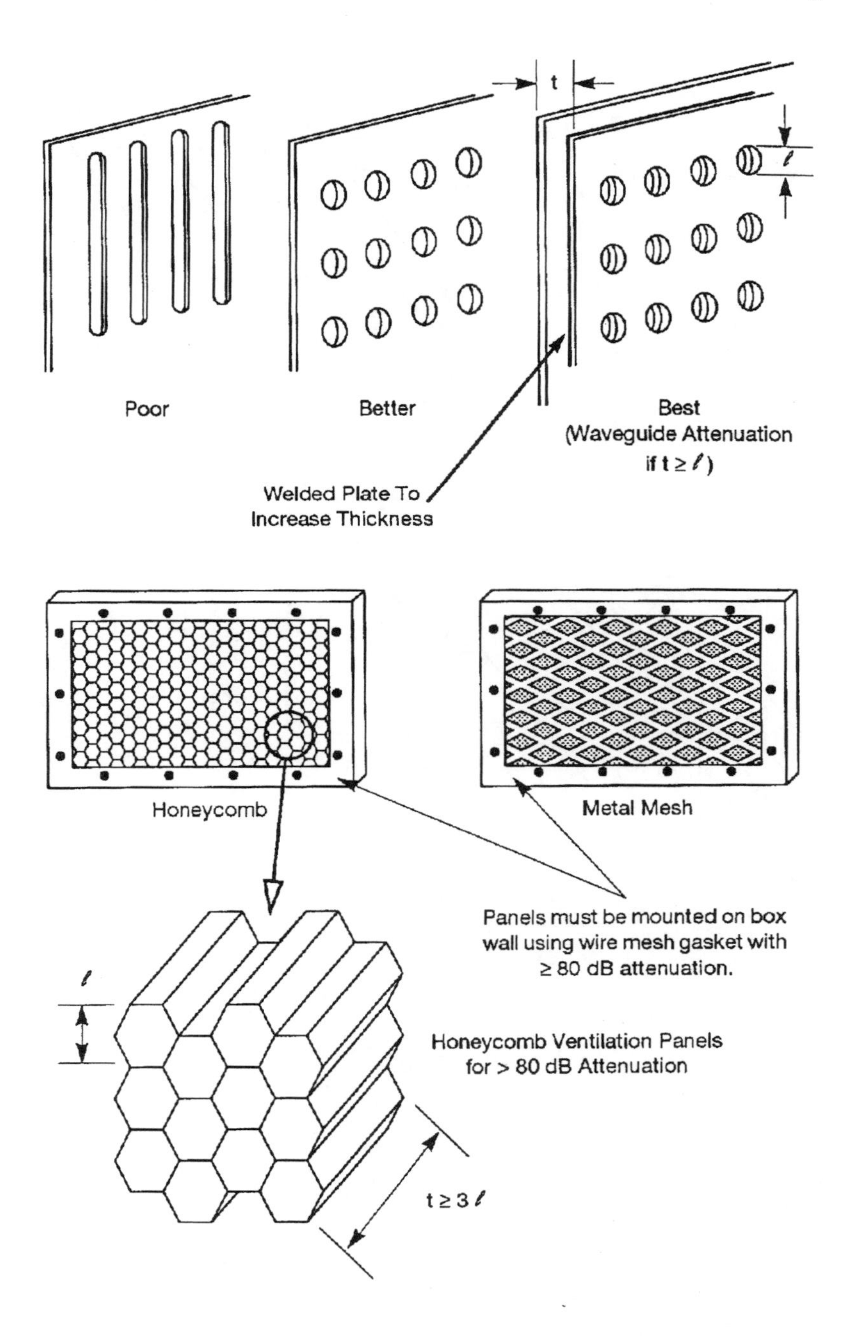

FIGURE 10.19 Methods of Shielding Cooling Apertures

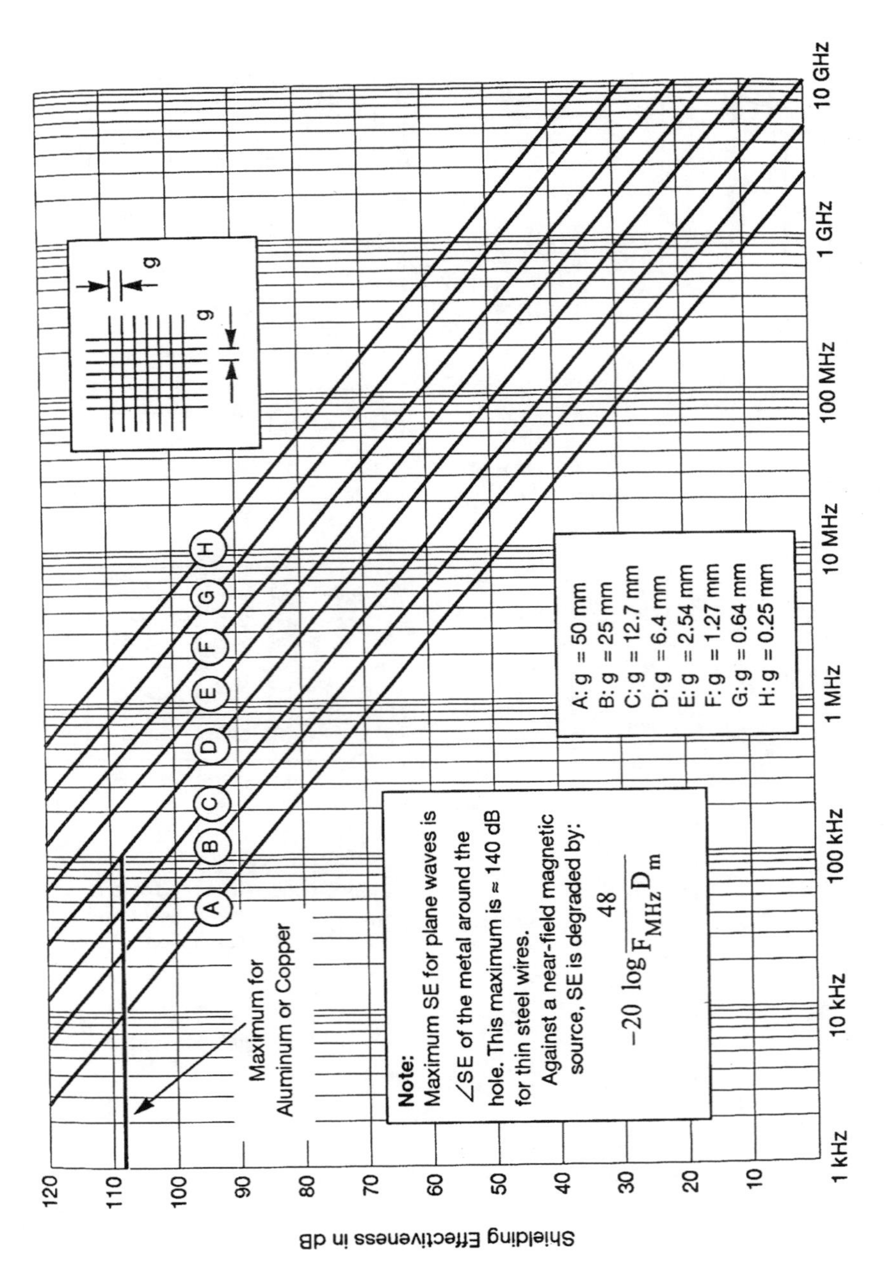

FIGURE 10.20 Shielding Effectiveness of Screen Wire Shields, in Far-field Conditions (Distance in meters > 48 /FMHz)

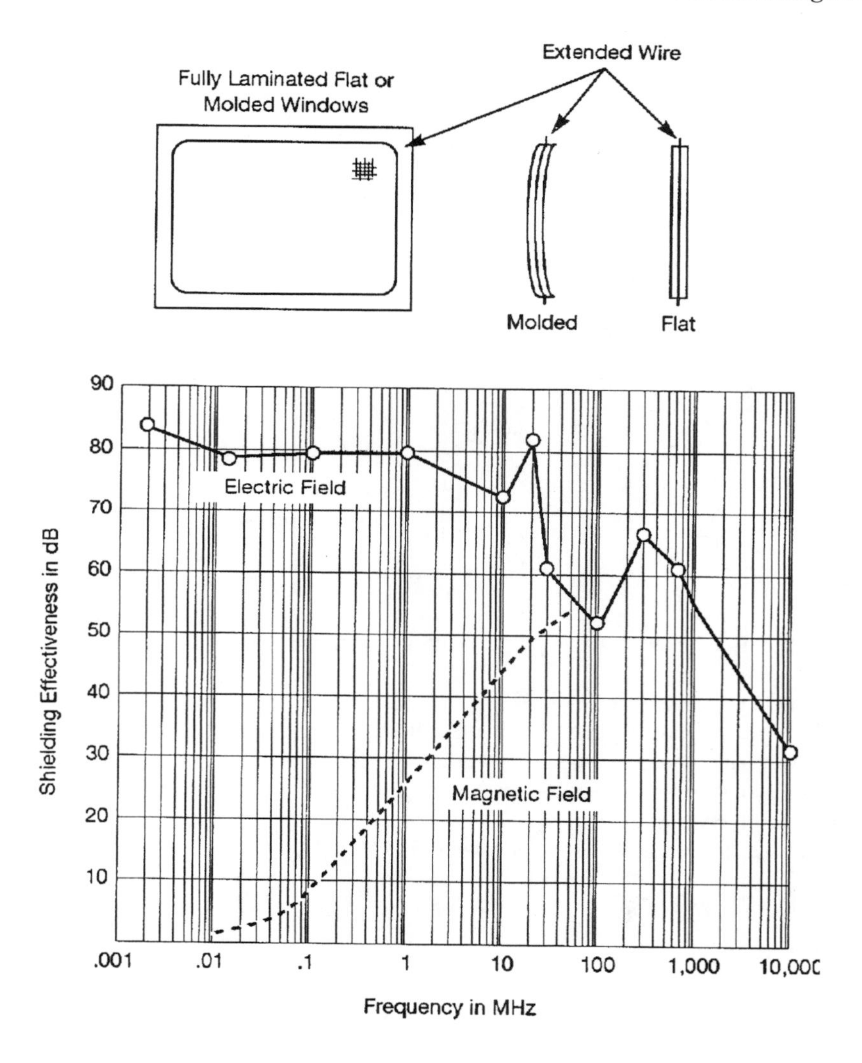

FIGURE 10.21 Shielded Transparent Windows for CRTs, Thin-Mesh Version, with 100 Wires Per Inch, Laminated and Anti-glare Treated (source: DonTech, U.S.A.)

The shaft, lever or fuse cartridge acts as a monopole, exiting via a coaxial line: just what is needed to transmit radio signals.

As far as FCC, CISPR and other civilian limits are concerned, component holes are seldom a problem because of the relatively small leakage. With MIL-STD-461 or TEMPEST emission tests, component holes can become significant contributors to EMI radiation.

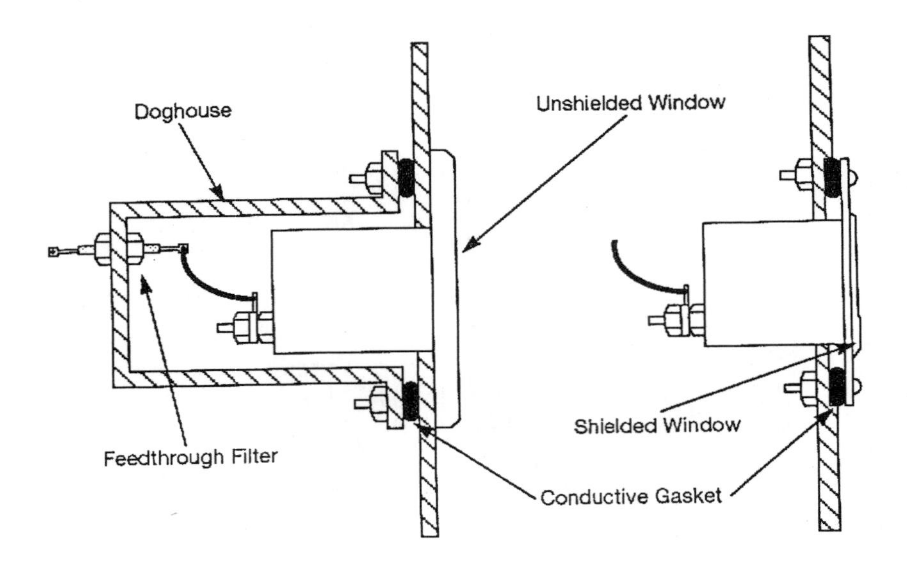

FIGURE 10.22 Shielding of Meter and Display Openings

The solutions are:
1. Use nonconductive shafts or levers, and increase the hole depth with a piece of metal tube to create a waveguide attenuation.
2. Use grounding washers or circular contact springs to make electrical contact between the shaft and panel.
3. Use shielded versions of the components.

10.6.5 Shielding of Cable Entries, Connectors and Nonconductive Penetrations

Last but not least, this breach in box skin integrity is a serious concern, since cables are the largest potential RF carriers in the entire system. The shielding (or no shielding) of the cable penetrations depends on the decision chart of Fig. 10.23.
From the first decision (evaluation of cable entry hole), the designer will follow one of two paths:

1. Calculation of box SE shows that the cable exit hole is tolerable (answer "Yes" at step 1). However, even if the box is O.K., the cable can behave as a radiator. If the cable needs to be shielded for radiation (and/or susceptibility), its shield must properly terminate at the barrier crossing with a 360° clamp, ultra-short strap or, best of all, a metallic connector shell.
If the cable is not shielded but still is a threat, each of its conductors must have been filtered (see Chapter 9). There is no point in shielding the hole.

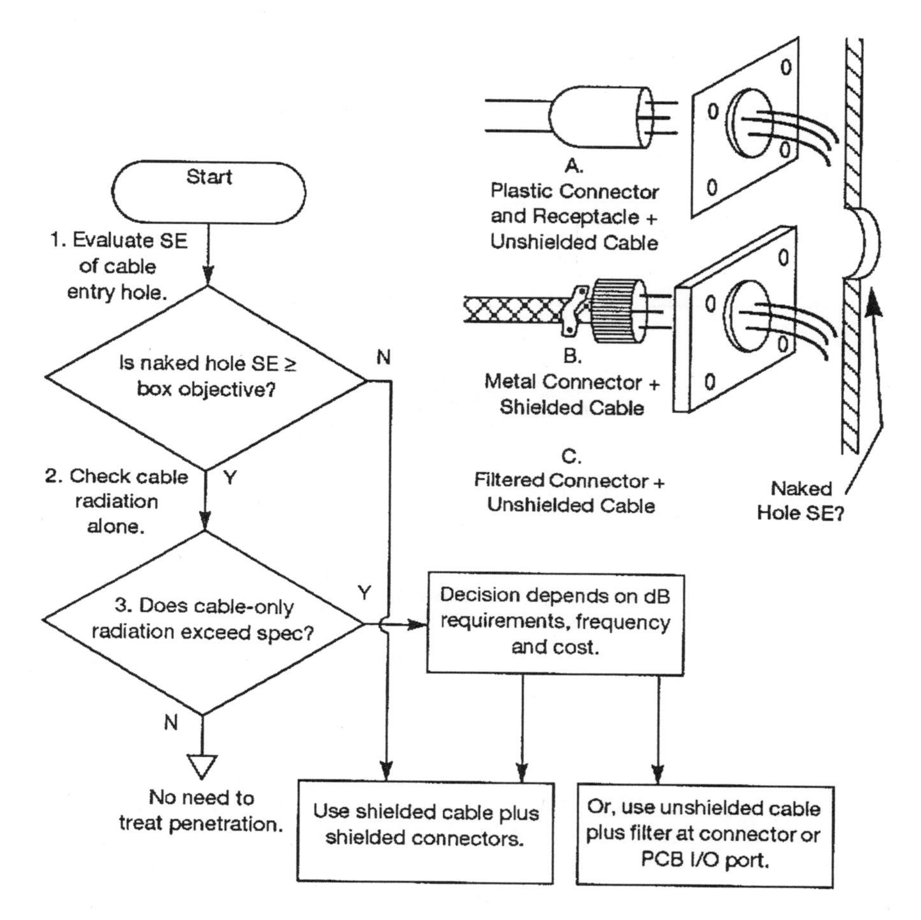

FIGURE 10.23 Decision Chart for Shielding of Cable Penetrations

2. Calculation of box SE show that the cable exit hole is not tolerable (answer "No" at step 1). In this case, it becomes imperative to use shielded cable and shielded connectors, creating a sort of shielded enclosure for the entire interconnect cabling system.

A compromise version would be to use unshielded cable but to block aperture leakage with a shielded and filtered connector receptacle. This would recreate a recessed shield barrier behind the cable hole.
The interface of cable shields at box penetration is a topic inseparable from cable shielding. This matter is addressed in the forthcoming Chapter 11.
Some other exit/entry ports exist for nonconductive lines such as pressure sensors, fluid lines, fiber optics and so forth. If the tube is nonconductive and the SE of the naked hole is insufficient, this type of leakage is easily reduced by using the waveguide effect. For fiber optics, transmitters, and receivers, metallic packages are available with appropriate tubular fittings.

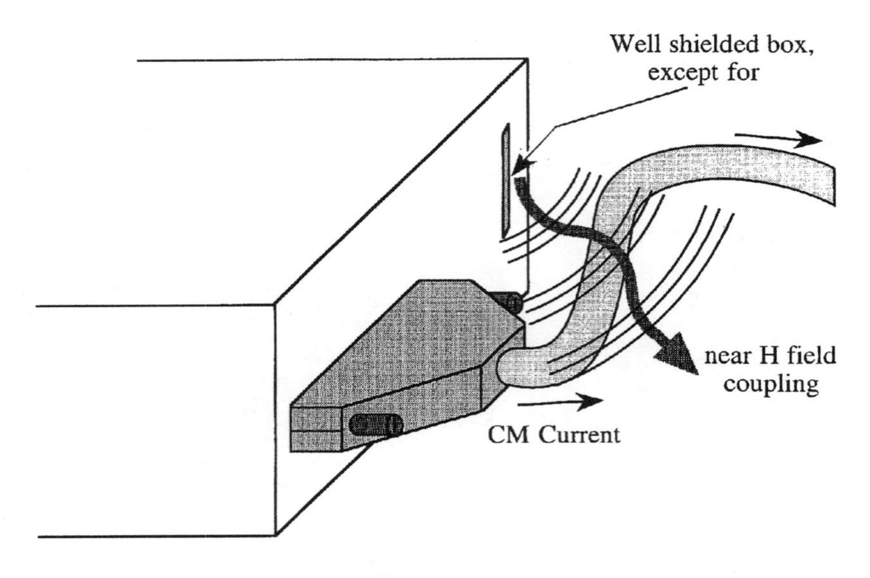

FIGURE 10.24 Excitation of I/O cables by a nearby slot.

10.6.6 Detrimental effect of box leakages near a cable penetration

When a cable exit is located in the proximity of a slot leakage (ex. open-edged slot, warped joint etc..), the slot attenuation is locally less than its theoretical far-field value. In fact the exciting source inside can couple to the first centimeters of the outside cable segment (Fig. 10.24) by a mechanism which is closer to magnetic or capacitive crosstalk than actual radiation. CM currents, then, will be found on the cable, e.g. using a current probe, even after I/O filters or ferrites have been installed, turning the cable into a secondary antenna. Such leakages in a "hot plate" area must be controlled very carefully.

10.7 SPECIALLY HARDENED EQUIPMENT HOUSINGS

Several vendors of ready-to-use racks and cabinets offer EMI shielded versions of their products. Even a standard steel or aluminium cabinet with some simple precautions (paint-free and zinc- or tin-plated contact areas, metal-mesh air filters) provides some degree of shielding.

Equipped with EMI gaskets and shielded air vents, 100 percent welded frame joints and piano-hinged doors for better seam tolerances, shielded cabinets offer valuable SE performance, as shown on Fig. 10.25, at a cost increase of $300 - 350 (1995 prices) as compared to the standard version. Be careful, however, when dealing with emission problems: many SE values reported on shielded cabinet specification sheets are measured by the MIL-STD-285 method, with a radiating source outside,

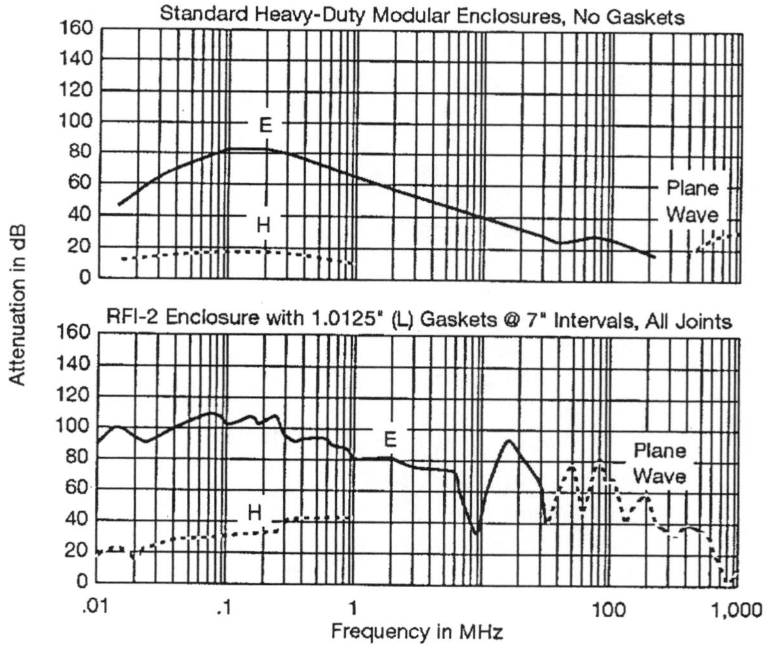

FIGURE 10.25 Example of a Commercially Available Shielded Cabinet (Source: Equipto, Aurora IL, USA)

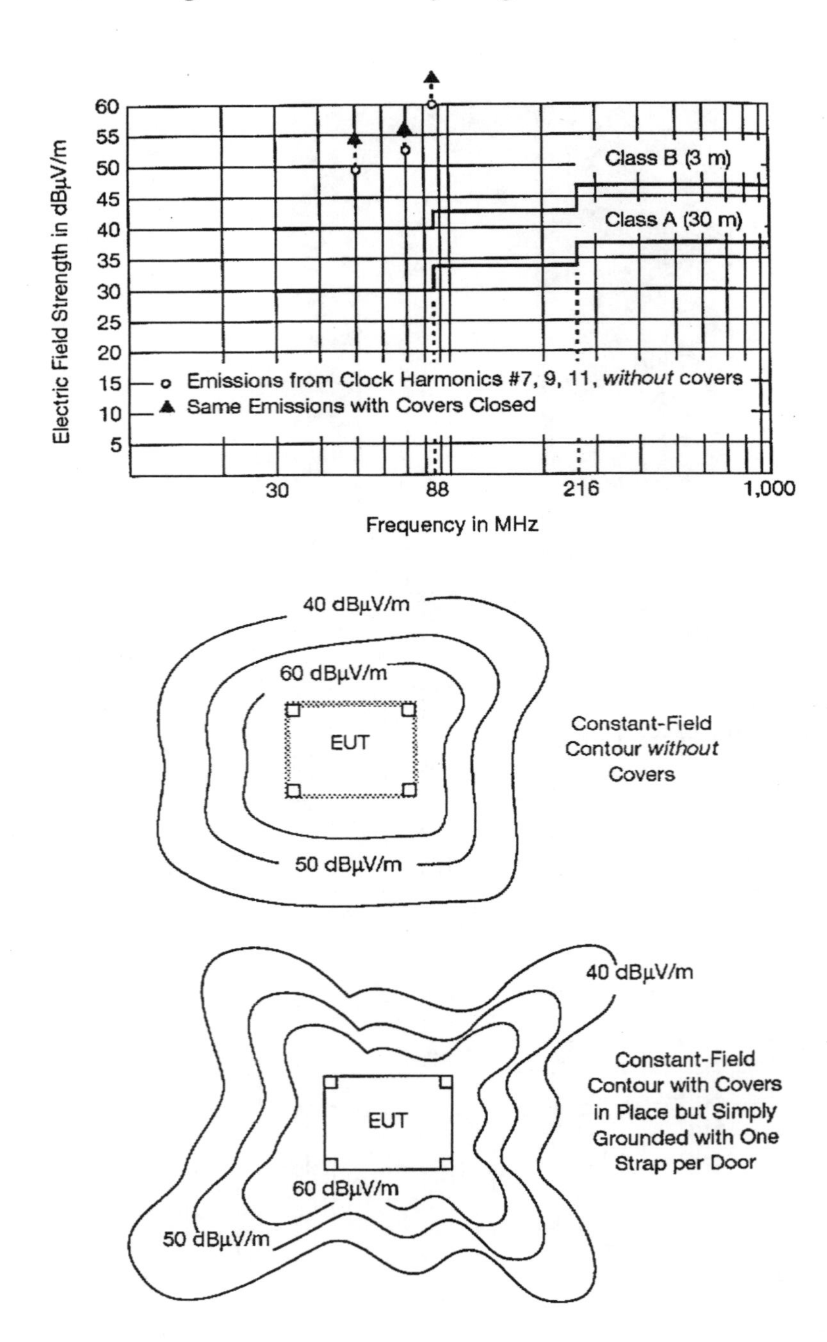

FIGURE 10.26 Apparent Increase in Radiated Field Due to Improper Bonding of Covers

at 30 cm from the doors. These values may not be applicable to a situation where the emitter is inside the cabinet.
Proximity effects can cause lower than expected values for SE, especially when housed electronics and cables are near the cover seams.

Figure 10.26, taken from our "real-cases story book", illustrates this point. A telecommunication electronic system was exceeding FCC Class A limits. To understand what was happening, a preliminary E-field scan was plotted, 360° around the frame, with the original, ungrounded covers removed. Then, another plot was made, with each steel cover grounded at one point by 12 cm, round wire straps. The engineers expected, based on SE of 1 mm thick steel, a significant attenuation (theoretical figure being above 300 dB). To the engineers' chagrin, the attenuation was in fact negative: in some azimuths the metal-covered cabinet radiated 2 to 4 dB more than the open-sided case. The answer, of course, is that long, ungasketed slits (here about 1.75 m long) behave as radiating dipoles with directional gain in some specific angles. The total radiated power had not changed, but its spatial distribution had changed.

10.8 APPLICATION EXAMPLE: BOX DESIGN FOR A GIVEN SE OBJECTIVE

We will use, as our example, the metal box of Fig. 10.10, with aperture leakages as estimated in Example 10.1. The principal radiating source is assumed to be the mother board/daughter cards assembly of Chapter 3, Example 3.2. The radiation profile of the 15 MHz clock harmonics is the same as shown in Fig. 3 4 (FCC/Class B compliance).

1. Calculate the existing box SE. Compare the result to the design objective and decide where modifications are necessary.
2. Repeat exercise 1 with MIL-STD-461, RE02, as the specification limit.

In both cases, an $\approx$ 6 dB margin is desired to cover manufacturing and installation variations.
Solutions
The general routine is:
1. Evaluate SE requirements.
2. Evaluate the SE of the box skin material as if it were a perfectly homogeneous barrier. Here, we have 2 mm-thick aluminium.
3. Evaluate SE for all apertures listed, per Example 10.1.
4. Retain the worst SE in each column, and compare with SE requirements to decide if improvements are needed.

Solutions for FCC Class B Specification (from Example 3.2)

Freq. in MHz	*15*	*45*	*75*	*150*	*300*	*500*
A. SE Objective*	NA	20	30	30	20	14
B. Metal SE, in dB	>300	>300	>300	>300	>300	>300
C. Openings' SE, in dB (from Ex. 10.1)						
Display	46	36	32	**26**	**20**	**16**
Cooling Fan	36	26	**22**	**16**	**10**	**6**
Cover Seams	44	34	30	**24**	**18**	**0**
Cooling Slots	49	39	45	29	23	19
Open-Sided Slot	48	38	34	28	**22**	18
D. Overall SE†	<36	<26	<22	<16	<10	0
E. Improvement Needed (D – A) in dB	0	0	>8	>14	>10	>14

*Specification violation plus ≈ 6 dB margin
†Worst of each individual SE

Comments:

• The exact calculation of overall SE could be done more precisely (Ref. 15) by combining the antilog of all leakages, assumed (worst case) in phase:

$$SEtotal = -20 \log [10^{-B/20} + 10^{-C1/20} + 10^{-C2/20}, \text{etc..}]$$

• Taking an average of 15 cm for center-to-sides distance, the box walls are at near-field from inside sources up to a frequency such that 0.15 m > 48/F; i.e., 320 MHz. Therefore, Eq. (10.6) for proximity effects should apply. However, because the radiating circuit impedance is not very different from 377 Ω, no correction has been applied.
• At any frequency, the aluminium SE is way above the requirements. Therefore, the metal poses no problem.
• Figures appearing in bold in the table designate insufficient attenuation.
• The needed improvement (8 to 15 dB in the 75 to 500 MHz range) is rather modest. In fact, leaving the box as is, except for the cover seams where the number of screws needs to be doubled, the field reduction could be obtained by changing the mother and daughter PCBs to multilayer ones, at much less cost. Notice that with a more precise calculation, actual mother-bd circuit impedance of 300Ω, then 200Ω should have been taken into account, resulting in a SE figure correction of - 5dB (at 150MHz) dropping progressinvely to 0dB at 320MHz.

If internal redesign is not feasible, the following changes are necessary:

1. Reduce cover seam lengths by doubling the number of screws. This would shift λ/2 resonance up to 1,000 MHz, and SE at 500 MHz would become 20 dB.

2. Subdivide the cooling aperture in a 10 mm x 10 mm grid. The SE at 500 MHz would become >26 dB. A thin wire mesh is not necessary; this could be done via cutouts in the sheet metal.
3. Subdivide the display opening into three 20 x 20 mm openings, leaving two thin metal ribs between. The SE at 500 MHz will become > 26 dB.
4. Replace the side-cut slots with oval holes. This will reduce the edge radiation. The SE will be >30dB at 500 MHz.

Solution for MIL-STD-461, RE02

First, to evaluate SE requirements, the field envelope should be recalculated for 1 m distance, and compared to RE02 limit. This produces:

Freq. in MHz	*15*	*45*	*75*	*150*	*300*	*500*
$E_{dB\mu V/m}$	84	74	74	78	69	64
RE02 Specification	21	24	28	32	37	40
Off Specification by:	63	50	46	46	32	24
A. New SE Objective (incl. 6 dB margin)	69	56	52	52	36	30
D. SE at Present	36	26	22	16	10	0
E. Improvement Needed	33	30	30	36	26	30

We see that the new requirement demands more hardware changes for frequencies up to 500 MHz. A fan hole grid should be employed, with 3 mm maximum spacing. The long cooling slots should be subdivided into shorter, 5 mm x 3 mm slots or, better yet, into an array of 3 mm diameter holes. The display window must be equipped with 3 mm maximum thin mesh. Cover screw spacing has to be reduced to 50 mm (or an EMI gasket installed).

10.9 SHIELDING COMPONENTS FOR MASS-PRODUCTION, CONSUMER PRODUCTS

In the late 1990's, the technical evolution has brought a huge number of miniature, popular electronic devices using high speed digital circuits and wireless RF techniques, operating at 0.9GHz, 1.8GHz or more. This has urged the development of new shielding hardware. These shielding components have to be economical, lend themselves to mass production techniques (one single manufacturer of portable telephones reported a year 1999 production of 60,000 devices per day!), and provide performances which were barely attainable by the expensive, sophisticated military electronics of the 1980's.
Such shielding hardware includes:

• Heat-formable, shrinkable films.
They are generally polymer-fiber films coated with a metal mesh with low fusion point (ex.tin, for the 3M #6100) or a conductive ink grid (G.E. Lexan). When

plastic housing that needs to be shielded. With total thickness of 0.2 to 0.8mm, foil resistance is in the range of 0.1-1Ω/sq., and the textured nature of the metal content prevents hi-Q resonant cavity effects.

• Thin, form-in-place gaskets.
Conductive caulking can be applied in a regular cord-gasket with diameter as small as 0.3 -1mm, acting both as EMI and weather gasket. They can be put with an automatic dispenser, or printed in a single operation like an ink, conforming to very intricate shapes.

• PCB component shields (see Chap. 5)
Five-sided cans, stamped from tin-plated steel or brass, are available off-the-shelf in standard shapes, with heights as low as 3mm. They can be wave-soldered to a printed ground belt around the specific component, or PCB section, which needs to be shielded.

10.10 SUMMARY OF RADIATION CONTROL VIA BOX SHIELDING

1. When the best affordable measures have been taken at PCB and internal wiring level, the equipment housing is the ultimate barrier against radiated emissions.

2. Until the last hole or slot is checked, the best metal box could appear to be useless as a shield.

3. For metal housings:
   a. Bond all metal parts (a floated item is a candidate for re-radiation).
   b. Avoid long seams and slots: a 30 cm seam is almost a total leak at the upper VHF frequencies.
   c. Use gaskets and waveguide effect.

4. For plastic housings:
   a. Use conductive coating, < 1 Ω/sq, then treat the box like a metal housing
   b. Avoid long, protruding screws inside.

5. Respect, or restore, shield integrity at:
   a. cooling holes
   b. viewing apertures
   c. component holes
   d. cable penetrations

6. Beware of noisy circuits or cables close to seams and slots: they degrade an otherwise sufficient SE.

Chapter 11

Controlling Radiation from External Cables

As soon as an equipment is fitted with external cables whose length exceeds the largest box dimension, it is highly probable that these cables will become the largest contributors to radiated emissions (and susceptibility, as well). We have seen (in sections 2.4 and 2.5) that two types of excitation drive these cables as radiating antennas:

1. Differential-mode (DM) excitation, where the currents are balanced (equal and opposite) in the wire pairs. Unless the outgoing and return carriers are spaced very far apart and untwisted, this mode is a minor contributor, though measurable.

2. Common-mode (CM) excitation, where the unbalanced portion of the current flows in the whole cable-to-ground loop. Due to the huge size of the antenna, this mode largely dominates DM excitation.

11.1 ADVANTAGES OF BALANCED INTERFACES

Compared with ordinary unbalanced driver/receiver links where each signal is simply transmitted or received between a single wire and a common return (e.g., RS-232), balanced links offer the advantage of pairing each signal with its associated return wire. Although primarily intended to improve EMI immunity, balanced links also reduce emissions.

The true balanced scheme (also known as true differential or bipolar) consists of transmitting a signal with equal positive and negative amplitudes relative to ground via a wire pair or twinax. In this manner, 100 percent of the outgoing current on one wire of the pair normally returns by the other wire, and no net current flows via the chassis and earth grounds (see Fig. 11.1).

In reality, CM cancellation is not perfect because the symmetry of drivers and receivers is never perfect, nor is the symmetry of the wire pair distributed RLC parameters. The symmetry generally deteriorates at high frequencies. True differential drivers and receiver pairs normally need a double power supply to provide +Vcc and -Vcc versus ground. However, more and more devices are

available where the double polarity is generated internally, so they can be supplied from a single voltage source.
The pseudo-balanced link simultaneously delivers a logic pulse A and its complement, $\bar{A}$. The signal is still referenced to a ground, as with an ordinary, unbalanced link. In reality, dynamically, when one output goes up, the other goes down. In theory, this scheme should provide a CM decoupling similar to that of the true symmetrical one. However, the positive signal edge and its complement are not perfectly in phase, due to different delays inside the device and, subsequently, on the PCB or backplane. The small shift causes both wires of a same pair to be at a positive (or negative) state at the same time for few nanoseconds or less. This corresponds to the circulation of short spikes in the CM loop (see Fig. 11.2). Therefore, in summary, balanced links will tend to radiate less than unbalanced ones, but no precise figure can be given for this reduction, particularly above 30 MHz.

11.2 LINE BALANCING DEVICES

With ordinary, unbalanced, drivers and receivers, it is still possible to transfer data in a balanced mode by using line balancing devices such as:

- signal/pulse isolation transformers
- longitudinal (no isolation) transformers

In general, such devices are installed within the equipment, near the I/O ports, so in a physical sense, they more properly belong in Chapter 9. However, because they are aimed at reducing susceptibility to and emissions from external cabling, they will be discussed here.

11.2.1 Signal Isolation Transformers

Signal or pulse transformers are used in many EMI suppression applications, from audio to video frequency. Uses include breaking ground-loops, creating galvanic isolation, line impedance matching and balance-to-unbalance conversion. The latter is our prime concern here.
The balancing transformers used in the Integrated Services Digital Network (ISDN), local area networks (LANs), MIL-STD-1553 and other communication links allow the rejection of CM currents in external loops while processing DM signals. Therefore, with proper mounting precautions to reduce primary-to-secondary couplings, the CM voltages developed on PCB traces and ground references (with respect to chassis) do not drive the external pairs (see Fig. 11.3).

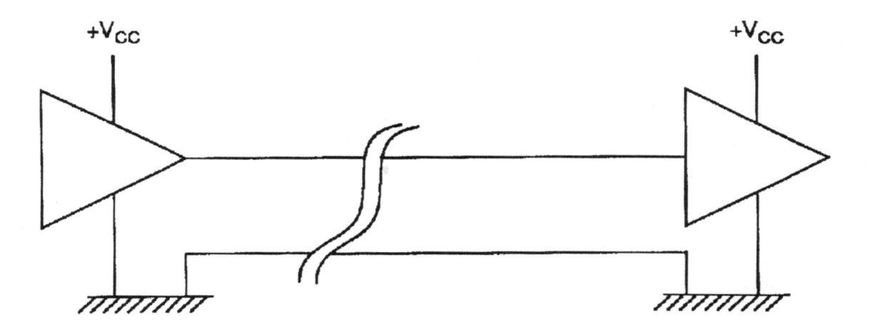

a. Unbalanced

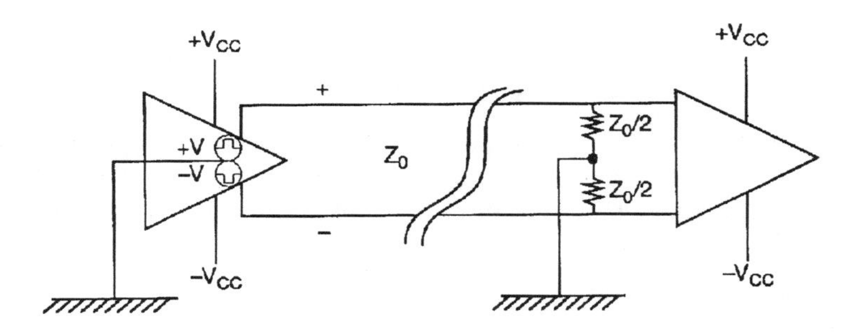

b. True Balanced (Differential or Bipolar)

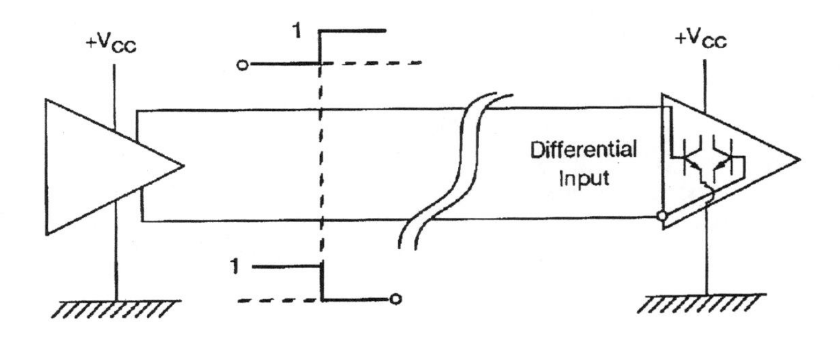

c. Pseudo-balanced (Unipolar Differential)

FIGURE 11.1 Different Types of Digital Line Balancing

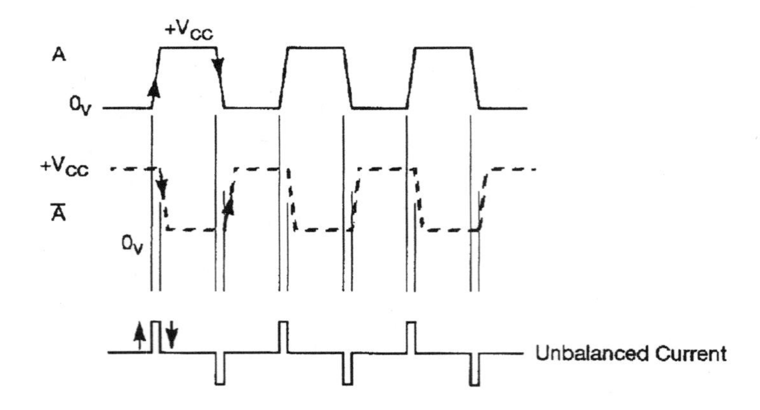

FIGURE 11.2 Unbalance Problem with Pseudo-Balanced Drivers. Due to a small shift between the A and Å transitions, short current spikes are driven into the system ground.

This, of course, is an idealistic view. In reality, several secondary effects that are not visible at low frequencies do show up above the bandwidth of the useful transmitted signals, right into the critical frequency domain for spurious radiation These effects are:

1. Mode conversion

Due to imperfect balancing of transformer windings, a certain amount of the CM voltage impressed on primary side appears as a DM signal on the line. For the best transformers, this fraction is only one to three percent, at frequencies above 1 MHz. However, the effect can be aggravated at higher frequencies The consequences on radiated emissions are moderate, since DM radiation is seldom the dominant mode.

2. Primary-to-secondary capacitance

This capacitance will let an increasing portion of high-frequency internal CM voltages to appear as CM voltages driving the wire pair. This is of serious concern because CM excitation is the most efficient contributor to cable radiation. An electrostatic (Faraday) shield between the primary and secondary can reduce this effect.

For the user, the useful information is the practical 3 dB bandwidth, corresponding to:

• the longest pulse width that can be processed without excessive amplitude droop
• the fastest rise time that can be processed without excessive distortion

For example, a pulse transformer with a 3 dB bandwidth of 0.15 to 100 MHz can process pulses as long as 2 μs and rise times as short as 3.2 ns. The following table shows some typical features of signal/pulse transformers.

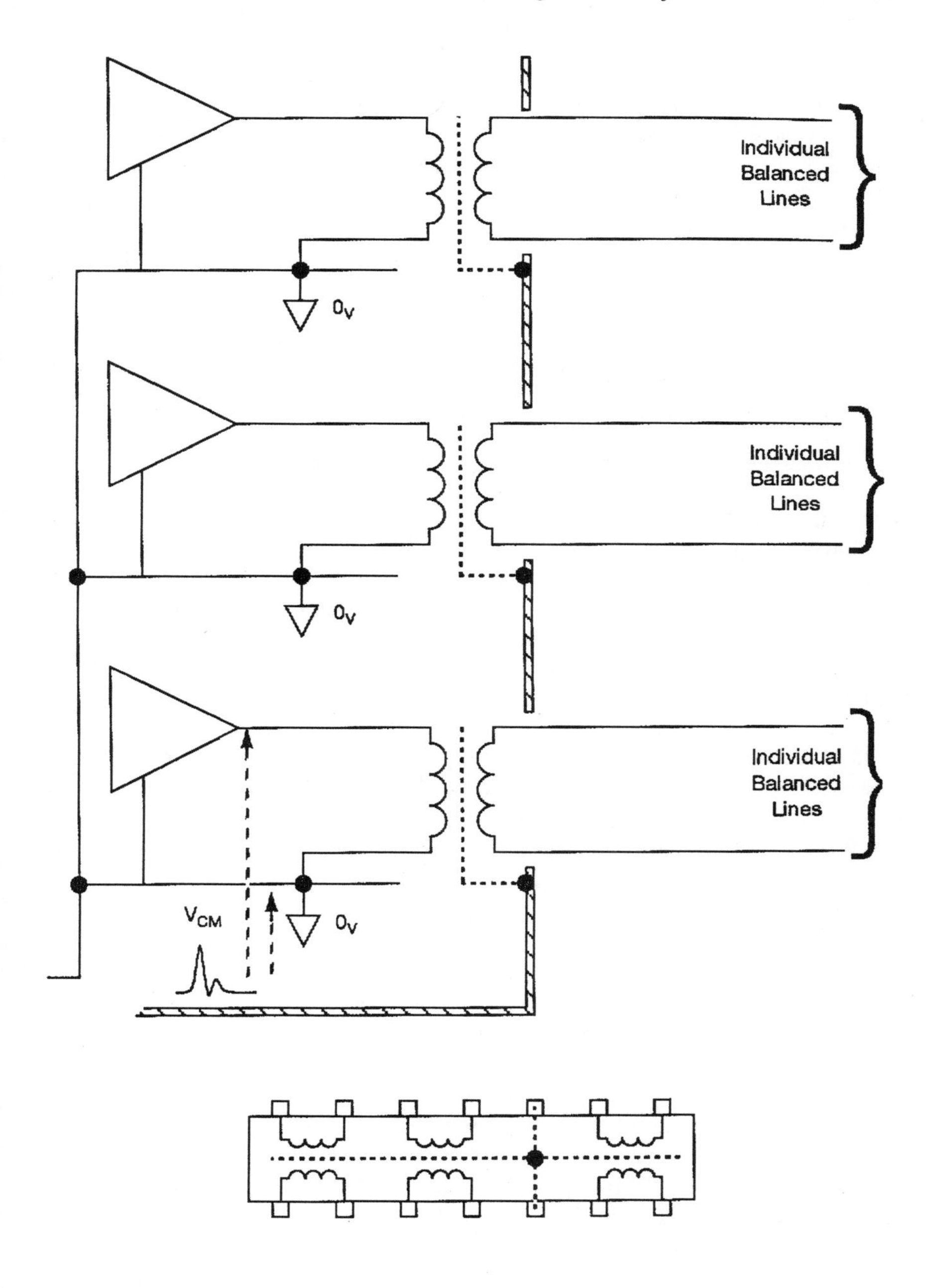

FIGURE 11.3 Signal Isolation Transformers for DM Coupling. These also can be obtained in dual in-line package, with a Faraday shield.

Type	*Bandwidth* (F_{min})	(F_{max})	*Parasitic Capacitance, Primary to Secondary*	*CM Rejection Ratio CMRR*
Audio Transformer	300 Hz	3,300 Hz	60–100 pF	60 dB (in band)
Signal/Pulse (DIP)	10 kHz	100 MHz	15–40 pF (unshielded) 3–5 pF (shielded)	40 dB @ 1 MHz 20 dB @ 50 MHz
Video-Wideband	20 Hz	30 MHz	50 pF	120 dB @ 60 Hz 30 dB @ 1 MHz

As a summary, isolation transformers behave as high-pass couplers for CM noise; their blocking effect against cable emission rates from excellent at LF to mediocre at VHF and above. If employed, they need to be complemented by CM capacitive decoupling (see Chapter 9) or CM ferrite beads.

11.2.2 Longitudinal, Non-isolating Transformers

Based on the principle of mutual cancellation of equal CM currents, these bifilar chokes have the opposite features of isolation transformers (see Fig. 11.4). They do not block low frequencies (in fact, they are transparent down to dc) but have excellent CM attenuation in the VHF and low UHF ranges. The mutual inductance between the coupled windings increases the CM loop impedance by as much as 3 to 20 times while having little effect on the useful, DM signal, because DM fluxes are cancelled.

Such transformers are available in DIP, SMC and discrete packages. One can predict that cable emission levels will be reduced by the same amount that CM current is reduced, and actual test results confirm this prediction. Isolation transformers and CM bifilar chokes are sometimes combined in a single miniature component (Fig. 11.4A).

11.3 REDUCING CM RADIATION BY FERRITE LOADING

Using a principle similar to that of the longitudinal transformers already discussed, ferrite beads are hollow cores or toroids that can be slipped over a wire or cable. They behave as a lossy inductance with one or a few turns. Although they are popular among EMC specialists as quick “last chance” fixes, they can be incorporated in original designs to achieve remarkable EMI reduction, provided their theory is properly understood (Ref. 28, 29).
In contrast to ferrites used for coupling between circuits in microwave applications (couplers, circulators, and so forth) where low loss and best efficiency are the aims, EMI ferrites are made of lossy materials that have a good magnetic permeability

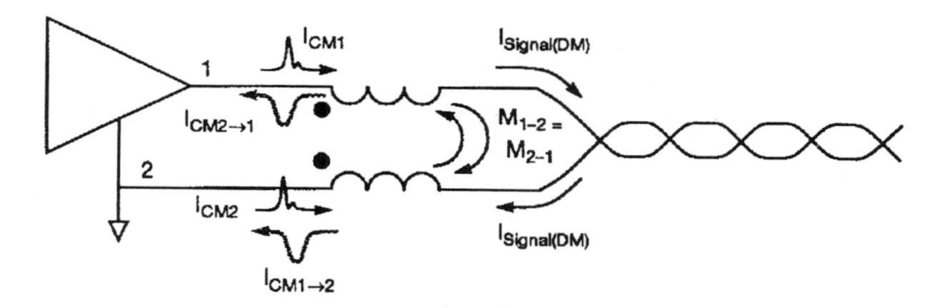

- For CM currents, fluxes are in phase in the coil. Each I_{CM} induces in the other wire a counter EMF that drives a cancelling current.
- For DM currents, fluxes are equal and in opposite directions; net flux = 0.

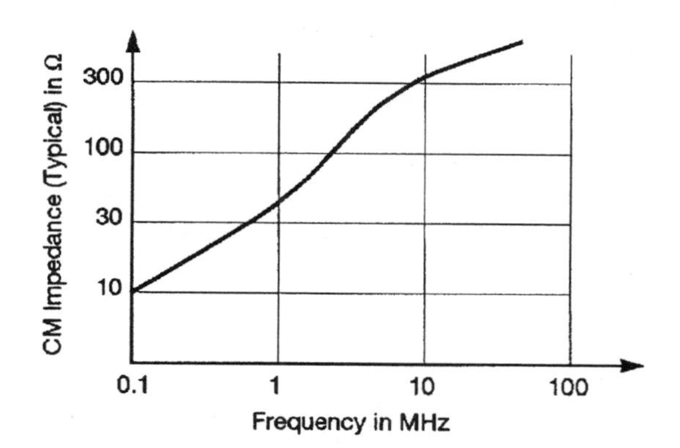

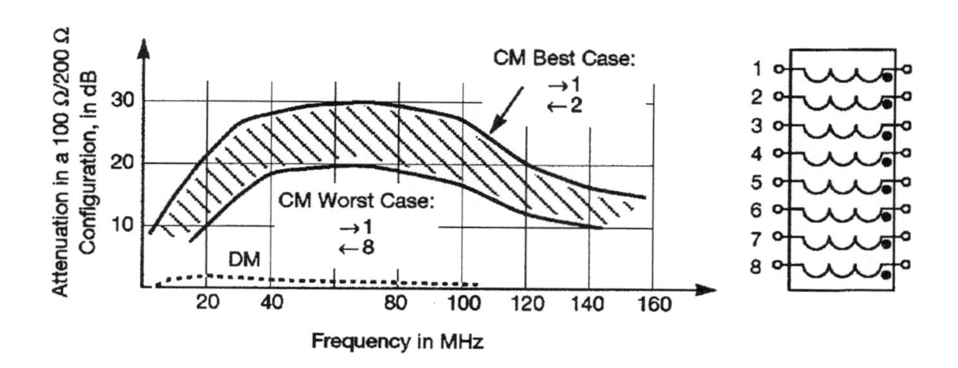

FIGURE 11.4 Longitudinal, Non isolating Transformer. Attenuation is shown for typical dual in-line signal transformers. Notice that DM Attenuation for normal signal stays below 2 dB.

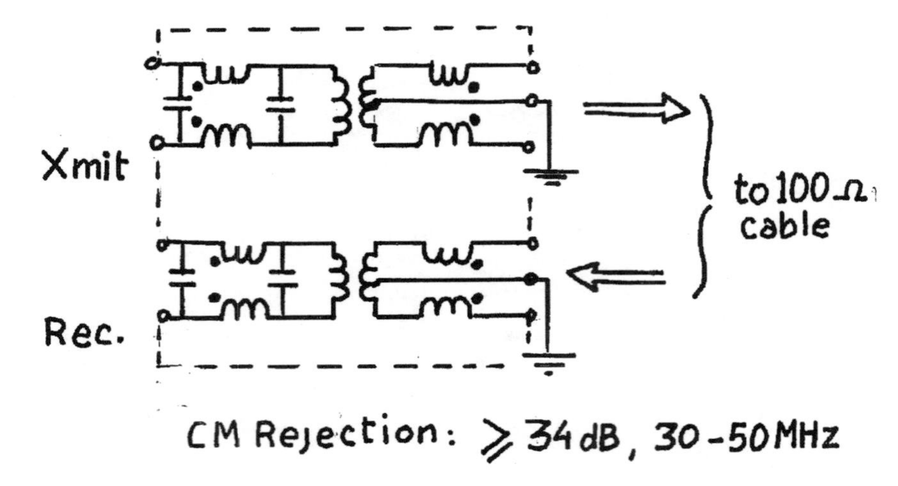

FIGURE 11.4A. Example of miniature signal transformer, combining isolation, Bandwidth filter and CM attenuation in a 25x10mm size (Source: Valor /Pulse Engineering)

(preferably, with μ_r being flat over a wide frequency span, and with typical values of 30 to 3,000). They also display an equivalent resistance of 10 to a few hundred ohms. Although ferrite beads are generally thought of as inductors, they are in fact lossy transformers. Therefore, the impedance of the ferrite is more like:

$$Z = \sqrt{R^2 + L^2\omega^2} \qquad (11.1)$$

The L term depends on the relative permeability, μ_r, which generally is in the range of a few hundreds for a typical ferrite. It can be demonstrated, that with L_o being the self-inductance of the segment of conductor prior to the ferrite mounting, the increase in inductance caused by the ferrite can be predicted from:

$$\frac{L}{L_o} = \mu_r \qquad (11.2)$$

An example is a ferrite 1cm in length, with $\mu r = 300$, slipped over a 1 m wire. As the inductance for an ordinary wire is about 1 μH/m, the initial value of Lo for 1 cm was:

$$L_o = 1\ \mu H \times 0.01 m = 10 nH$$

Once the ferrite is installed, L becomes:

$$L = 10\ nH \times 300 = 3\ \mu H$$

Compared to the initial 1 μH inductance of the 1 m wire, the increase in L caused by the ferrite is 3 μH/1 μH; i.e., three times.

Due to the generally small size of ferrite beads, they can easily saturate for the normal current and become inefficient against the EMI current. The amount of current a bead can handle without significant decrease of μ, is given by the manufacturer. It is related to:

$$\text{logn}\,(r_2/r_1)$$

where r_2 and r_1 are the bead's outside and inside diameters. Therefore, beads with proportionally small holes will behave better.
The permeability is also affected by frequency. Some beads are optimized to work below 10 MHz, and others are suitable from 10 to 100 or even 1,000 MHz.

Figure 11.5 shows the resistive and inductive parts of a typical bead impedance. Because of their low Q, beads are especially efficient for damping high-frequency contents of switching transients, clock harmonics and parasitic resonances. Since they add series impedance, they are an inexpensive way to create EMI losses without affecting dc or low-frequency intentional currents. If a circuit or cable is exposed to a high-frequency EMI coupling, the beads will prevent the circulation of induced currents.
Conversely, if a signal source contains undesired spurious noise, ferrites will prevent these currents from propagating and making the wire a radiating antenna.
To gainfully use ferrite beads, it must be understood that they work by series insertion loss. Therefore, the attenuation provided by the ferrite will be:

$$A_{dB} = 20 \log \frac{V_o \text{ without ferrite}}{V_o \text{ with ferrite}} \tag{11.3}$$

$$= 20 \log \frac{\dfrac{Z_L}{Z_g + Z_w + Z_L}}{\dfrac{Z_L}{Z_g + Z_w + Z_L + Z_b}}$$

$$= 20 \log \frac{Z_g + Z_w + Z_L + Z_b}{Z_g + Z_w + Z_L} \tag{11.4}$$

where.

AdB = attenuation in dB
Z_g, Z_L = circuit source and load impedances,
Z_w = wire impedance
Z_b = ferrite bead impedance
V_0 = voltage across load

This equation reveals two things:

1. Ferrites will not work efficiently in high-impedance circuits. Although significant progress was made by manufacturers in the 1980s, the best ferrites today achieve values of Z_b in the 300 to 600 Ω range, above 50 MHz. Neglecting the wire impedance, the best attenuation they can provide in a 100 Ω/100 Ω configuration is:

$$AdB = 20 \log [(100 + 100 + 600) / (100 + 100)] = 12 \text{ dB for 1 turn}$$

Conversely, ferrites will be extremely efficient in low-impedance circuits such as power distribution, power supplies or radio-type circuits where impedances are 50 or 75 Ω.

2. If the wire impedance itself is significant, the ferrite performance may be disappointing because of the presence of Z_w in the equation.

An extremely useful application of ferrite is in the blockage of common-mode currents. As explained in Section 11.2.2, with regard to inductors, if the two wires of a signal pair are threaded into the bead, the ferrite will affect only the undesired EMI currents and will have no effect on the intentional differential-mode current. The same is true when a ferrite is slipped over a coaxial cable.

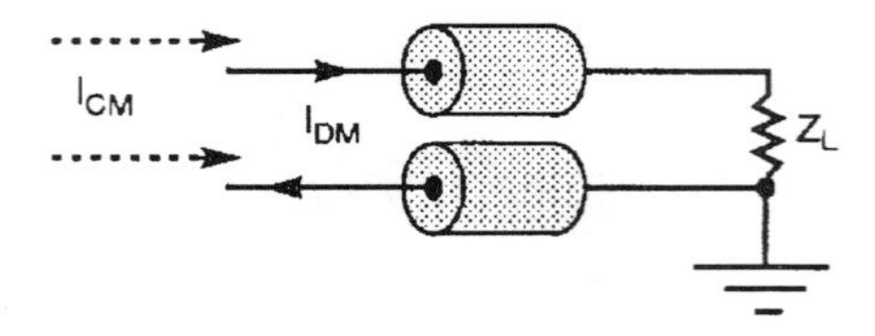

One ferrite per wire attenuates *DM and CM* currents (might affect useful signal)

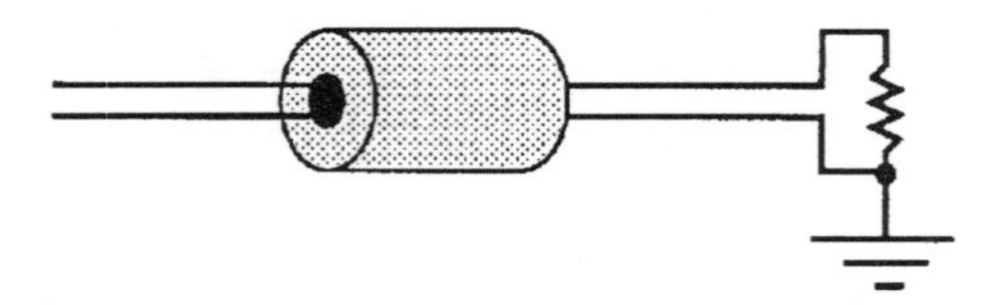

One ferrite around two wires attenuates *CM only.*

FIGURE 11.5 Common-Mode vs. Differential-Mode Operation of Ferrite Beads

In addition to their limited impedances, ferrites have the following limitations:

1. When bead length approaches λ/2, the bead becomes inefficient if it gets into a current "null". This can be somewhat overcome by always placing the ferrite near to the cable end.

2. The end-to-end parasitic capacitance of the ferrite (typically 1 to 3 pF) may bypass its resistance above a certain frequency and cause its attenuation to collapse.
3. Beyond about 1,500 to 2,000 G, saturation occurs and efficiency decreases.
4. When slipped over multipair cables, ferrites may increase inductive crosstalk between adjacent pairs.

The saturation problem can be controlled by checking the flux density $B=\phi/S$. Figure 11.6 shows that ferrite bead flux is given by $B \times l(r_2 - r_1)$. However, if B is too large, the core will saturate and μ_r will decrease.

Example 11.1
A ferrite has the following dimensions:

length $l = 1$ cm
outside radius $r_2 = 0.5$ cm
inside radius $r_1 = 0.2$ cm

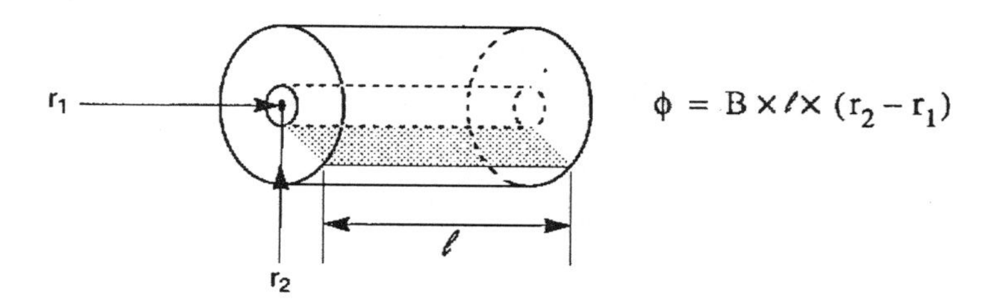

FIGURE 11.6 Ferrite bead cross-section and Flux Density

According to manufacturer's data, the equivalent inductance L_b is 10^{-6} H, provided that B does not exceed 1,500 G. What is the maximum current this bead can handle without losing efficiency?

$$I = \phi/L_b$$

$$= \phi/10^{-6} \tag{11.5}$$

Also,

$$\phi = B \times S = B \times l \times (r_2 - r_1) = B \times 0.01\ (0.5 - 0.2).10^{-2}$$

$$\phi = B \times 0.3.10^{-4}$$

Given, $B_{max} = 1{,}500\text{G} = 0.15\text{T}$

$$\phi_{max} = 0.15 \times 0.3.10^{-4} = 0.45.10^{-5}$$

Then, from (11.5)

$$I_{max} = \phi_{max} / 10^{-6} = 4.5\text{A}$$

If the attenuation with a ferrite bead is not sufficient, this can be improved in several ways. One method is to make more than one turn of the wire in the bead hole, using two or three turns. However, this may rapidly bring the ferrite into saturation. Also, the turn-to-turn capacitance may ruin the inductance improvement.

The increase in impedance is (theoretically) proportional to N^2, so two turns will give a 4 times larger impedance (and, hence, attenuation). Three turns produces 9 times the impedance and attenuation, and so forth. Generally, if one turn gives insufficient results, two or three may be enough. Additional turns won't help or will simply shift the problem toward a higher frequency. Putting several beads in a end to end installation is another method, but less efficient; it would take nine beads to duplicate the result of three turns into a single bead. Multihole beads are not suitable for fitting on external cables but can be mounted in the PCB I/O area. Be careful that, in this case, the two wires of a line are not mutually coupled: a multihole bead will affect DM signal as well as CM.
Ferrite beads are available in cylindrical or flat core shapes (see Fig. 11.7). Split versions are available for quick installation during EMC troubleshooting.
Figure 11.8 shows a typical result of installing ferrites on unshielded cables connected to a small personal computer. The objective is to meet FCC Class B specifications. In this case, the cables were the major radiators, and the ferrites did wonders (Ref. 30). As discussed in Chapter 13, this may not be sufficient if box radiation also violates the limits.

Ferrite-Loaded Cables and Tubing

Derived from the lossy ferrite principle, an interesting extension has attracted a growing interest since the late 1970s. Instead of inserting discrete ferrites over the cable, the whole cable length is made lossy by coating all the wires in the cable with a ferrite-loaded jacket (Ref. 31). This provides basically CM attenuation, with very little DM attenuation. To maintain flexibility, the ferrite powder is embedded in a soft binder, so the ferrite percentage in volume is rather low, corresponding to a relative permeability of a few tens. By avoiding the ringing due to impedance discontinuities that a cascade of beads would create, ferrite-loaded wires have a more even attenuation performance above 100 MHz. The impedance of typical lossy cables is shown in Fig. 11.9.

11.4 REDUCING DM RADIATION BY TWISTING

Twisting the two wires of a pair has virtually no effect on CM radiation (although a slight improvement in the symmetry of each wire-to-ground distributed inductance and capacitance may result). It has a strong effect on DM radiation. Since this mode is seldom a problem, twisting does not create a major reduction in overall radiated field.

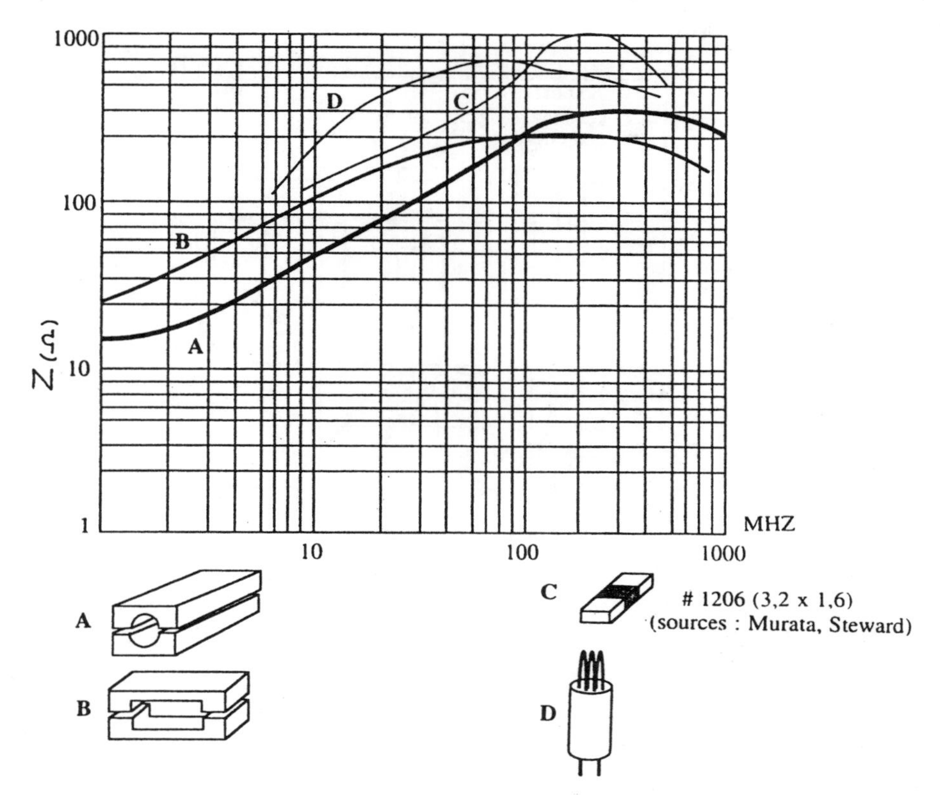

FIGURE 11.7 EMI Suppression Performance of Sample Small and Large Ferrite Beads
* For A, B styles, impedance is shown for one pass (N=1).

- For N = 2 or 3, Z is multiplied by N^2 up to ≈150 MHz

However, since each contributor plays its part, when efficient CM reduction techniques have been applied (such as reducing CM pollution of I/O cables by 25 or 30 dB) the DM contributor may resurface. In fact, if the DM contribution of the untwisted pair is only 5 percent of the CM one, reducing this latter by 30 dB (a 30 times factor) will unveil the DM radiation.
The reduction in a radiated field produced by twisting a wire pair is expressed (from Ref 16) by:

$$A_{dB} = 20\log\frac{\text{E field w/o twisting}}{\text{E field after twisting}} \qquad (11.6)$$

$$\approx 20\ \log\frac{1+2n\ell}{1+2n\ell\ \sin\dfrac{F_{MHz}}{100\times n}} \qquad (11.7)$$

where,

n = number of twists/meter
l = total twisted length

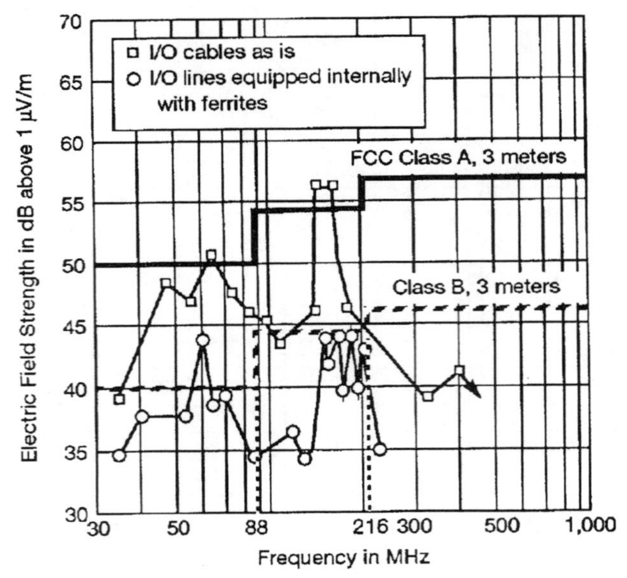

FIGURE 11.8 Effect of Long Ferrite Beads on Radiation from I/O Cables (From Ref. 30)

When $F_{MHz} << 100 \times n$, the attenuation becomes: $20 \log (1 + 2 n l)$
When $(F / 100 \times n)$ approaches $\pi/2$, A = 0 dB

This formula assumes that:

1. There is always an odd (uncancelled) loop, or a small untwisted segment at the end of the wire, whose length equals approximately l/n meters.
2. The entire radiating wire length is twisted. Twisted flat cables with untwisted sections will provide less reduction.

11.5 REDUCING CABLE RADIATION BY SHIELDING

Although shielding a cable may seem to be the obvious "catch-all" barrier to radiated emissions, application may not be so easy. Throwing in shielded cables at the last minute may give disappointing or nonexistent results. The author has even seen weird cases where shielded cables increased radiated levels at some frequencies. There are explanations for this, of course, as will be seen.

The basic principle for a shield to work against all types of EMI, with the widest coverage of situations (E field and H field, LF and HF, DM and CM, etc.), is to create a continuous barrier that encloses the conductors and is 360° bonded to the conductive boxes at both ends. No matter which theory is employed to model this shield (reflection loss, absorption loss, skin depth, Faraday cage, mutual inductance, ad infinitum), calculations and experiments show that when the entire system is enclosed in a continuous barrier, its radiated EMI is reduced. This principle works whether the barrier is earthed or not (see Fig.11.10).

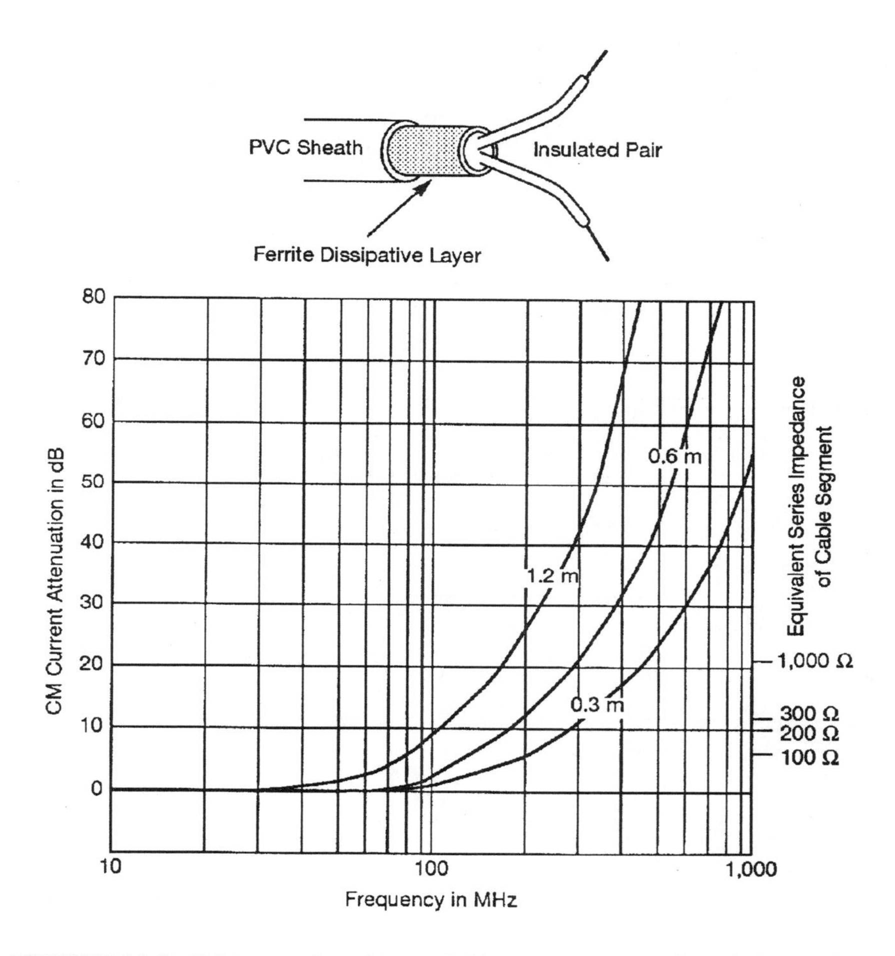

FIGURE 11.9 CM Attenuation of Lossy Cable, Measured in 50 Ω/50 Ω, with Cable 5 cm above Ground

If the box is not a six-face metallic cubicle, the principle still can work, provided there is at least one large metal face or ground plane on both ends and that it terminates the shields and closes the cable-to-shield return path for CM currents. Otherwise, if there are no ground plane references for terminating the shields (as in the case of solid plastic boxes), a cable shield will not be efficient in reducing radiated emissions in the VHF range. Adequate I/O port decoupling and ferrite loading are more appropriate for such a situation, if less than 20 to 30 dB reduction needed.

If, for a well understood reason (e.g., LF ground loops caused by large noise voltages between distant chassis, upsetting a low level analog input), a cable shield has to be grounded at one end only, it is only effective in reducing LF electric fields and capacitive crosstalk. It has virtually no effect on CM immunity or radiation, as the CM loop current does not return by the shield but, rather, by the chassis and

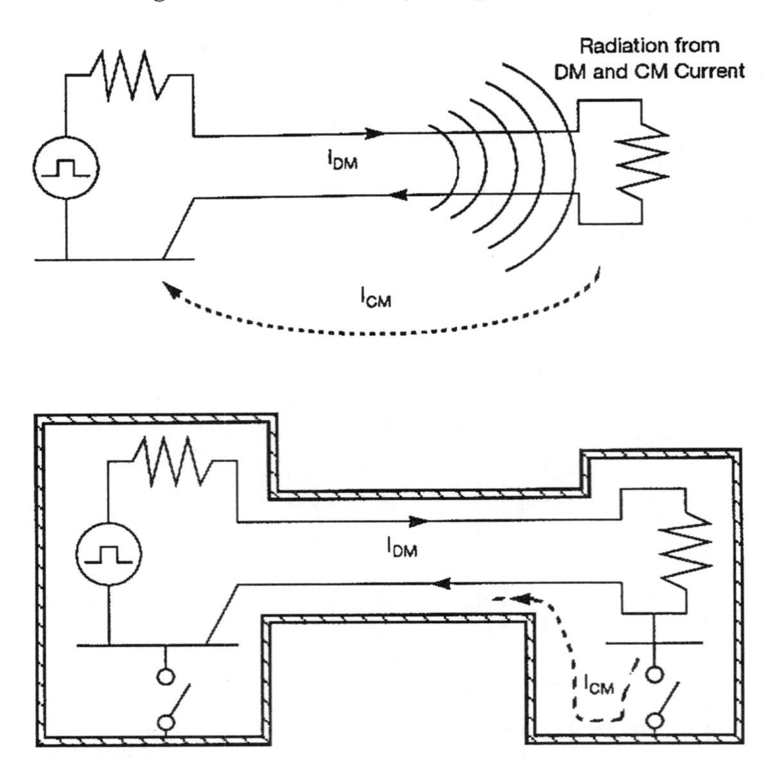

FIGURE 11.10 Ideal Shielded System. Provided the metal barrier is uninterrupted and homogeneous, radiation is strongly reduced, whether or not the circuit is grounded to the shield, or the shield is connected to earth.

and ground plane, as if there were no shield. However, as frequency increases, some percentage of the CM current returns by the internal wire-to- shield capacitance. But the shield impedance also increases with frequency (see Fig. 11.11). Therefore, if we designate :

V_{CM} = CM noise voltage driving the pair
C_p = pair-to-shield distributed capacitance
Z_{sh} = shield impedance
$= R_{sh} + j\omega L_{sh}$

then, we can express the voltage versus ground of the shield's floated end:

$$V_{CM} = \Sigma I_{sh} \times Z_{sh} = V_{CM} C_p \omega \times (R_{sh} + \omega L_{sh}) \qquad (11.8)$$

Therefore, above few kilohertz,

$$V_{sh} / V_{CM} = L_{sh}\, C_p\, \omega^2$$

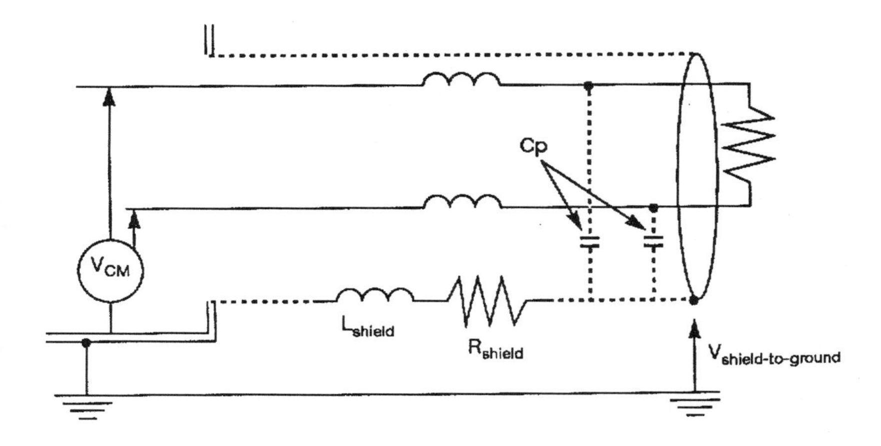

FIGURE 11.11 Radiated Emission with Floated-End Shield

The shield voltage versus ground increases with the square of frequency and becomes a significant fraction of V_{CM}. The floated end of the shield becomes the"hot" tip of a radiating monopole, and we have just replaced a radiating pair by a radiating shield. So, exceptions acknowledged, a cable shield must be connected at both ends to the boxes, whether these are grounded or not.

The exceptions are:

- low-level analog instrumentation (strain gages, thermocouples, etc.)
- audio interface cables

In these applications, only an electrostatic shield is needed. Ground loops are addressed by galvanic isolation amplifiers, differential amplifiers, and so forth, and grounding a shield at both ends could inject LF noise into the cable. A few millivolts injected this way are harmless for digital interfaces but can constitute strong interference for low-level analog signals.
Notice that it is unlikely that such analog cables would radiate RFI. If this were to happen, at least the floated end of the cable shield should be grounded at RF frequencies through a capacitor of a few nanofarads value.

11.5.1 Fields Radiated by Coaxial Cable

RF signals, baseband video, some LAN links and other high-frequency signals are carried on coaxial cable. When a coaxial cable carries a signal, very little current (typically 0.3 to 0.1 percent, above a few megahertz) returns by paths other than the shield itself. This assumes that the shield is at least correctly tied to the ground references at both ends, and preferably also to the chassis by the coaxial connectors.

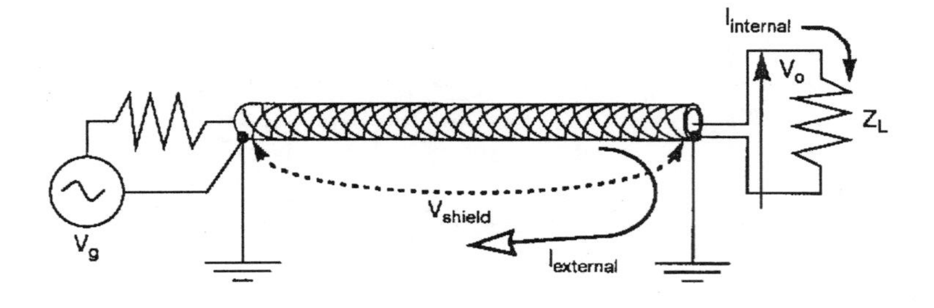

FIGURE 11.12 Field Radiation by the External Shield Current

This external current radiates a small electromagnetic field (see Fig.11.12) that can be associated with the quality of the shield and its installation. It is related to the voltage appearing along the shield due to the transfer impedance of the braid.

Transfer impedance, Z_t, is a convenient way of characterizing the merit of a cable shield (Ref. 27,32). The transfer impedance relates the current flowing on a shield surface to the voltage it develops on the other side of this surface. This voltage is due to a diffusion current through the shield thickness (if the shield is a solid tube, this diffusion rapidly becomes unmeasurable, due to skin effect, as frequency increases) and to the leakage inductance through the braid's holes. The better the quality of the braid, the less the longitudinal shield's voltage.

Initially, Z_t was conceived for susceptibility calculations and defined as:

$$Z_t \text{ in } \Omega/\text{m} = \frac{V_i}{\ell_m \times I_o} \tag{11.9}$$

where.

V_i = longitudinal voltage induced inside the shield, causing a noise current to circulate in the center conductor

I_o = external current injected into the shield by the EMI source

However, the principle is perfectly reciprocal and can be applied to emissions as well (see Fig. 11. 13). The internal, intentional current I_o returning by the shield's inner side causes an EMI voltage that appears along the outer side of the shield; it appears along the outer side of the shield and can be expressed as:

$$V_{ext} = Z_t\ \Omega/\text{m} \times \ell_m \times I_o \tag{11.10}$$

$$= Z_t\ \Omega/\text{m} \times \ell_m \times V_o/Z_L \tag{11.11}$$

This voltage excites the antenna formed by the external cable-to-ground loop.

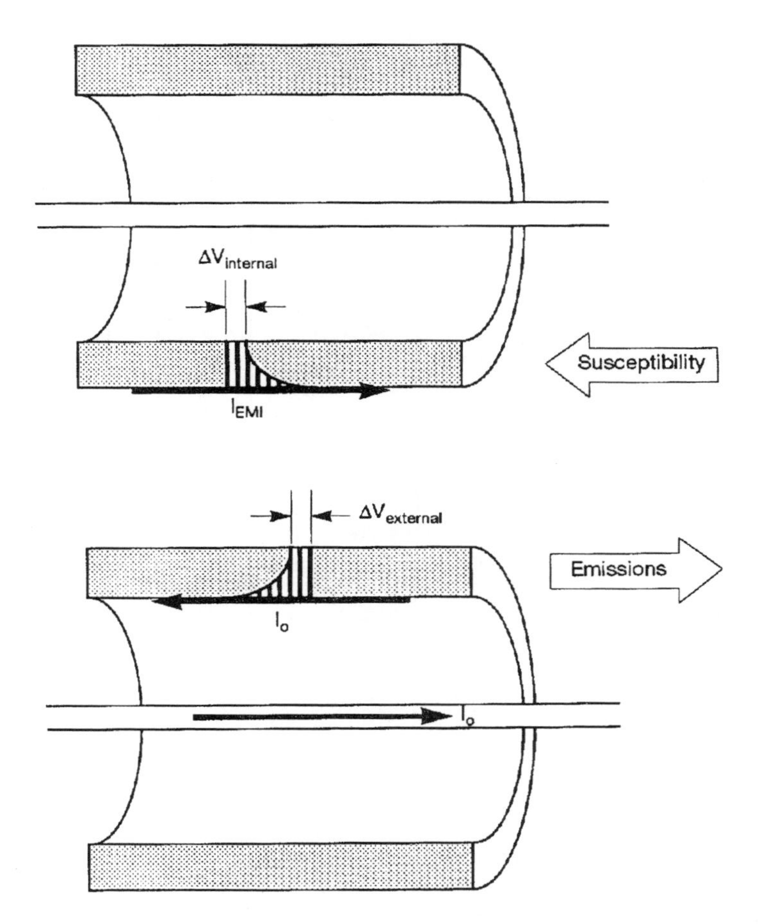

FIGURE 11.13 The Transfer Impedance Concept: Reciprocity of Susceptibility and Emissions

Typical values of Z_t for various coaxial cables are shown in Fig. 11.14. If the shield is grounded by pigtails (a poor practice), the pigtails and other impedances must be added to Z_t and loop impedance calculations.

An approximate value of the external loop impedance for a single-braid coaxial, with an outer diameter in the 6 to 15 mm range, and at a height of 50 to 500 mm above ground is:

$$Z_{ext} = (\ 10\ \text{m}\Omega + j5\ \Omega \times F_{MHz})\ \text{per meter length}$$

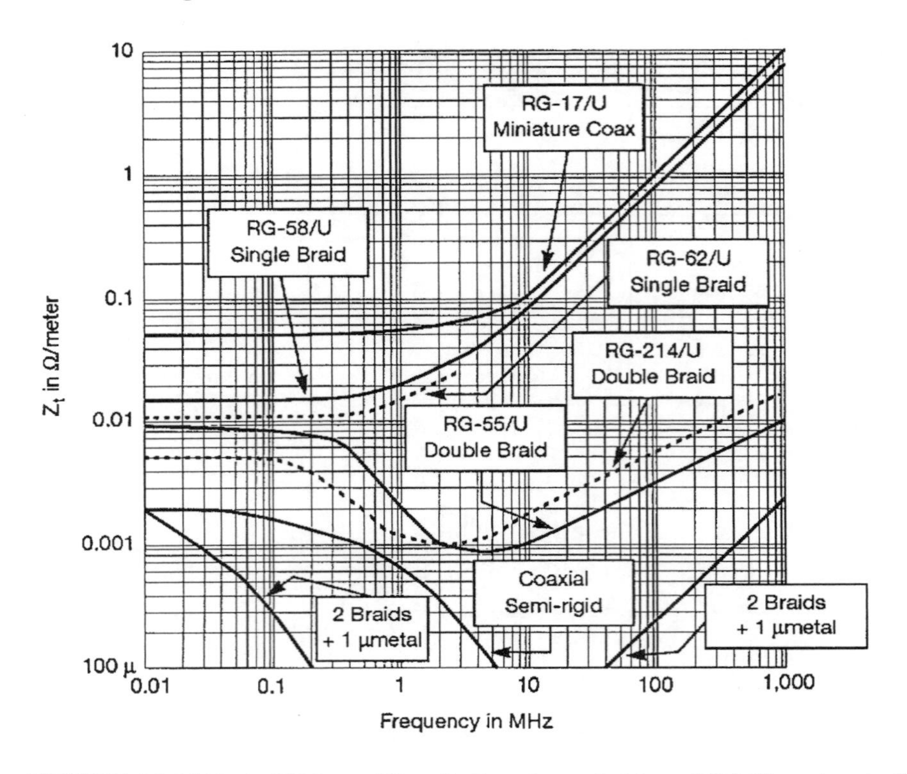

FIGURE 11.14 Typical Values of Transfer Impedance, Z_t (above 150 MHz, values in Ω/m are indicative only, since $l \geqslant \lambda/2$)

Eventually, pigtail or connector impedances have to be incorporated into Z_{ext}, although their contribution is usually minimal. *In contrast, their contribution to Z_t is very important*, as Z_t must be hundreds or thousands of times smaller than Z_{ext}, for a good shield.)

Then, to estimate E and H field from this low-impedance loop (see Fig. 11.15), the external shield current can be calculated by:

$$I_{ext} = V_{ext} / Z_{ext}$$

If the shield is floated from the chassis, the coax becomes an electrically driven radiator, and Fig. 2.6 can be used with V_{ext}, as an input.
When the cable becomes electrically long, Z_t (Ω/m) no longer can be multiplied by the length, since the current is not uniform along the cable shield. A default approximation is to consider that the maximum amplitudes of the shield voltages distributed along the shield are:

$$V_{ext\,(max)} = Z_t\,(\Omega/m) \times \lambda/2(m)$$

So, as Z_t increases with F, the effective length, which multiplies Z_t, decreases with F. At the same time, the cable-to-ground external impedance needs to be replaced by Z_0, the corresponding characteristic impedance, using the formula shown earlier as Eq. (2.25):

$$Z_0 = 60 \log n\,(4h/d)$$

Example 11.2

A 2 m piece of RG-58 coax is connecting two cabinets. The electrical parameters are:

Useful signal:	15 MHz video
Load resistance:	75Ω
V_o amplitudes:	fundamental = 10 V
	harmonic #3 = 3 V
	harmonic #10 = 0.3 V

The geometry is:

cable diameter= 0.5 cm
height above ground = 30 cm

Good-quality BNC connectors are used at both ends. Calculate the radiated field at 3 m for these three frequencies (fundamental and harmonics #3 and # 10). First, we need to determine the area, A, of the radiating loop:

$$A = 2\text{ m} \times 0.3\text{ m} = 0.6\text{ m}^2 = 6{,}000\text{ cm}^2 \text{ or } 76\text{ dBcm}^2$$

Above 37.5 MHz (λ/4 for 2 m), the cable and its image versus ground plane become a λ/2 folded dipole, and the effective radiating area is reduced accordingly. Zext will be calculated by:

$$Z = (0.01 + j5 \times F_{MHz}) \times 2\text{ m}$$

Above 75 MHz, this shield-to-ground loop impedance will be replaced by:

$$Z_0 = 60 \log n\,(4 \times 30/0.5) = 330\ \Omega$$

The internal current is: $I_o = V_o/75\ \Omega$, i.e., $I_o(\text{dBA}) = V_o - 38$ dB.

The radiated field at 3 m is calculated with the following results:

F	15 MHz	45 MHz	150 MHz
V_o, dBV	20	10	–10
Z_{load}	(75 Ω)	(75 Ω)	(75 Ω)
(1) I_o, dBA = V_o – 38	–18	–28	–48
(2) Z_t, dBΩ/m	–18	–8	+2
(3) $L_{dBmeter}$ (clamp to λ/2 above 75 MHz)	+6	+6	0
(4) V_{ext} = (1) + (2) + (3)	–30	–30	–46
Z_{ext}	(150 Ω)	(450 Ω)	(400 Ω)
(5) Area correction, dBcm^2	76	74	64
(6) E_o for 1 V – 1 cm^2 (Fig. 2.6) dBμV/m	–4	10	30
(7) E_{tot} = 4 + 5 + 6 dBμV/m	42	54	50
FCC Class B	NA	40	43

Although these radiated levels are about 50 dB lower than if a bare wire were carrying the same currents with a return by the ground plane, the FCC limit is exceeded by 14 dB* @ 75MHz and 7dB @ 150MHz. Several possibilities exist to reduce the radiated field:

1. Select a coaxial cable with a lower Z_t; i.e., $Z_t \leqslant$ -30 dB Ω/m at 45 MHz. Such performance can be achieved with "optimized" braided shields (thicker, denser braid) or more easily with double-braid shields.

2. Slip a large ferrite bead over the cable shield. It will take an added series impedance of about 1,200 Ω to achieve the required attenuation. Passing the cable twice into a large bead hole will provide such impedance (see Fig. 11.8).

3. Decrease cable height above ground.

11.5.2 Fields Radiated by Shielded Pairs or Multiconductor Shielded Cables

The concept of transfer impedance, used for radiated emission modeling of a coaxial cable, is transposable to shielded pairs. However, there is a noticeable difference: the shield is no longer an active return conductor (see Fig. 11. 16).

* It has been explained in Chap. 2 sect.2.5 that the 3dB which should be subtracted for peak-to-rms conversion are ≈ offset by the ground plane reflection of the CISPR/FCC test set-up.

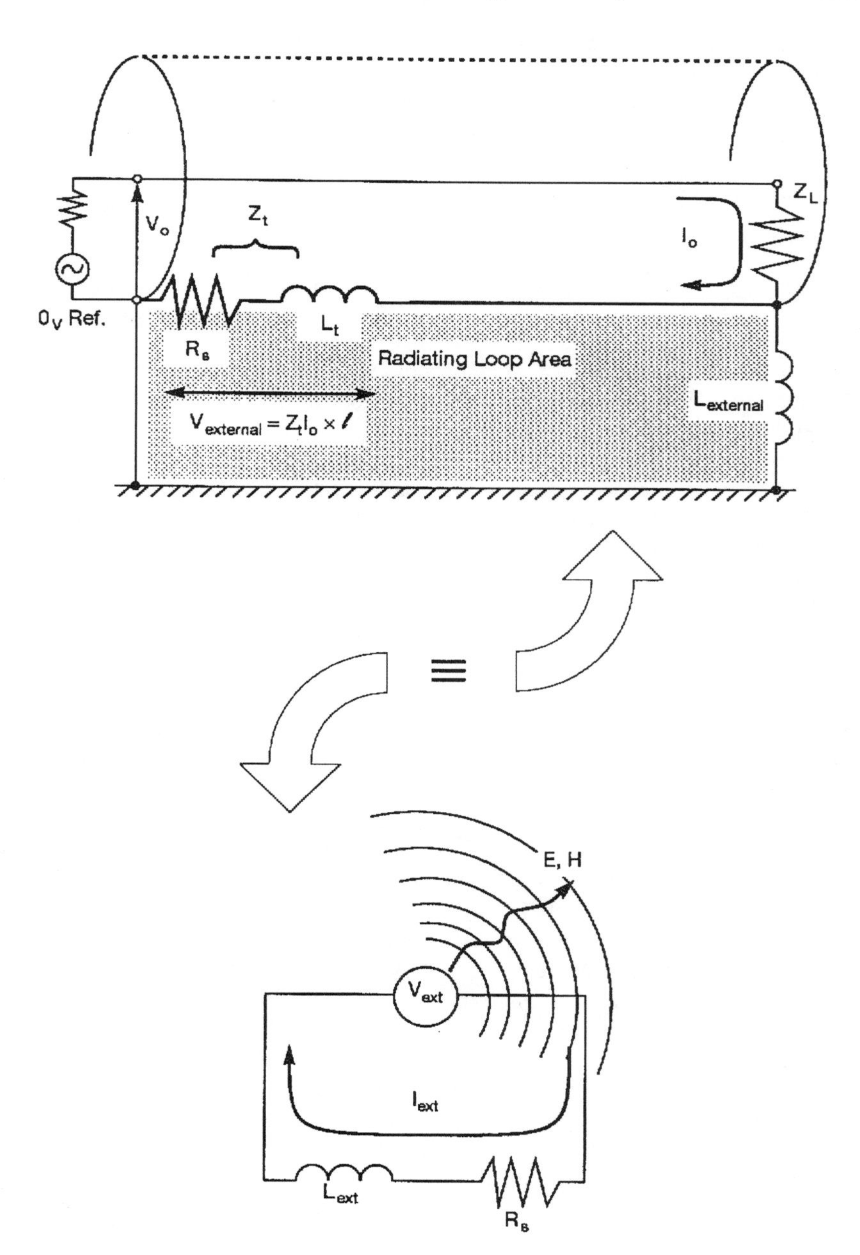

FIGURE 11.15 Equivalent Circuit to Predict Coaxial Cable Radiation

With balanced interfaces and wire pairs, the current returning by the shield is only prorated to the percentage of asymmetry in the pair. If the transmission link is balanced with X percent tolerance, the current returning by the shield is, for the worst possible combination of tolerances, only X percent of the total current.

In this case, Eq. (11.11) becomes:

$$V_{ext} = X\% \left(Z_t \, \Omega/m \times \ell \times \frac{V_o}{Z_L} \right) \quad (11.12)$$

The radiated field is reduced by a factor equal to X percent, compared to a coaxial cable situation. Depending on the quality of the balanced link, X may range from 1 to 10 percent, with typical (default) value being 5 percent.
If the wire pairs are interfacing circuits that are not balanced (e.g., the signal references being grounded at both ends), a larger portion of the signal current will use the shield as a fortuitous return. This portion is difficult to predict. At worst, this unbalanced configuration cannot radiate more than the coaxial case.

11.5.3 Shielded Flat Cables

One specific case of shielded multiwire cable is the shielded flat cable. A few typical versions are shown in Fig 11. 17. Version (a), sometimes described as "shielded", is in reality a flat cable with ground plane. Although offering some advantages, in general it has insufficient reduction properties because CM current can still flow on the single-side foil edges and radiate.

The (b) version, also marketed as "shielded" is leaky at HF due to the long, unclosed seam that runs over the entire length. The drain wire is sometimes acceptable as a low frequency shield connection, but it is absolutely inadequate at HF.
Versions (c) and (d) deserve to be called "shielded" as the shield totally encircles the wires. However, with (c), because there is no access to outer metal surface, 360° bonding is not easily made, and the drain wire is still there.

Figure 11.18 (Ref. 33) shows some results of radiated EMI measurements from flat cables carrying high-speed digital pulses. It is clear that no cable carrying clock pulses faster than 10 MHz and 10 ns rise time can meet Class B limits without having a 360° shield (unless the signal is cleaned up by other methods, such as ferrites, I/O filtering, balancing and so on).

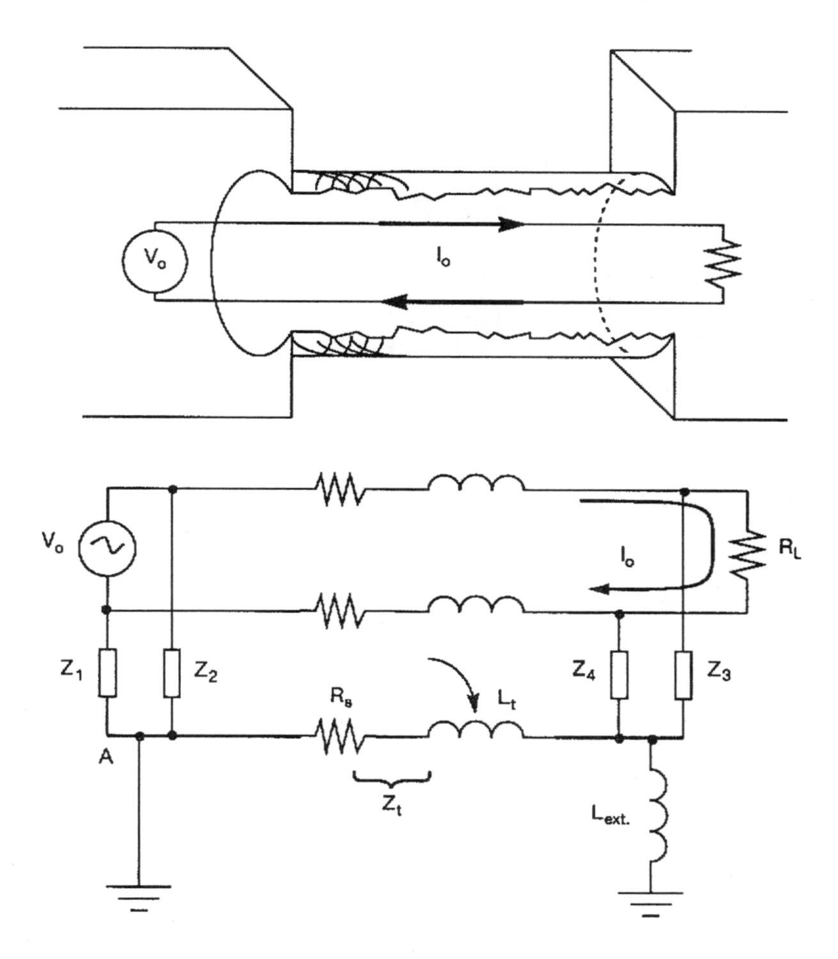

FIGURE 11.16 Radiation from a Balanced Shielded Pair

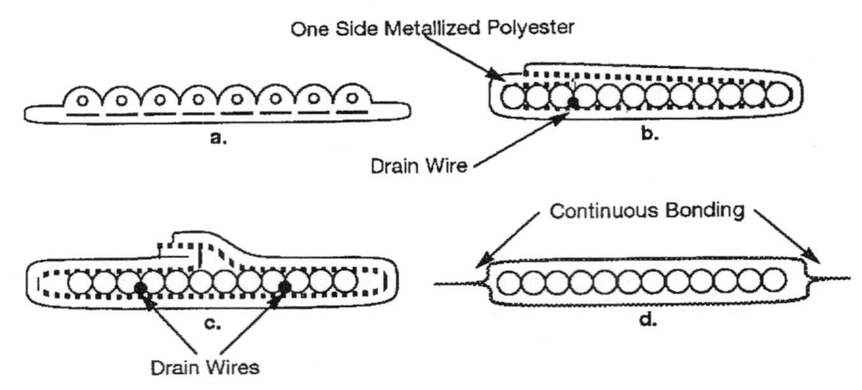

FIGURE 11.17 Shielded Flat Cable Configurations

11.5.4 Importance of the Shield Connection

As important as a good shield with low Z_t is its low-impedance termination to the equipment metal boxes. Figure 11.19 shows that connection impedance Z_{ct} is directly in the signal current return path, in series with Z_t. Therefore, Z_{ct} can increase seriously the voltage V_{ext}, which excites the cable-to-ground radiating loop.

The following values can be taken for typical impedances of one shield connection:

	DC to 10 MHz	*100 MHz*	*1,000 MHz*
BNC connector	1 to 3 mΩ	10 mΩ	100 mΩ
N connector	< 0.1 mΩ	1 mΩ	10 mΩ
Ordinary Multicontact Connector (metallic case but just pluggable or non-threaded bayonet style)	10 to 50 mΩ	10 to 50 mΩ	300 mΩ
Pigtail, 5 cm	$Z = 3\ \text{m}\Omega + j\ 0.3\ \Omega \times F_{MHz}$		

Example 11.3
Referring back to Example 11.2, find the radiation increase if the cable shield were terminated by 2.5 cm pigtails, one at each end.
The new value of Z_t to use in the calculations would be:

$$(Z_{t\ cable}\ \Omega/\text{m} \times l) + 2\ Z_{ct}$$

Voltage: 5.3 V
Duty Cycle: 50%
Length: 3.0 m

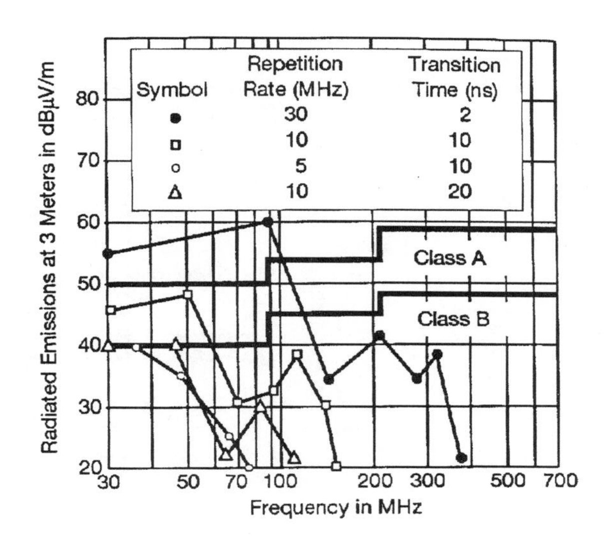

a. Effect of Clock Rate and Transition Time on Radiated Field (cable #3517, edge connector)

Repetition Rate: 5 MHz
Transition Time: 10 ns
Voltage: 5.3 V
Duty Cycle: 50%
Cable Length: 3.0 m

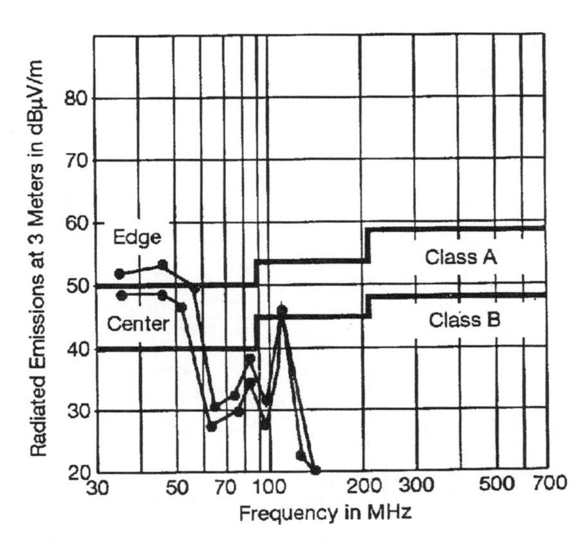

b. Effect of Conductor Location (cable #3469)

FIGURE 11.18 Comparison of Radiated Levels from Different Flat Cables (Source: 3M, Ref. 32) (continued next page)

Repetition Rate: 10 MHz
Transition Time: 10 ns
Voltage: 5.3 V
Duty Cycle: 50%
Cable Length: 3.0 m

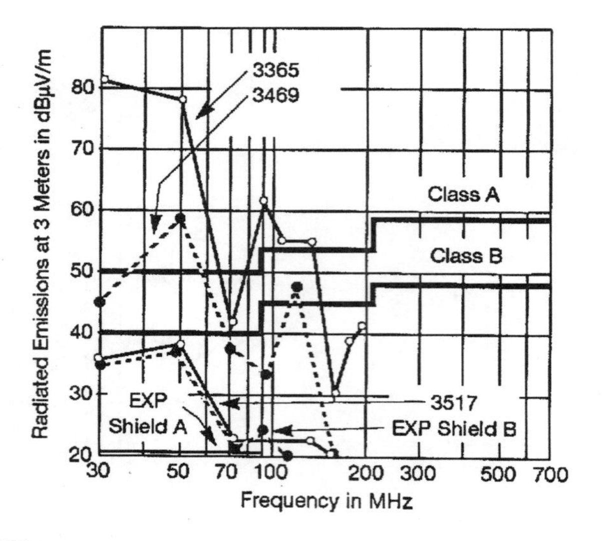

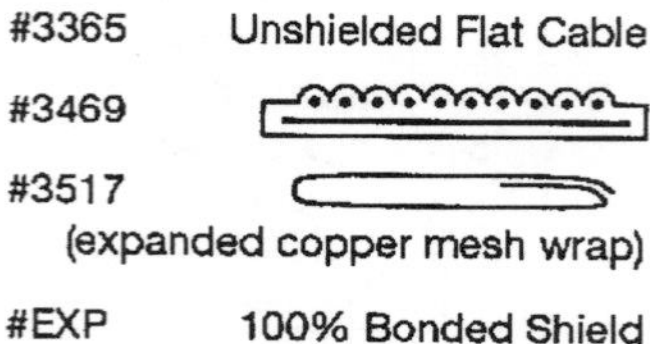

c. Comparison of Radiated Levels from Different Cables for a Constant Signal (center conductor)

FIGURE 11.18 (continued)

The calculations show:

	15 MHz	*45 MHz*	*150 MHz*
Z_t, 2 m Cable Alone	0.125 Ω	0.4 Ω	1.2 Ω
Z_{ct}, 2 Pigtails	4.5 Ω	13.5 Ω	45 Ω
New Z_t (total dBΩ)	13.3	23	33
Old Z_t, dBΩ	–12	–2	+2
Z_t Increase, dB	25	25	31

The pigtail impedance alone represents 30 times the cable shield transfer impedance. The radiated field will increase by the same amount as the Zt increase. Any attempt to solve the problem by using a better shield will be for naught, as the pigtails are the problem.

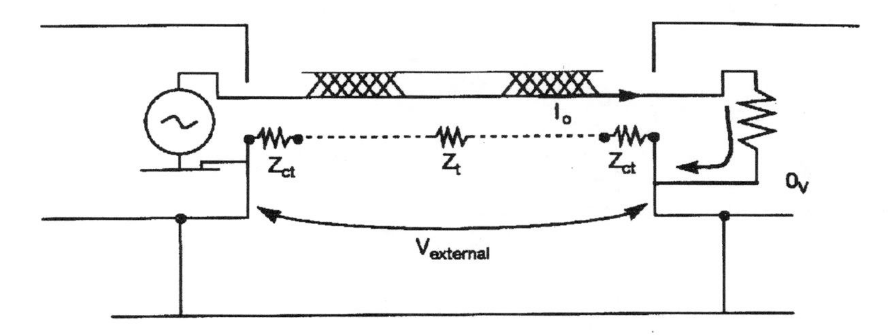

FIGURE 11.19 Contribution of Shield Termination Impedance

Figure 11.20 shows another frequent cause of cable shield inefficiency: an internal pigtail picks up PCB radiation and drives the resulting current over the shields, causing them to radiate. Fixing this simple detail can produce a 20 dB reduction in the radiated emission level.

Obviously, it is vital that a cable shield be terminated by a low-impedance connection (lower than l x Z_t of the cable itself). Most connector styles are now available in shielded versions, allowing a 360° contact on braid (see Fig. 11.21).

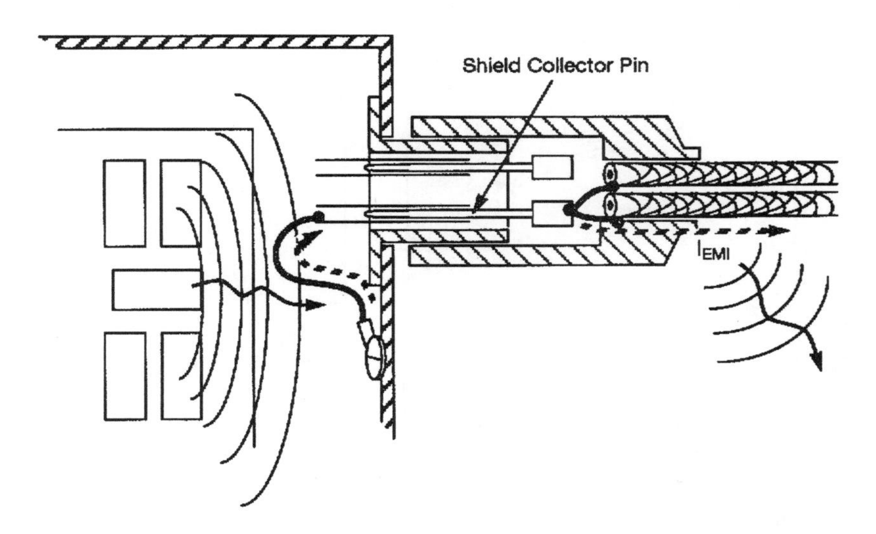

FIGURE 11.20 External Radiation of Cable Shields Caused by Internal Pigtail

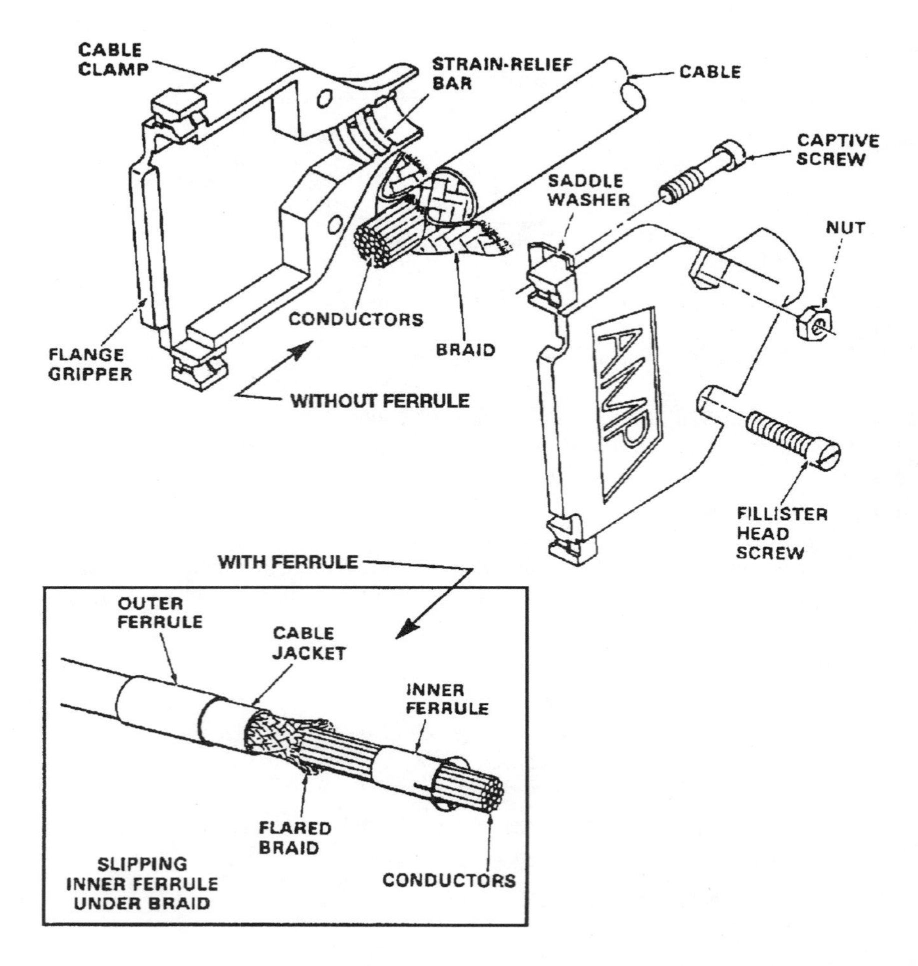

FIGURE 11.21 Commercial Shielded Connectors with Low-lmpedance Shield Connection (courtesy of AMP, Harrisburg, PA)

11.5.5 Specifying Zt from SE Objectives

At the first gross estimate of EMC requirements for a system, hardening is often presented in terms of shielding effectiveness (SE) requirements. Therefore, it would be interesting, as an example, to be able to grade the cable hardening in the same terms as the box/enclosure hardening. Although correlation is not rigorously accurate, several authors (Ref. 32) have derived and validated practical relationships to correlate Zt and SE (see Fig. 11.22).

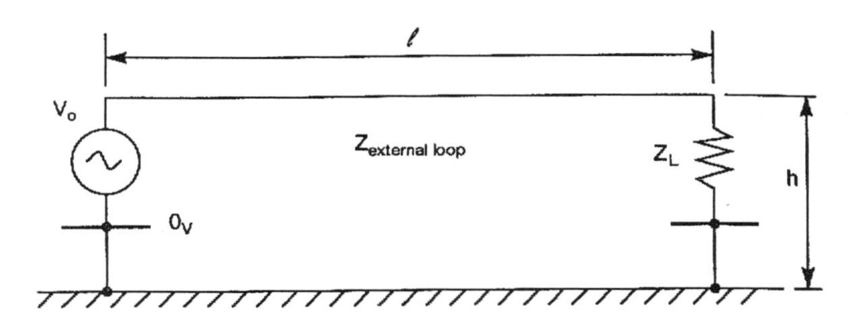

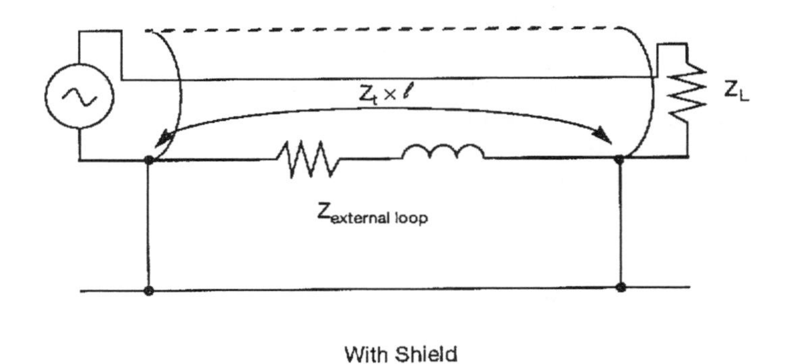

FIGURE 11.22 Comparison of Radiating Circuit with and without Shield, to Correlate Shielding Effectiveness with Z_t

The effectiveness of a cable shield in reducing a radiated field can be defined by :

$$SE_{dB} = 20 \log \left[\frac{E_1 \text{(field radiated by the bare conductor)}}{E_2 \text{ (field radiated by the same conductor, shielded)}} \right]$$

Looking at Fig. 11.22, E_1 can be expressed by:

$$E_2 = \frac{K}{D} \times \frac{V_{ext}}{Z_{ext}} \times \ell \times h \times F^n$$

where K is the proportionality factor depending on the near- or far-field conditions, which will be the same for E_1 and E_2. The latter is calculated as follows:

$$E_2 = \frac{K}{D} \times \frac{V_{ext}}{Z_{ext}} \times \ell \times h \times F^n$$

Thus, the ratio of E_1/E_2 simplifies as:

$$\frac{E_1}{E_2} = \frac{V_o}{Z_{loop}} \times \frac{Z_{ext}}{V_{ext}}$$

The total impedance of the "bare wire" comparison circuit, Z_{loop}, can be expressed as:

$$Z_{loop} = Z_L + Z_{ext}$$

The impedance of the shield above ground, Zext, can be expressed as:

$$Z_{loop} = Z_L + Z_{ext}$$

V_{ext} is derived from Eq. (11.11). Integrating all these terms, and remarking that V_0 eliminates itself:

$$Z_{ext} = R_{shield} + j\omega L_{ext} \approx (R_{sh}\ \Omega/m + j5\Omega \times F_{MHz}) \times \ell_m$$

Example 11.4

The following SE requirements have been set for system radiation containment:

SE = 30 dB at 10 MHz
SE = 40 dB at 50 MHz
Cable length = 2 m
$Z_L = 50\ \Omega$

Using Eq. (11.13):

$$SE = 20 \log(50\ \Omega) + 20 \log Z_{ext} - 20 \log(Z_t \times l) - 20 \log(Z_{ext} + 50\ \Omega)$$

At 10 MHz:

$$30\ dB = 34 + 20 \log(5\Omega \times 10\ MHz \times 2m) - 20 \log(Z_t \times l) - 20 \log(50\Omega + j5 \times 10 \times 2)$$

$$20 \log(Z_t \times l) = 34 - 30 + 40 - 41 = 3\ dB$$

$$Z_t \leq -3\ dB\Omega/m \text{ or } 0.7\ \Omega/m$$

At 50 MHz:, where $Z_{ext} = 5\Omega \times 50\ MHz \times 2m = 500\Omega$

$$40\ dB = 34 + 20 \log(500) - 20 \log(Z_t \times l) - 20 \log(50 + j500)$$

$20 \log(Z_t \times l) = 34\text{-}40 = -6\text{dB}$

$Z_t \lesssim -12$ dBΩ/m or 0.25 Ω/m

These data can serve as a cable shield selection guideline. Notice that connection impedance is incorporated in this Z_t objective.

When cable length exceeds λ/2, Z_{ext} must be replaced by Z_0, the shield-above-ground characteristic impedance, and *l* must be replaced by λ/2 as Z_t multiplier.

For an even quicker approximation above a few megahertz (where, in general, Z_{ext} is $> Z_L$), Eq (11.3) can be replaced by:

$$SE_{dB} = 20 \log Z_L - 20 \log(Z_t \times \ell) \tag{11.14}$$

For $Z_L = 50\ \Omega$

$SEdB = 34 - 20\log Z_t \times l$

11.6 DISCUSSION REGARDING SHIELDED VS. UNSHIELDED TWISTED PAIRS

There has been a continuous controversy regarding the possibility of using unshielded twisted pairs (UTP) for high speed data links inside buildings, without resorting to more expensive shielded twisted pairs (STP).
Based on measured data, as well as cable CM radiation models of Chap. 2, we saw that in the 30-200 MHz range, a 20-25 dB reduction factor was needed for an ordinary wire pair to satisfy FCC class B. If, instead of the typical 10-30% unbalance of an ordinary single-ended link, a differential link with a high grade UTP is used, a total 2% unbalance can be achieved, meeting the 20 dB reduction goal. The same goal could also be achieved with the help of balancing transformers (seesect.11.2).

Furthermore, it is true that the normal % symmetry with a STP cable is slightly inferior to that of the same cable in UTP version; this is due to the fact that a perfect geometry of two wires twisted inside a shield is more difficult to control: a 0.05mm un-eveness in a 0.5mm dielectric thickness results in 10% assymetry of each wire-to-shield capacitance and inductance (see impedances Z_1/Z_2 and Z_3/Z_4 in Fig.11.16). High grade, class 5 UTP with $\leqslant$ 2% unbalance are available, while the corresponding STP version exhibits ≈ 3%. In addition, symmetry impairments are frequently aggravated by mediocre practices for shield continuity in building wiring.

Therefore, many articles are written, suggesting substantial savings by not using STP. But radiated emission is not the only EMC constraint. In industrial sites, hospitals or high-rise commercial buildings, immunity to RF fields of 10V/m and to

Electrical Fast Transients require 30-40 dB of added CM protection, which even higher grade UTP cannot provide. Hence, STP is often preferred in harsh environments, on the basis of immunity.

11.7 ELIMINATING CABLE RADIATION BY FIBER OPTICS

Optical fibers, besides their low loss, smaller cross section, and wider bandwidth, offer total EMI isolation, since they neither emit nor pick up electromagnetic fields. Replacement kits are available for most standard (RS 232, RS 422, etc.) digital interfaces. So, when affordable, fiber optics are the ultimate answer to I/O cable emissions.

Chapter 12

Principal Radiated Emission Specifications and Test Methods

This chapter is intended to make readily available only the essential radiated EMI limits as a complement to the design information provided in this book. These excerpts by no means constitute an acceptable substitute to the official texts, and any reader who is actually involved in testing should use the formal specification documents instead.
All specifications share several characteristics in that they define:

- a receiver bandwidth(s) and detection mode
- a test distance and a typical arrangement of the test instrumentation and equipment under test (EUT)

12.1 MIL-STANDARD 461-C, D AND 462

MIL-STD-461 C, still widely in use as of the year 2000, is divided into 10 parts. Part 1 specifies the general requirements of the standard, and Parts 2 to 10 give the requirements for the various categories of equipment, as listed below :

Part 1: general requirements
Part 2: requirements for equipment and subsystems aboard aircrafts, including associated ground support equipment (GSE)
Part 3: requirements for equipment and subsystems installed spacecraft and launch vehicles, including associated (GSE)
Part 4: requirements for equipment and subsystems installed in ground facilities (fixed and mobile)
Part 5: requirements for equipment and subsystems in surface ships
Part 6: requirements for equipment and subsystems in submarines
Part 7: requirements for ancillary or support equipment and subsystems installed in noncritical ground areas
Part 8: requirements for tactical and special-purpose vehicles and engine-driven equipment
Part 9: requirements for engine generators and associated components, uninterruptible power sets and mobile electric power equipment

used in critical areas
Part 10: requirements for commercial and electromechanical equipment

Different radiated limits are specified according to the EUT type. Figure 12.1 displays the most often quoted radiated limits, corresponding to the most severe categories in Parts 2, 4, 5, 6.
The various test methods are described in MIL-STD-462. These are broken down into conducted emissions (CE), conducted susceptibility (CS), radiated emissions (RE) and radiated susceptibility (RS) tests. For illustrative purposes, this book provides a brief description of the specifications and limits for two of the most widely applied emissions tests. Again, the reader should refer to the original documents for practical applications.

RE01: Magnetic Field, Narrowband (NB), 30 Hz to 30 kHz

Test distance: 7 cm (in some variations of this specification, a 1 m distance is specified)
Antenna: magnetic loop
Suggested receiver bandwidth (3 dB):
10 to 50 Hz for up to 1 kHz
100 Hz for 1 kHz < F < 10 kHz
1 kHz for above 10 kHz

Test practice requires that loop be oriented for maximum reception on each of all EUT faces. Only equipment intended for use in critical magnetic ambients (e.g., Navy, antisubmarine warfare) are subject to this very severe specification.

RE02: Electric Field (NB and BB), 14 kHz to 10 GHz

Test distance: 1 m
Antennas: vertical monopole (14 kHz to 30 MHz)
biconical (20/30 MHz to 200/300 MHz)
log periodic or log conical above 300 MHz~

Suggested receiver bandwidth (3 dB):
100 Hz to 1 kHz for 14 kHz < F < 250 kHz
1 kHz to 10 kHz for 0.25 MHz < F < 30 MHz
10 kHz to 100 kHz for 30 MHz < F < 1 GHz
1 MHz for 1 GHz < F < 10 GHz

These bandwidths are suggested simply as a good tradeoff between test duration and receiver sensitivity. Any other bandwidth can be used, especially for making the NB/BB discrimination.
The test requires both vertical and horizontal polarization to be tested above 30 MHz, with the highest readings retained. The EUT must meet:

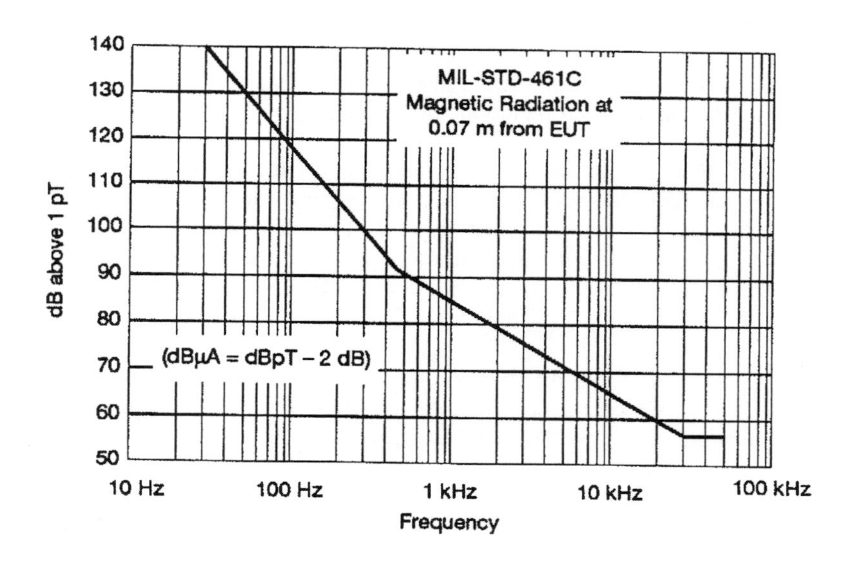

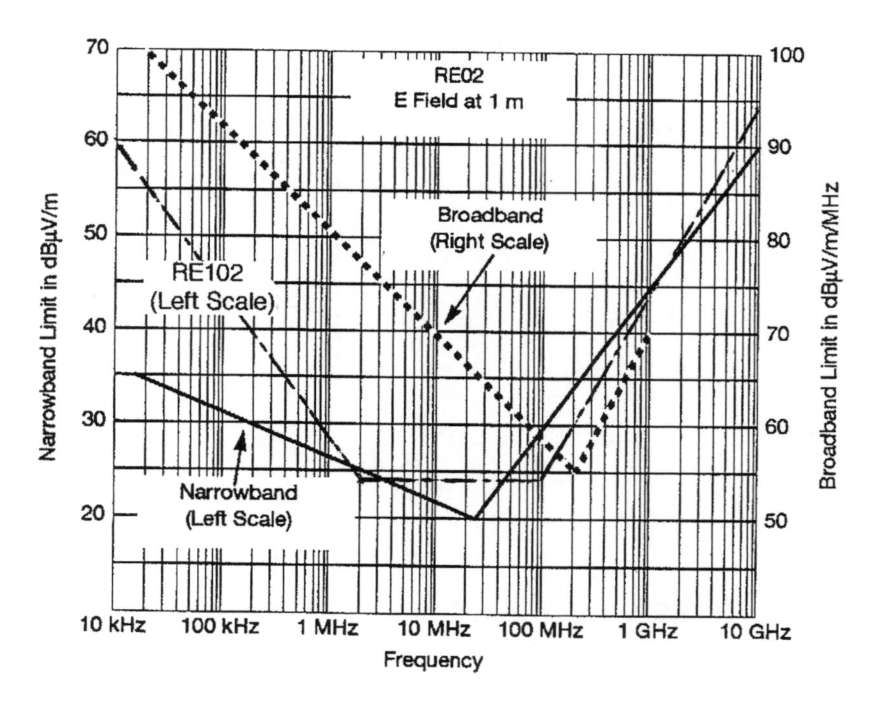

FIGURE 12.1 Radiated Emission Limits of MIL-STD-461C (only the most severe curves shown) and 461D.

1. the dBμV/m NB limit for its NB emissions (signals that increase by <3 dB when receiver bandwidth is doubled)
2. the dBμV/m/MHz BB limit for its BB emissions (signals that decrease by >3 dB when receiver bandwidth is divided by two)

The EUT is installed as shown in Fig.12.2. One important feature of MIL-ST-461/462 testing is that cables are laid at a fixed height (5 cm) above a copper ground plane. This is a good practice for test repeatability, but it can give optimistic results when compared with an actual equipment installation where one might encounter, for example, cables higher than 5 cm above a metallic structure or no metallic structure at all.

Mil-Std 461 D (released in 1993) introduced significant changes, a major one being the suppression of the NB/BB discrimination, which resulted in a single set of emission limits. Instead of the ten independent chapters repeating numerous limits and curves for each equipment category, Rev."D" is regrouping them into generic curves. The limits are generally identical to, or less severe than the former "C" version up to ≈ 1GHz, becoming progressively more severe above. The new radiated limits (aircraft category) have been overlaid on Fig. 12.1, for comparison. A quick C vs D cross-reference is shown below:

	461-C	461-D
H field	RE-01 @ 7cm	RE-101 @ 7 and 50cm
E field	RE-02 @1m up to10GHz	RE-102 @ 1m up to 18 GHz

Mil-Std 461D/462D defines also specific bandwidths. In contrast with the 461C where bandwidths were simply recommended, these new ones are mandatory, since the single NB/BB limit implies a strict respect of a standard bandwidth :

Frequency range	Mil 461D /462 D Bandwidth
1 - 10 kHz	100 Hz
10 - 250 kHz	1 kHz
0.25 - 30 MHz	10 kHz
30 - 1000 MHz	100 kHz
> 1 GHz	1 MHz

12.2 CISPR INTERNATIONAL LIMITS, TEST INSTRUMENTATION AND METHODS

Outside the military and aerospace domains, the radiated emissions specifications, no matter if they are:

- recommended in a given industry.
- legally enforced in a given industry.
- legally enforced in a group of nations, such as the European Community.

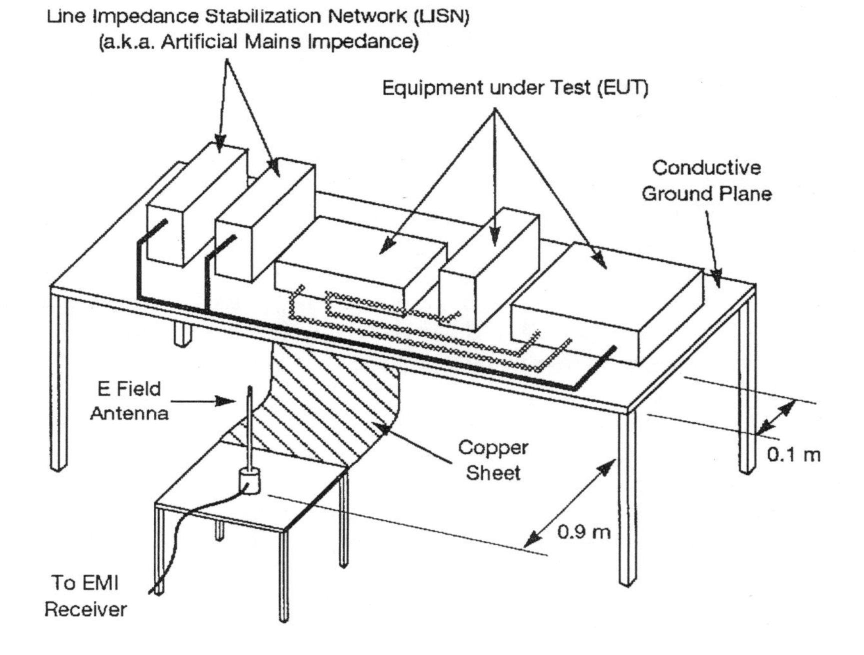

FIGURE 12.2 Setup for Radiated Emission Testing, per MIL-STD-462 or RTCA/DO-160

generally correspond to the CISPR recommendations (see Chapter 2), endorsed internationally by the International Electrotechnical Commission.

This is particularly the case with the few limits described herein for FCC regulations, European Norms (ENs), and Japanese VCCI standards. Although there can be some variation in the limit values, the measurement methods correspond to those of CISPR publications # 16 and 22, which can include the following methodology:

- use of a 10 m measuring distance (or 3 m or 30 m in some cases, with a 1/D correction factor)
- use of an open-area test site or, as a substitute, an anechoic chamber whose result can be correlated to open area. New proposals are under study for the use of a fully-anechoic chamber, recreating free-space. This would eliminate the tedious vertical antenna scanning, provided a corresponding -5dB re-inforcement of the limit.
- laying the I/O cables in a typical customer arrangement but trying to reproduce a realistic worst-case scenario for the first 1.50 m of cable - e.g., a vertical cable drop of 80 cm for tabletop equipment.

- searching for maximum field reception by:
    a. using the horizontal and vertical polarization of antennas
    b. shifting antenna height from 1 to 4 m to search for maximum ground reflection (see proposal above)
    c. rotating the EUT 360° to find worst emission pattern (or moving the antenna around the EUT)
- receiver conformity to CISPR Publication 16, including a bandwidth (6 dB) of 9 kHz for $0.15 < f < 30$MHz, or 120 kHz for $30 < f < 1{,}000$ MHz
- use of a quasi-peak detector

The quasi-peak detector, plus the fixed bandwidth, allows for no distinction between BB and NB signals. The limit is designed to protect against both types of interference. Also, the Class A/B dichotomy (residential versus industrial and commercial use) is common to most CISPR-derived civilian standards.

12.3 FCC PART 15 sub-part B

FCC Part 15-B applies to electronic data processing equipment. The radiated emission limits for Class A and B computing devices are shown in Fig. 12.3. Notice that the FCC has anticipated a trend toward increasing clock speeds by extending the limit up to 5,000 MHz and even beyond. However, the 1,000 to 5,000 MHz band is explored only if EUT uses clock frequencies >108 MHz.
The EUT must be operated using the maximum configurations of hardware and software options, and at least one sample of each attachable peripheral device.

12.4 EUROPEAN NORM (EN) 55022

Equivalent to CISPR 22, the radiated emission limits for Class A and B digital devices (termed "information technology equipment" in typical bureaucratic technical jargon) are shown in Fig. 12.4. There is one-limit curve, and the stronger severity of class B is expressed by a closer measurement distance. These limits are close (within 2 dB, accounting for proper 1/D distance factor) to FCC 15-B. The test setup is similar. While FCC had foreseen the extension of limits above 1GHz, to protect satellite and cellular telephone bands, CISPR 22, as of the year 2000 had not published limits above 1GHz, yet.

12.5 GERMAN VDE 871

Germany had a long history of efficient RFI control by law, based on the premise that "everything must have a permit unless specifically exempted". VDE 871 was encompassing all intentional RF sources operating above 10 kHz. This included industrial, scientific and medical (ISM) equipment, computers and so forth. Telecommunication equipment (e.g., modems, PBXs) were excluded, as they were covered by VDE 878.
This norm has been phased out, as the EC Directive has inforced CISPR Standards among Member Countries. However, there are still equipments in existence, and certain industries, where procurement traditions still quote the VDE 871, because it

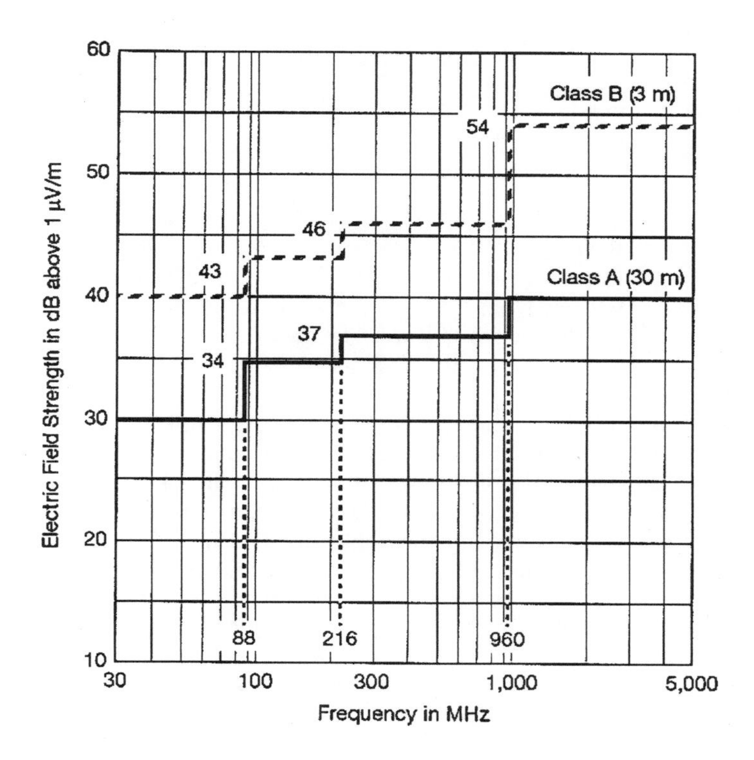

FIGURE 12.3 Radiated Limits of FCC Part 15-B, for EDP Equipment. Receiver (6 dB) bandwidth is 120 kHz, with quasi-peak detection. The limit extension above 1,000 MHz applies only to equipments using clock frequencies above 108 MHz.

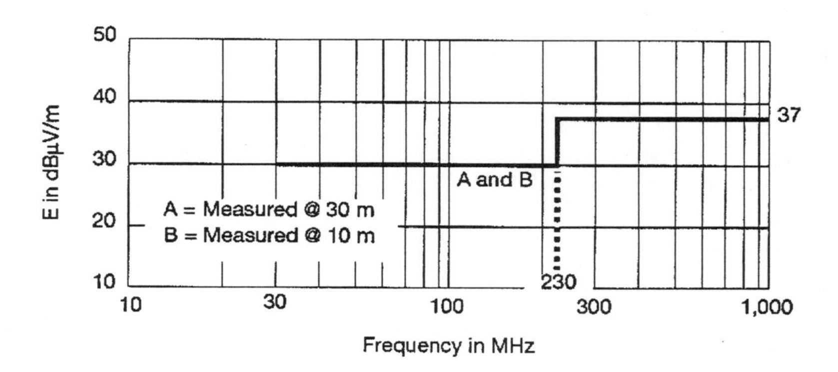

FIGURE 12.4 Radiated Limits of CISPR 22 (or EN 55 022)

is the only surviving norm having a H field emission limit, below 30MHz (Fig. 12.5). Although it is measured with a magnetic loop, the results are artificially converted from dBμA/m into "equivalent" dBμV/m by using a 52 dB (377 Ω) factor.

12.6 EN 55014 / CISPR 14

These limits are mandatory for "non-intentional RF sources", defined as electric/electronic devices not using operating frequencies above 10 kHz. They cover household appliances with motors, dimmer switches, electromechanical relays, heating resistances, igniters and so forth—basically BB sources (as perceived with a 9 kHz or 120 kHz receiver BW). For the normal category, the limit is 40 dBμV/m at 3 meters. In Germany, this norm has replaced its former national equivalent VDE 875.

For small equipment (i.e., box size less than 1 m), a simpler method for radiated emission assessment is used. Instead of measuring actual field strength, the equivalent radiated power is measured using a special current probe called the "absorbing clamp". The limit is, accordingly, expressed in decibels above a picowatt (dBpW). Interestingly, there is an approximate relationship between the dBpW and the actual field at 3 m. Between 30 and 300 MHz, this relationship is:

$$E\,(\mathrm{dB\mu V/m}) = P(\mathrm{dBpW}) + 20 \log (\,F_{MHz}/200)$$

12.7 JAPAN VOLUNTARY COUNCIL FOR THE CONTROL OF INTERFERENCE (VCCI)

The Japanese VCCI limits for EDP equipment are exactly those of CISPR 22, for Classes A and B.

12.8 FCC PART 18

This regulation covers ISM high-frequency generators (see discussion in Chapter 2) such as:

- industrial heating systems
- medical diathermy equipment
- ultrasonic devices
- microwave ovens
- plasma generators

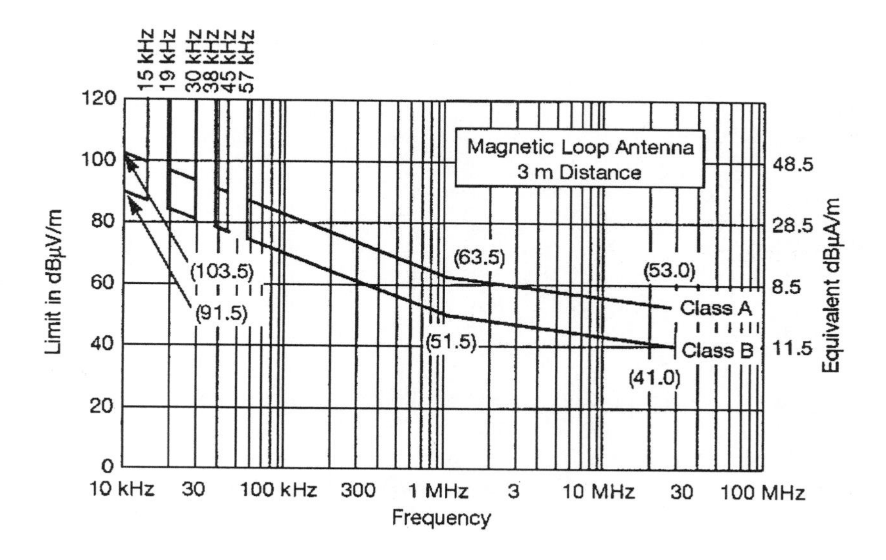

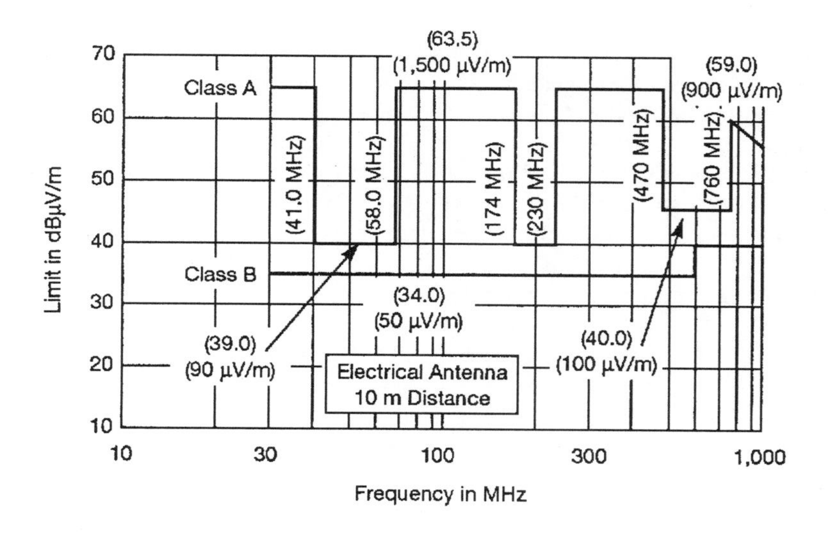

FIGURE 12.5 VDE 871 Radiated Emission Limits for EDP Equipment. The magnetic limit corresponds to the 1986 relaxed curve, translated to a 3 m distance. These limits have been progressively superseded by the EN 55022 standard as EEC directives came into force.

Due to the operational nature of such devices, a certain number of frequencies have been allocated to them, within which they are permitted to radiate without any restrictions. These are:

6.78 MHz	(± 0.015)
13.56 MHz	(± 0.007)
27.12 MHz	(± 0.16)
40.68 MHz	(± 0.02)
915 MHz	(± 13)
2,450 MHz	(± 50)
5,800 MHz	(± 75)
24,125 MHz	(± 125)

Outside of these authorized fixed frequencies, the emission limits are:

1. General requirement, for RF power
   P < 500W   250 µV/m at 30m(48dBµV/m)
   P ≥ 500W: 250 µV/m x V Pw/500 at 30m distance
   or (21 + 10 log Pw) dBµV/m
2. Industrial heaters: 530 µV/m (55 dBµV/m)

12.9 CISPR 25, AUTOMOBILE ELECTRONICS

In order to prevent RF interference from on-board electronics to the car, bus, or truck radio receivers, very stringent limits are recommended by CISPR 25 (See Chapt.1). Depending on the location of the culprit equipment in the vehicle (under the hood, on the dashboard, in the trunk etc..), different severities are specified. Contrasting with the evolution of Mil-Std 461, this rather recent standard calls for compliance with both NB and BB limits. Table. 12.1 shows the levels for severity #4, the one which is most used for passengers cars.

Table 12.1 CISPR 25 limits, severity #4. Field measured at 1m.

Freq. MHz	0.15-0.3	0.5-2.00	5.9-6.2	30-54	70-108	144-172	420-512	820-960
EdBµV/m								
NB	31	26	28	28	18	18	18	18
BB	66	59	42	42	31	31	31	31

Rcvr BW: - - - - - - 9kHz - - - - - - - >< - - - - - - - - - 120kHz - - - - - - - - - - - - >

BB limits are peak values. When measured in Quasi-peak, limit is reduced by 13dB

12.10 RTCA /DO-160.

This standard, published by the U.S. Radio Technical Commission for Aeronautics, applies to equipment used aboard civilian aircraft. Section 21, "Emission of RF energy", describes test methods (similar to MIL-.STD-462) and limits. In the B and C versions, the limits include three severity grades, according to the location of the equipment. These are:

Z: areas where it is vital to have no interference
A: areas where it is desirable to have no interference
B: areas with moderate interference control needs

The distinction between "vital" and "desirable" is rather subjective and, in fact, limits shown in Fig. 12.7 are identical for Z and A classes. As in the case of MIL-STD-461 RE02, there are NB and BB field limits. Notice that the limit is about 15 dB more permissive for civilian aircraft than for military ones.

DO-160 Rev.D was issued in 1997. In a manner similar to the Mil.Std 461 evolution, this revision has suppressed the double NB/BB limits, replaced by a one-limit / one-bandwidth approach. The categories have been changed as well:

Severity B : general
Severity L: reinforced category, for equipment in the electronic bay of the aircraft. Similar to the NB curve A of previous version.
Severity H: super-reinforced category; equipment in direct view of receiving antennas, typically outside the aircraft fuselage, or in non-metallic hulls (e.g. helicopters.).
Similar to "L", with re-inforced severity in certain slots:

F_{MHz}	E dBμV/m
100-150	25
20-340	38
1020-1100	45
1525-1680	48
5020-5100(GPS)	57

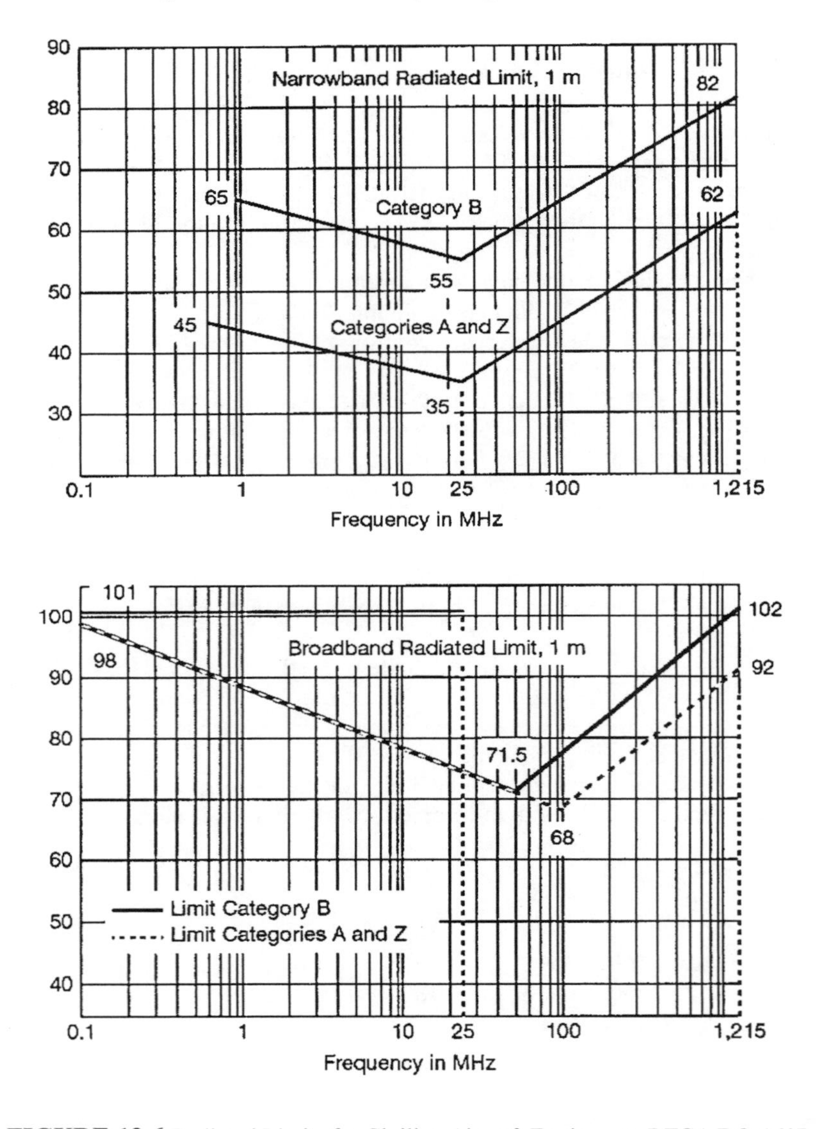

FIGURE 12.6 Radiated Limits for Civilian Aircraft Equipment, RTCA DO-160B

Chapter 13

Troubleshooting Radiated EMI Problems

Compliance with radiated emission limits of FCC or EEC regulations, or the stiffer levels of MIL-STD-461 or TEMPEST, has always been a nightmare for the EMI engineer or, more generally the electronic engineer. Too many times, using a blend of rules-of-thumb, tradition and his company's homegrown recipes, the designer does his best to have a prototype working unplagued by internal noise problems, then brings it to an EMI test site to "see if it passes". In many cases it does not, and several redesign and retest iterations are necessary to reduce sufficiently the radiated spectrum. These test iterations are generally costly, even if they are covered in the general "hidden costs" that appear in preliminary early financial projections. The method described in this chapter allows the designer to identify and reduce out-of-spec radiations without necessarily having to return to an RFI test site for each trial fix.

13.1 CABLE RADIATION VS. BOX RADIATION IN SPECIFICATION COMPLIANCE

In Chapters 2 and 3, we saw that external cables, via CM excitation and antenna size, generally cause higher field amplitudes than box radiation, at least up to ≈200MHz, for boxes whose largest dimension is less than one meter. The circuit areas formed inside the equipment by the PCB traces, the IC modules, power supply wiring and other internal wiring are several orders of magnitude below those formed by the external cables. Instead of square meters, we are dealing with tens or hundreds of square centimeters. However, due to their smaller length, these components reach their first quarter-wavelength resonance at a higher frequency and may cause specification violations if the box is poorly shielded or unshielded. In contrast with I/O cables, these internal circuits are generally neither twisted nor balanced, so their excitation is basically DM. This means that, in some critical frequency ranges (typically at mid-VHF and above), box radiation levels can be close to (only 10 or 20 dB below) cable radiation.

A difficulty arising from this situation is that after having struggled very hard to reduce CM cable radiation by shielding, ferrite loading, balancing and so forth, the engineer returning to the test range does not see all the improvement he was expecting. This frustration is even aggravated by the fact that box radiation can illuminate cables that have been "cleaned up" from CM emissions, again turning them into secondary radiators.
The method explained next has been developed and proven over time to avoid fruitless cut-and-try iterations (ref. 37). Because schedules are usually tight by the time compliance testing takes place, this is particularly useful.

13.2 STRATEGY WHEN A PRODUCT FAILS RADIATED EMISSION TESTS: QUANTIFYING THE dB REDUCTION

The philosophy behind this method is that one must first try to identify whether the most significant coupling is caused by the external cable(s) or by the equipment box alone. The routine is described in the flow chart of Fig. 13.1. (Notice that the method is equally applicable to susceptibility problems.)
Referring to the first box in that figure, try to gather as much information as possible while the equipment is still on the test site (shielded room, open-field site, anechoic chamber, or whatever). The reason is that test sites and labs are generally busy and are not convenient places to rework PC boards, cables or mechanical packaging. The steps are as described in the following paragraphs.

Step 1
While on the test site, disconnect all external cables from the EUT.

Step 2
Once all the external cables have been disconnected (excluding the power cord, unless the EUT can be powered from an internal battery), rerun the test and see if the unit passes.
This implies, of course, that the unit can be set up to run in a stand-alone mode, using self-diagnostics, self-looping or dummy loads of some sort, such that the EUT is exercised exactly like in its actual operating environment.

Step 3
If the answer is "yes" in decision box A (diamond shaped), this clearly indicates that these cables were the RF carriers. At this point, if necessary, we can take the equipment out of the test area and bring it back to the engineering lab for a more efficient application of EMI fixes.

Step 4
If the answer from box A is "no" we can conclude one of two things:

1. The only cable left (the power cord) is radiating.
2. The box itself is radiating.

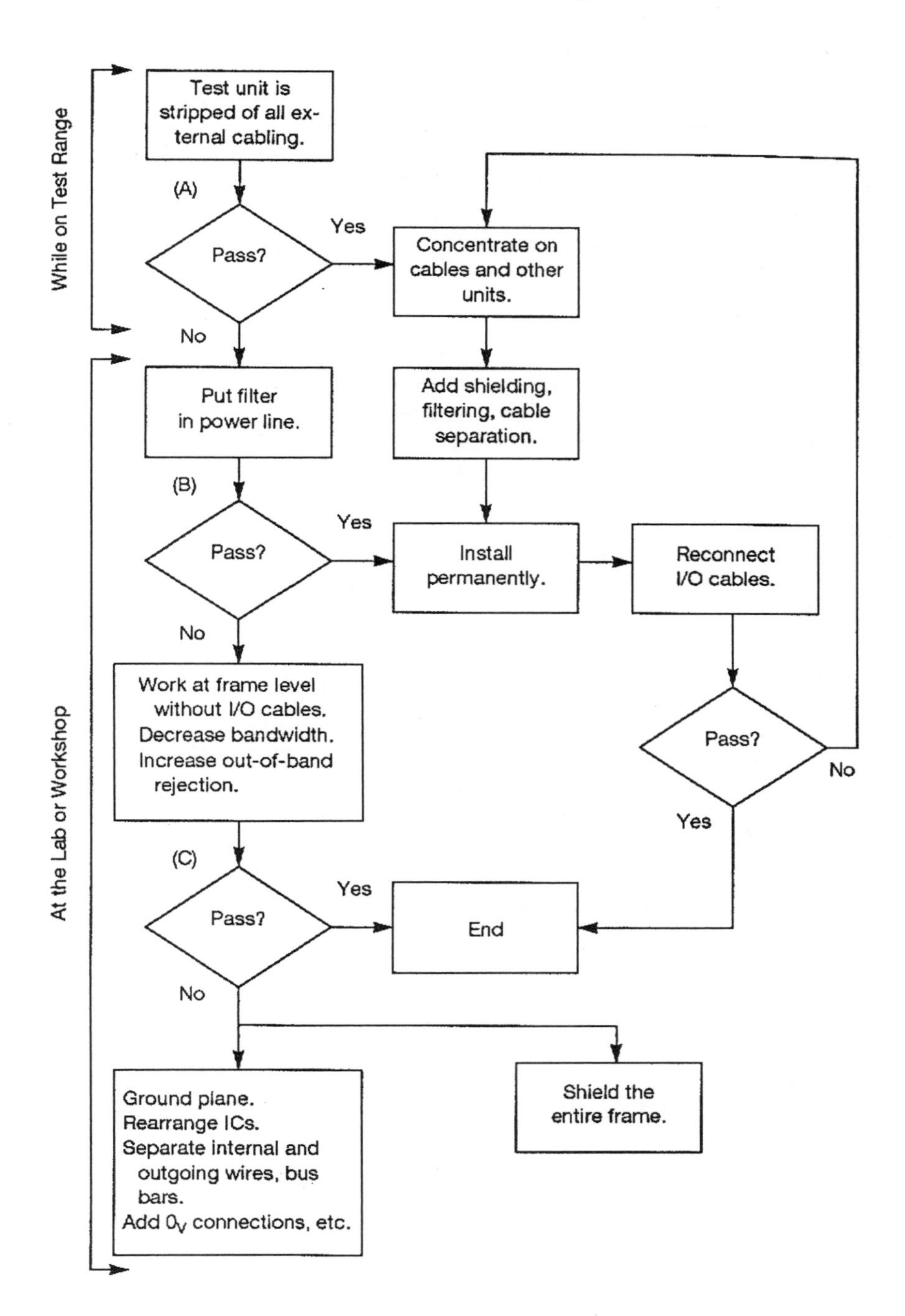

FIGURE 13.1 Strategy for Radiated EMI Diagnostic and Prototype Improvements

To decide which of (1) or (2) is true, we need to install temporarily, on the test range, a good-quality* filter [CM + DM] at the power cable port on the machine. One could also shield this cable, but the results could be misleading unless this shielding were made perfect by a solid conduit, bonded 360° to the EUT and extending beyond the test area. This is generally difficult to improvise on the spot. By default, if the specification violation occurs beyond about 10 MHz, we can insert one or two large ferrites on the power cable, with several turns on each. Select ferrites with the best possible impedance vs. frequency curve, providing at least 150Ω @ 30MHz, for one turn.

Step 5

After doing this, if answer from box B is "yes" it is prudent to reconnect the other cables and again check them for radiation while still on the test range. It is possible that the power cable radiation, being the dominant coupling mode, was masking some radiation from the I/O cables. In this case, some rework will be required in this area as well (as in the "yes" path from box A).

If the external cables are the culprits (refer to the box in the upper right corner of the chart), once the EUT is back in the engineering lab, we can work efficiently at the workbench by using an EMC current probe. We will try to filter the I/O ports using filtered connectors, feedthrough filters or ferrites. Shielding the cables is an alternative, paying special attention to a good, integral bonding of the shield to the chassis. One must keep working until the current probe readout shows that, in the entire frequency range, the conducted spectrum has been reduced by the amount (in decibels) by which the radiated test limit was exceeded.

This is, of course, for a radiated emission case. If we wanted to test for susceptibility instead, we would need to use an inverse method whereby we inject, via a current probe, the same currents that were induced during the actual radiation of the EUT. In any case, it is almost guaranteed that the EUT will pass after such cable hardening has been effected.

If answer to box B is "no", we will become involved in a more difficult task; i.e., hardening the product itself. Depending on the stage of the EUT in the development or production cycles, one can work:

1. at circuit or internal packaging level. (This is more labor intensive but may be more cost-effective in the long run.)

2. at box shielding level. (This is the "brute-force" approach, using gaskets, screen meshes and seam tightening. It is often the only option left when the calendar is the driving force.)

---

*"Good-quality" of course, is vague. But the filter must have an additional attenuation that is commensurate with our specification violation in decibels, at least up to the highest frequency at which the unit failed.

Fig. 13.2 shows an example of an EUT violating the RE02 limit of MIL-STD-461. The initial test plot shows many spectral lines above the limit. When the I/O cables are removed, many of the narrowband emissions (related to a 4 MHz clock) have decreased, but a significant number are still out of spec (see Fig. 13.3).

These two plots will be our fiduciary references in the forthcoming investigations because they tell us what is contributed by the I/O cables and what is due to the box alone (plus, eventually, the power cord).

There is no need to go back to the test range every time to check our progress. A good deal of evaluation can be done right at the workbench, using a miniature field probe.

Back at the engineering lab, we will first concentrate on reducing the emissions coming through the box alone. It is important to do this first; otherwise, any future progress in cable EMI reduction will be masked by box emissions. The equipment is still stripped of its two I/O cables, and a set of heavy tubular ferrites is placed over the power cable, right at its box exit. Then:

Step 6

Before making any changes, the "sniffer" H-field probe (Ref.37) is brought to 2 cm from the case*. All the faces, especially around edges, seams and apertures, are explored. At each face, the leakage which produces the highest profile on the spectrum analyzer display is retained, taking a photograph or an X-Y plot (see Fig. 13.4).

Step 7

We know that these levels have no absolute meaning but can be related to the specified-distance test for each frequency that was significantly out of spec by ΔdB. Therefore, we simply subtract the ΔdB from the sniffer probe results. This will produce a kind of broken line which becomes our goal for reducing box emissions, as seen from the H-field probe. Below this line, actual test will be in spec, too. Therefore, PC boards, flat cables and boxes will be treated until the "sniffer" antenna reveals that we have decreased the emission level by at least the amount that the EUT failed the limit (ΔdB). There is no uncertainty: if a close-proximity probe indicates an appropriate reduction in decibels at each of the leaky spots, this reduction will show up in at least the same range of magnitude in the final test.

In our example, after putting planar bus decoupling capacitors underneath the clock oscillator and clock drivers, plus adding EMI gaskets to the cover lid, our probe readout was below our translated limit, even on the worst-case side of the EUT. In many cases, it is indiscernible from the noise floor.

---

*The absolute value of this distance is not critical. But once it is set, it must be kept rigorously constant across the whole procedure. A distance caliper made by a piece of stiff cardboard or plastic can be stuck on the probe edge to this intent.

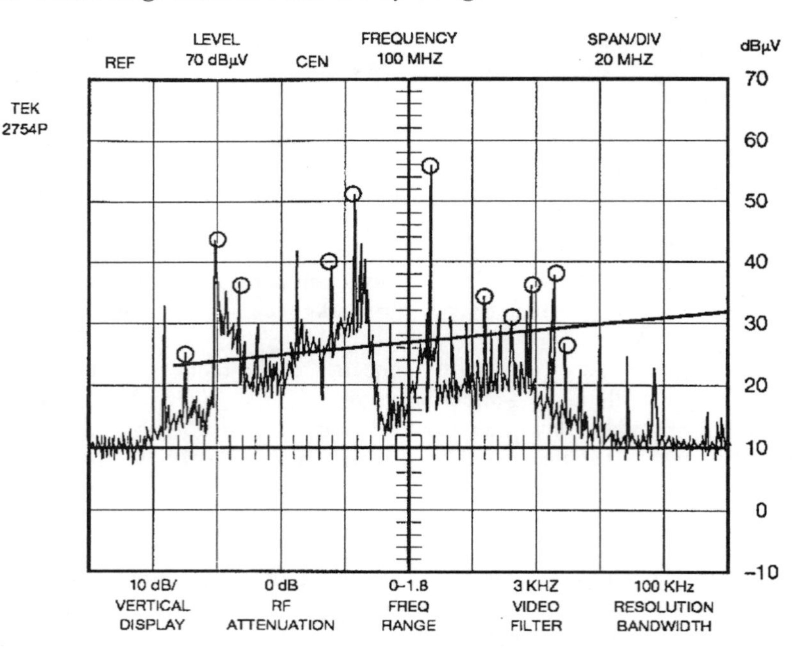

FIGURE 13.2 Actual Data—Initial Test Results with I/O Cables in Place. Circles indicate emissions that later will be traceable to the cables.

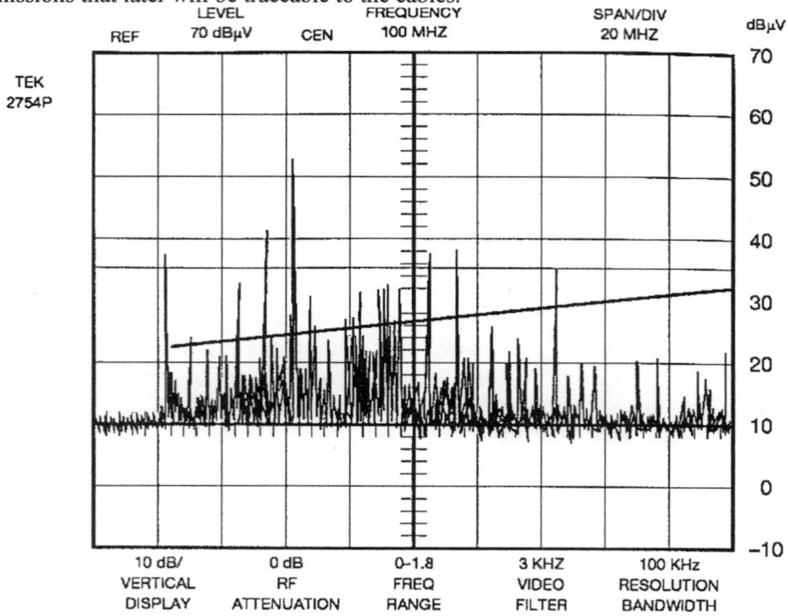

FIGURE 13.3 Actual Data—Test Results of Same EUT as in Fig. 13.2, with Cables Removed. Emissions are caused by box only.

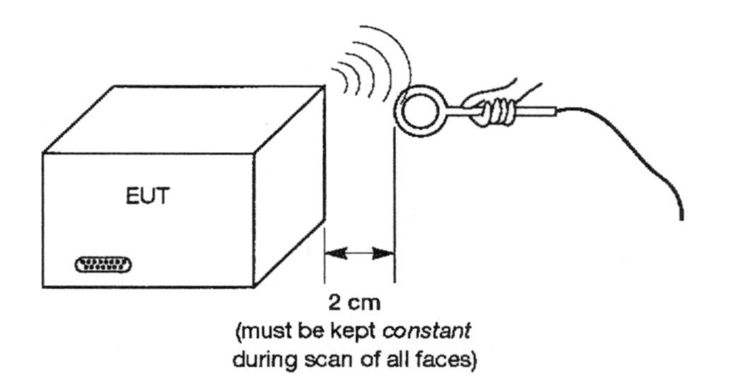

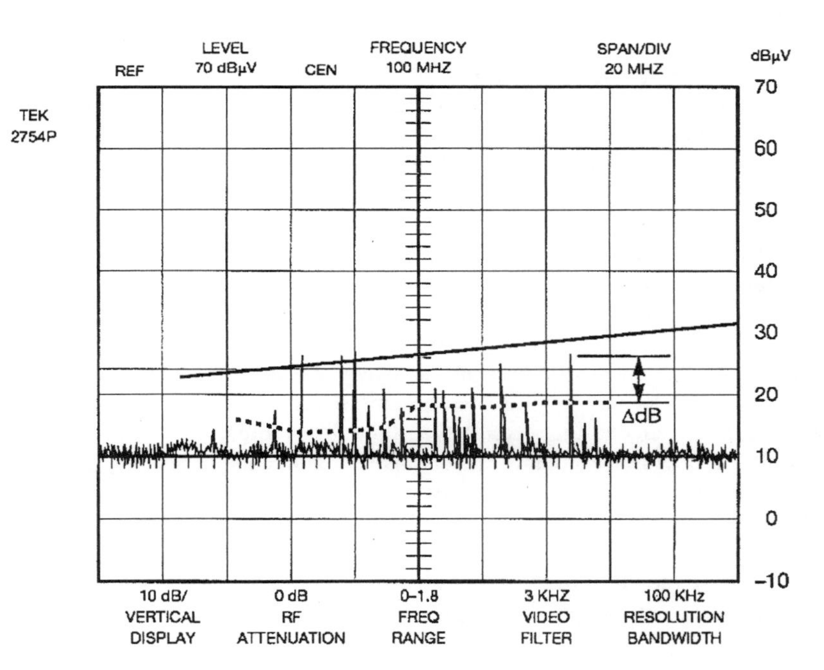

FIGURE 13.4 Actual Data—Readout from the Sniffer Probe, without Cables. (The broken line represents the objective.)

Step 8

The I/O cables are put back in place, and elevated ≈ 5 cm (2 in.) above our workbench ground plane (which can be a simple sheet of aluminium foil), over their entire length or at least 1.50m, whichever comes first.This standard height is an important detail for the reliability of the measurements, since it stabilizes the CM impedance.

We will now use the EMC current probe to control and reduce the cable contribution to total radiation. The probe readout displayed on the spectrum analyzer shows significant current levels at frequencies where radiated fields were found.

As in the case of the H-field probe, these probe readings do not represent actual E-field levels, but they can be related. So, for each frequency that was significantly out of spec by ΔdB, we simply subtract ΔdB from the current probe spectral results (see Fig. 13.5). The resulting line will become our goal for reducing cable CM currents.

Step 9

After installing feedthrough filters on the I/O connector pins, improving the power cord attenuation with CM ferrites, and using a shielded power cord, the current spectrum on the I/O cable is reduced by > ΔdB.

Step 10

We now have good expectations that the product will meet the specification' which is confirmed when the EUT is brought back to the shielded test room (see Fig. 13.6). Two important observations need to be remembered while doing these types of investigations:

1. Our measuring receiver or spectrum analyzer needs a sufficient reserve of dynamic range to let us see the decibel improvement. For instance, the IF bandwidth can be reduced to 10 kHz, even though it requires a slower scan rate, and a 20-30 dB low-noise preamplifier is also recommended. Beware that if the current probe reveals a BB current, a BW correction may be needed to adjust to the FCC or Mil-461 bandwidth.

2. While evaluating the merit of the various fixes, we must ensure that they have not simply "shifted" the problem; i.e., compressed some harmonics but increased some others.

13.3 APPROXIMATION OF RADIATED RFI LEVELS FROM I/O CABLE CM CURRENTS (VHF REGION)

In Chapter 2, Eq. (2.27), we gave the far field, free-space radiation from a cable acting as a dipole below resonance:

$$E_{\mu V/m} = \frac{0.63 I_{\mu A}}{D_m} \times \ell_m \times F_{MHz}$$

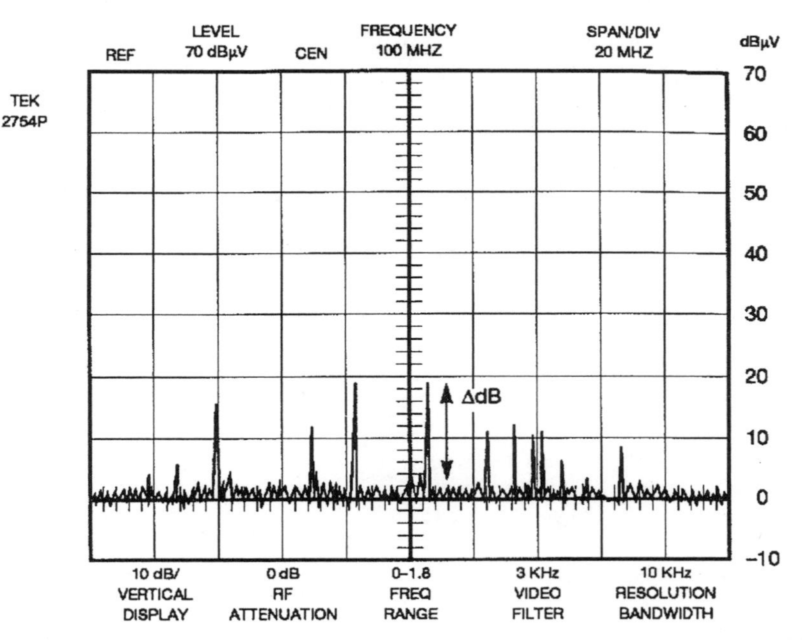

FIGURE 13.5 Readout from the Current Probe on I/O Cables After the Box Alone Has Been Fixed. Each prominent current harmonic relates to a certain field level being ΔdB off specification.

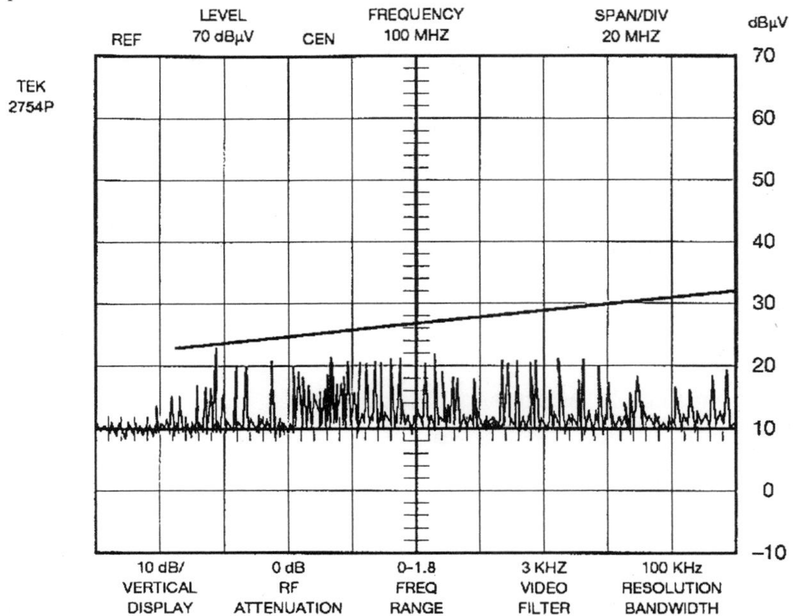

FIGURE 13.6 Actual Test Data—Field Radiated by EUT with Cables in Place, after Box and I/O Cables Have Been Treated. The unit is now within specifications.

where,

D_m = distance from source to receiving antenna

For a base-driven monopole (the equivalent model of a typical I/O cable carrying CM currents), when the cable length exceeds $\lambda/4$ and the height above ground exceeds $\lambda/8$, the length l in the equation must be replaced by $\lambda/4$ (the antenna is electrically "shrinking"): in Eq. (2.28), *E becomes independent of F* and relates only to I and D.
Therefore, in practical situations of FCC /CISPR 22 radiated testing, under the following test conditions:

- frequencies above 50 MHz
- cable length greater than 1.50 m
- cable height greater than 0.75m

A simple current criteria for pass/fail prognostic can be set:

$$I_{cm}\ (\mu A) \leqslant E_{\mu V/m}\ (\text{spec.limit}) \times D_m / 60 \qquad (13.1)$$

For FCC/CISPR, electronic data processing devices, simply measuring Icm with a current probe on all I/O cables, and taking into account a ≈ 5dB margin for ground reflection, if we find, on every spectral line,

	50 to 230MHz	230 to 400MHz
• I_{cm}, for class A:	**≤ 10 μA (20dBμA)**	**≤ 20 μA (26dBμA)**
• I_{cm}, for class B:	**≤ 3 μA (10dBμA)**	**≤ 6 μA (16dBμA)**

the equipment can be brought to the test site with a high confidence that it will pass (400MHz is the upper practical range of validity for this method).The current probe must be moved along the 1.50 m cable section that is closer to the EUT box to make sure you do not miss a current standing wave maximum.

If, to the contrary, it is found for Class A:

- $I_{cm} > 30\ \mu A$ from 50 to 230 MHz, and $> 60 \mu A$ above

And for Class B:

- $I_{cm} > 10\ \mu A$ from 216 to 230 MHz, and 18 μA above

Do not waste your time bringing the EUT to a test site: it is almost certain to exceed the limits.

Between the two criteria is some latitude for trying our luck, depending on:
• How costly is engineering effort versus test lab cost ?
• How critical is our $/dB optimization (e.g., the cost penalty of a slight over-design with mass-produced equipment) ?

With MIL-Std-461C,RE02, or 461D, RE102 , two more factors are coming into play:

1. Cables are laid at 5 cm above the ground plane, which reduces E by a 10 x h/λ factor (Ref. chap.2, sec.2.5).
2. The limit relaxes progressively above 20MHz (for 461C) or 100MHz (461D).

After entering these variables, and considering the 1m test distance for far field conditions, the criterion for I_{cm} max becomes:

$$I_{cm}\ (\mu A) \leqslant E\ \mu V/m\ \text{(spec limit)} \times D \times \lambda / (60 \times 10 \times h) \qquad (13.2)$$

i.e., $I_{cm} \leqslant E\mu V/m\ \text{(limit)} / (0.1 \times F_{MHz})$

or, $I_{cm}(dB\mu A) \leqslant EdB\mu V/m + 20 - 20 \log F_{MHz}$

This translates into the following table:

FMHz	50	75	100	200	300
Δ(dB)	-14	-17	-20	-26	-30
Mil.461C/RE02					
EdB(μV/m):	25	28	30	34	37
I_{cm}(dBμA) = E - Δ :	**11**	**11**	**10**	**8**	**7**
Mil.461D/RE102					
EdB(μV/m)	24	24	24	30	34
I_{cm}(dBμA) = E - Δ :	**10**	7	**4**	**4**	**4**

We see that, up to ≈ 75 MHz and regarding cable radiation only, the criteria for Mil.Std are quite close to what would be required for FCC & CISPR class B.

FIGURE 13.7 Example of EMI Investigation Routine on a class B computer. Here, the CM current is checked on the power cord alone, all signal cables being disconnected. As a variation of the routine step = (, the unit is tightly wrapped in aluminium foil, in order to prevent box leakages from obscuring our diagnosis (photo courtesy of Don Sweeney, D.L.S ELECTRONIC SYSTEMS Inc)

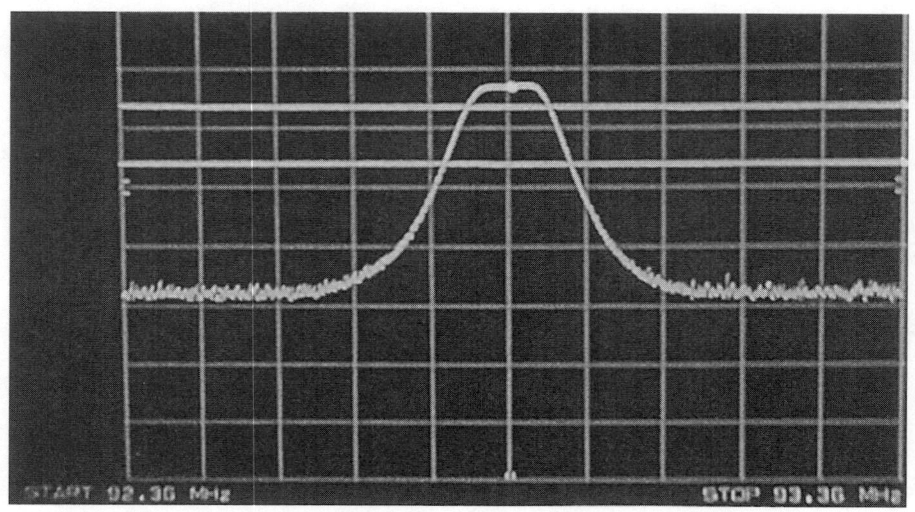

FIGURE 13.7 (cont'd). Here, at one spefific frequency where limit is violated, the CM current is exceeding our criteria by ≈ 14 dB.

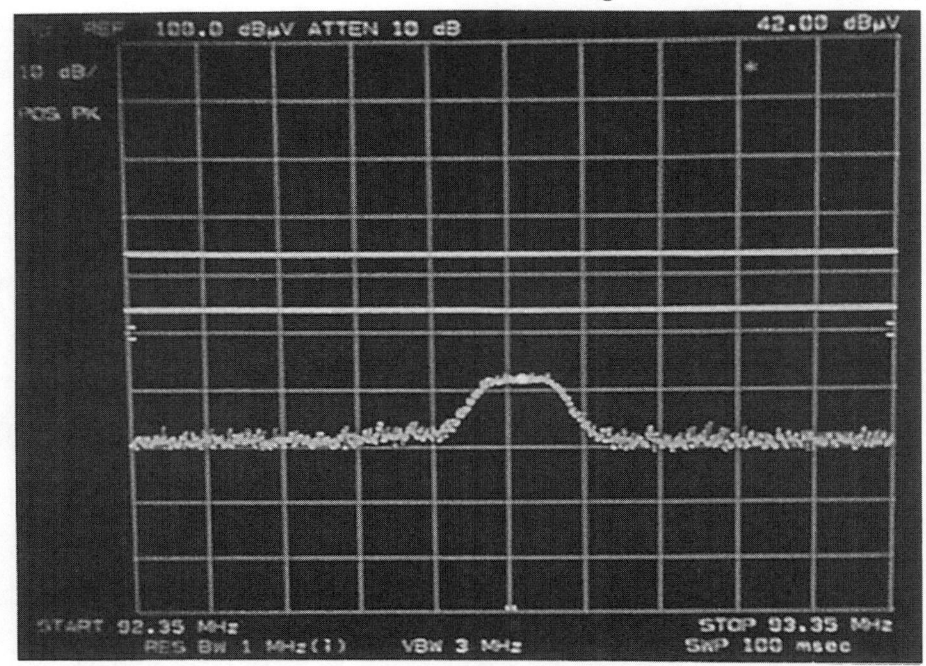

FIGURE 13.8 Following-on the procedure, additional CM filtering is installed on the power cord input. The CM current has been reduced by ≈ 20 dB. The radiated check confirms the product is now within limits (Photo courtesy of Don Sweeney, D.L.S ELECTRONIC SYSTEMS Inc).

Appendix A

The Modified Dipole Model

The practical radiation formulas presented in Chapter 2 and applied from that point onward have been established via the process described in this appendix.
When the source is at a distance $D < \lambda/2\pi$ (near-field conditions), the E/H ratio of an electromagnetic field departs from the free-space impedance Z_0, which is :

$$Z_0 = \sqrt{\mu_0 / \varepsilon_0} = \sqrt{\frac{4\pi.10^{-7}\,Henry\ /\ m}{1.10^{-9}\ /36\pi F\ /\ m}} = 120\pi,\ \text{or } 377\Omega$$

The near-field E/H ratio depends on the source impedance but can never exceed $377 \times \lambda/2\pi D$ nor be less than $377 \times 2\pi D/\lambda$. The question of how source-circuit and wave impedance are related in the near field is important because the estimation of E and H, and the shielding effectiveness of barriers, are dependent on this relation.

The development of a discrete relation between circuit impedance, Z_c, and wave impedance, Z_w, in the near field is beyond the scope of this handbook. However, the following mathematical relations are suggested for all conditions in which the circuit dimensions, $D \ll \lambda$:

For $Z_c \geq Z_0$ (high-Z source):

$$Z_w = Z_0\lambda/2\pi D,\ \text{for } Z_c > Z_0\lambda/2\pi D \geq Z_0$$
$$\approx Z_c,\ \text{for } Z_0\lambda/2\pi D > Z_c > Z_0$$
$$\approx Z0,\ \text{for } Zc = Z0$$

For $Z_c \leq Z_0$ (low-Z source):

$$Z_w = Z_c,\ \text{for } Z_0 > Z_c > Z_0 2\pi D/\lambda$$
$$\approx Z_0 2\pi D/\lambda,\ \text{for } Z_0 > Z_0 2\pi D/\lambda > Z_c$$

These equations are plotted in Fig. A.1 for several values of common circuit impedances of 50, 100, 300 and 600 Ω. To the extent that these conditions exist, the finite source circuit impedance, then, does not "permit" an infinitely high or null wave impedance E/H. Rewriting the above equations in more practical terms, the near-field wave impedance for *any* circuit is:

$$Z_w(\Omega) = 18{,}000 / D \times F_{MHz}\ ,\ \text{for } Z_c = 18{,}000 / D \times F$$

$$Z_w(\Omega) = Z_c,\ \text{for } 18{,}000 / D \times F \geq Z_c \geq 7.9\ D \times F$$

$$Z_w(\Omega) = 7.9\ D \times F,\ \text{for } 7.9\ D \times F \geq Z_c$$

Far-Field Values

The E field radiated by an isolated wire at a distance $D > \lambda/2\pi$ is:

$$EV/m = (1/D) \times 60\pi . I \times l/\lambda$$

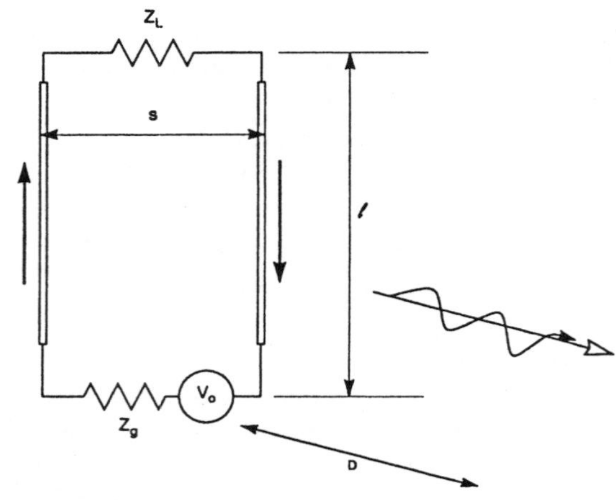

FIGURE A.1 The Two-Dipole Model

If, instead, we have two wires carrying equal but opposite currents, the radiated field in the plane of the two wires is calculated from the phase lag of the equal and opposite fields:

$$EV/m = [(l/D) \times 60\pi \times I \times l/\lambda \ (\sin 2\pi s/\lambda)]$$

Recognizing that, for small values of "x", $\sin x \approx x$, replacing λ by $300/F_{MHz}$ and expressing $l \times s$ in cm^2:

$$E\mu V/m = 1.3 \times V\ (l \times s)F^2 / (D \times Z_L)$$

V being the drive voltage, this is the same expression as the loop model, in the far field.

Case of two wires loaded by $Z_L > 377\Omega$

In section 2.3 and Fig. 2.6, we stated that when Z_L exceeds 377Ω, the radiated field no longer depends on the terminal load impedance. Using a different approach, Chatterton and Houlden (ref. 38) reach a similar conclusion, with a formula for the E field from any loop, with loads from few ohms to ∞ :

$$E_{V/m} = \frac{\pi . l . s}{D . \lambda^2} . V \left[\sqrt{1 + (Z_0 / Z)^2} \right]$$

The bracket term allows an easy entry for math modeling. When Z_0 is replaced by its 120π value:

$$Ev/m = 0.35.10^{-2}.F_{MHz}.V_{volt}.(l.s)_{cm^2}[\sqrt{1+(Z_0/Z_L)^2}]$$

If $Z_L >> 377\Omega$, the field value no longer depends on the load current, only on the voltage, as shown in curves of Fig. 2.6a and b. Numerous experiments have validated this model, confirming that when a loop is terminated by anything greater than ≈ 400Ω, it behaves as an open, folded dipole.

Values at Transition Distance

Replacing F by its corresponding value at the near-far transition distance, i.e., $F_{NF} = 300/2\pi D$, or 48/Dm:

$$E\mu V/m = \frac{1.3xV}{D.Z_L}.l.s.(\frac{48}{D})^2$$

$$= \frac{V.(l.s)cm^2}{Z_L} \times \frac{3000}{D^3}$$

This new formula is used as the reference value to calculate the near-field terms, since the near-field wave impedance will become asymptotic to the impedance of the source circuit, increasing from 377Ω to Z_c (if $Z_c > 377\ \Omega$) for high-impedance circuits, or decreasing to Z_c if $Z_c < 377\ \Omega$.

Near-Field Values (i.e., $F < F_{N\text{-}F}$)

$$E\mu V/m = \frac{V.(l.s)cm^2}{Z_L}.\frac{3000}{D^3}$$

multiplied by:

F/F_{NF}, if $Z<377 \times (F/F_{NF})$, or by : $Z/377$, if $Z > 377 \times (F/F_{NF})$

Therefore,

1) if $Z < 377 \times F/F_{NF}$ (low-Z circuit), or $Z < 7.9\ F \times D$:

$$E\mu v/m = 62\ V\ (l \times s) \times F_{MHz} / (Z \times D^2)$$

2) if $Z > 377 \times F/F_{NF}$ (high-Z circuit), or $Z > 7.9\ F \times D$:

$$E\mu v/m = 7.9\ V\ (l \times s) / D^3$$

Quasi-static Values for E or H

In the near field, field prediction curves of Chapter 2 (Fig. 2.6) show that E becomes constant for a given drive voltage and distance. This raises the question: What happens to the associated H field? The previous equations, plus Figs. A2 and A3, provide the answer.

For a constant voltage excitation, the wave impedance increases when F decreases below F_{NF}, until it reaches Z_c (unless $Z_c = \infty$). This would meet the case of a monopole, or open loop excited in dc, creating a static E field but no H field.

Conversely, for a magnetic, low-Z circuit, the wave impedance decreases when F decreases below F_{NF}, until it reaches Z_c (unless $Z_c = 0$). Therefore, the associated E field decreases, but not down to zero, unless $Z_c = 0$. This would meet the case of a perfectly shorted loop at dc, having no E field and a static H field.

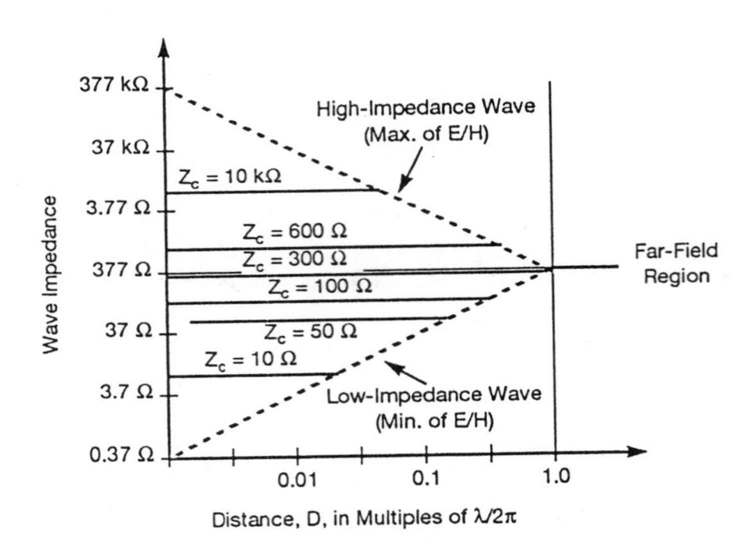

FIGURE A.2 Wave Impedance vs Circuit Impedance

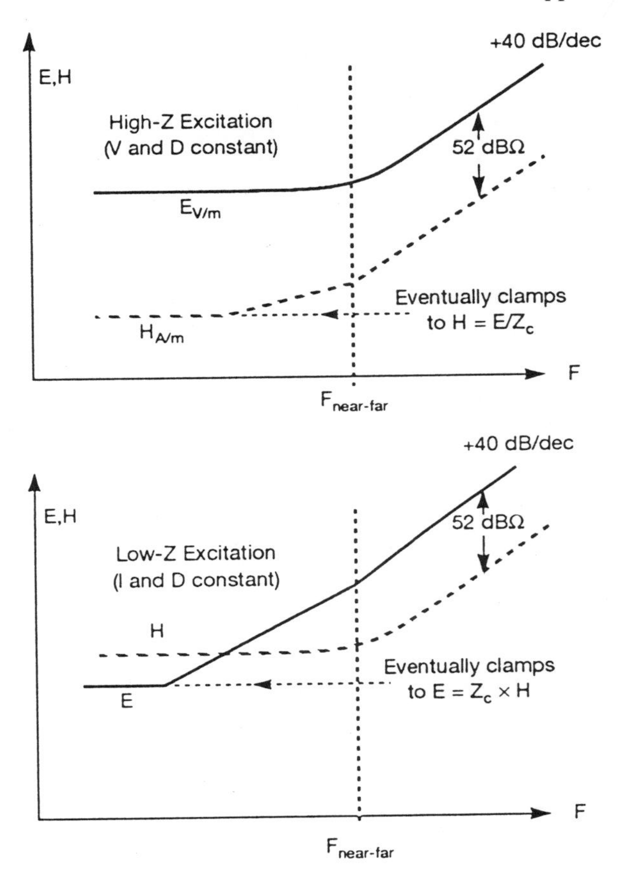

FIGURE A.3 Electric and Magnetic Field Trends at Very Low Frequencies (Quasi-static)

Appendix B

Some Validation Results Supporting the Simplified Radiation Model

Several validation measurements performed by the author on simple circuits, as well as other measurements reported in the literature, give an indication of the error margin incurred.

Figures B.1 and B.2 show the results for a personal computer single-layer board radiation and a backplane with 10 MHz clock runs, both measured on calibrated FCC test sites. Interestingly, in Fig. B.1, the influence of changing from a clock oscillator supplied by source A to one provided by source B, with slightly different rise times, is clearly visible.

The compilation of about 60 radiated test results, compared to the predicted results per this book's method, showed a mean of differences of 8.5 dB.

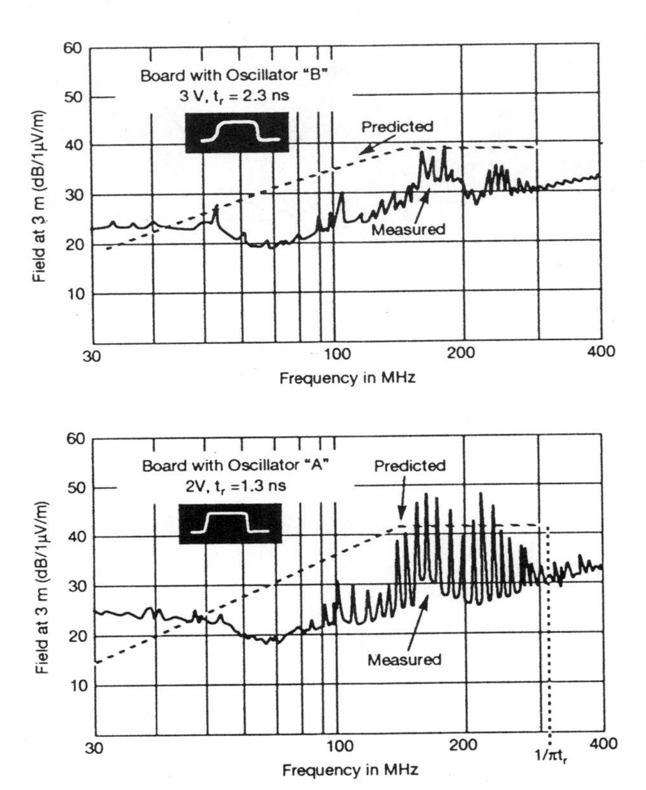

FIGURE B.1 Measured vs. Predicted Radiated Emissions from a PCB, 3 m Test Site per FCC Part 1 5-J (Ref 39) (continued next page)

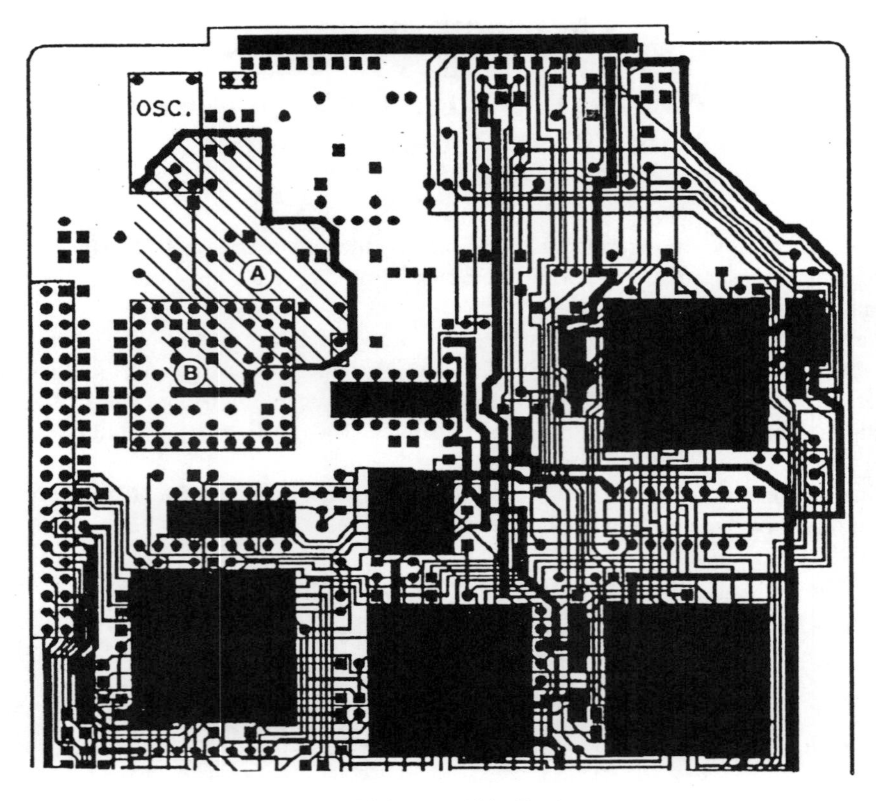

OSC. = 7.5 MHz Clock
A = Loop Area, 6.5 cm^2
B = Driver Module

FIGURE B.1.a. The Single Layer PCB responsible of Radiated Spectrum in Fig.B.1.

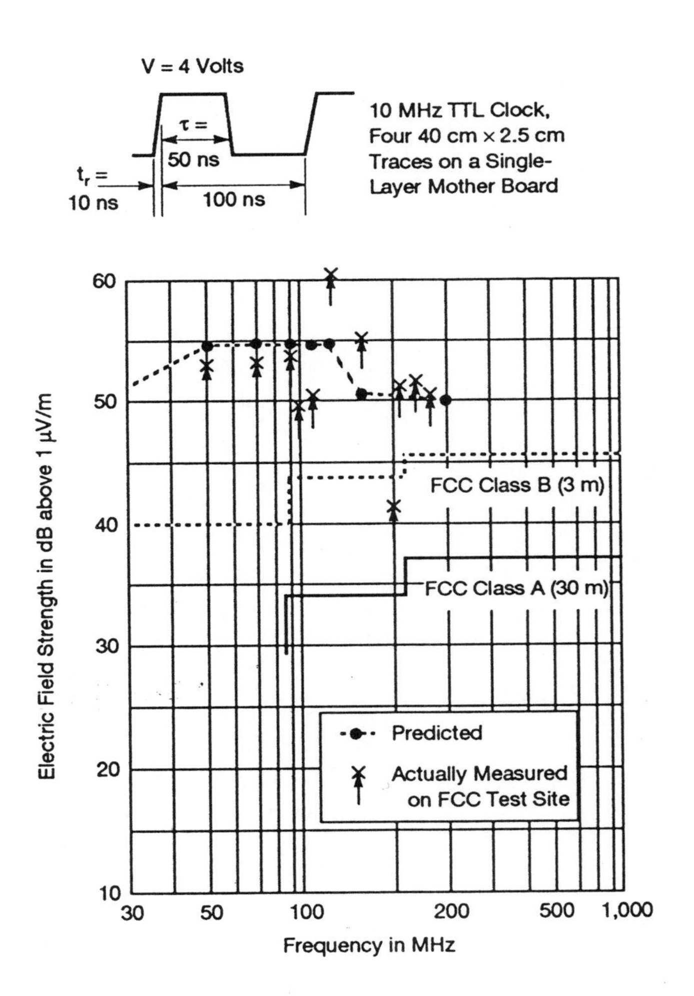

FIGURE B.2 Measured vs. Predicted Radiated Emissions from PCB Traces

Appendix C

Inductances and Capacitances of PCB Traces

The following formulas and curves, essentially derived from the remarkable work by C.Walker (Ref.40), are useful for most calculations regarding EMI coupling in PCBs. When it was necessary for simplification, formulas have been rounded within +/- 5%.

C.1 Isolated trace

Although a completely isolated trace is an academic configuration, the following formula gives the exact value of inductance L when the associated trace or plane is extremely far (in practice, at a distance greater than the length l). Notice that in this case, strictly speaking, no linear value in nH /cm can be given since L will depend on the l/w ratio. The 10nH/cm value we have been using in many calculations is a close, approximation, not an exact value.

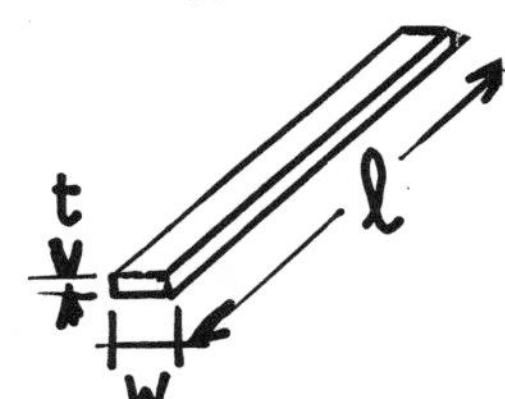

$$L\,\mathrm{nH/cm} = 2 \times l_{cm} \left[\log n \left(2l / (w + t) \right) + 0.5 + (2w + t) / l \right]$$

l, w and t in same units.

C.2 Trace above a ground plane (microstrip), epoxy dielectric (εr = 4.5)

$$L_0 nH / cm = 2 l_n \left(\frac{6h}{t + 0.8w} \right)$$

$$C_0 pF / cm = \frac{1.55}{\ln\left(\frac{6h}{t + 0.8w} \right)}$$

$$Z_0 = 35.8 \ln\left(\frac{6h}{t + 0.8w} \right)$$

C.3 Strip Line, epoxy dielectric (εr = 4.5)

$$L_0 nH / cm = 2 \log n \left[\frac{3.8h}{(0.8w + t)} \right]$$

$$C_0 pF / cm \cong 0.76(1 + w / h)$$

$$Z_0 \Omega = 28.3 \log n \left[\frac{3.8h}{(0.8w + t)} \right]$$

C.4 Mutual capacitance, microstrip traces

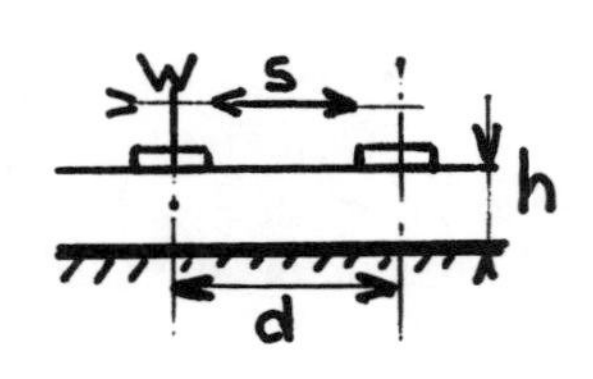

Exact formula:

$$C1\text{-}2(pF/m) = 0.7\varepsilon r.\ KLKc(w/h)^2 \log_n [1+(2h/d)^2]$$

in practice, for εr=4.5 and length in cm:

$$C1\text{-}2(pF/cm) = 0.031.KLKc\ (w/h)^2 \log_n[1+(2h/d)^2]$$

KL,Kc are the fringing terms given by complex formulas.
The product KL x Kc has the following values:

h / w	0.5	1	2	3	5	8	10
w / h	2	1	0.5	0.3	0.2	0.12	0.1
KL.Kc	3	6.5	13	24	48	96	130

C.5 Mutual inductance, microstrip traces

$$M1\text{-}2(nH/cm) = \log_n [1 + 2 (d - w) /h]^2$$

if "s" is used instead of "d":

$$M1\text{-}2 = \log_n [1 + (2s /h)]^2$$

C.6 Mutual capacitance, strip Line traces

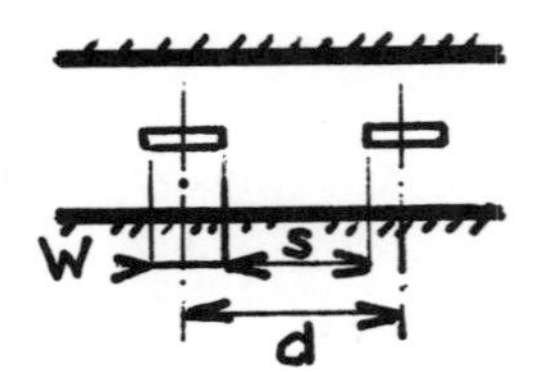

$$C1\text{-}2(pF/cm) = 0.126 [(h+w) /d]^2$$

if "s" is used instead of "d":

$$C1\text{-}2(pF/cm) = 0.126 [(h+w) / (s+w)]^2$$

C.7 Mutual inductance, strip Line traces

$$M1\text{-}2\ (nH/cm) = \log_n [1 +(d - w /h)^2]$$

if "s" is used instead of "d":

$$M1\text{-}2\ (nH/cm) = \log_n [1 +(s /h)^2]$$

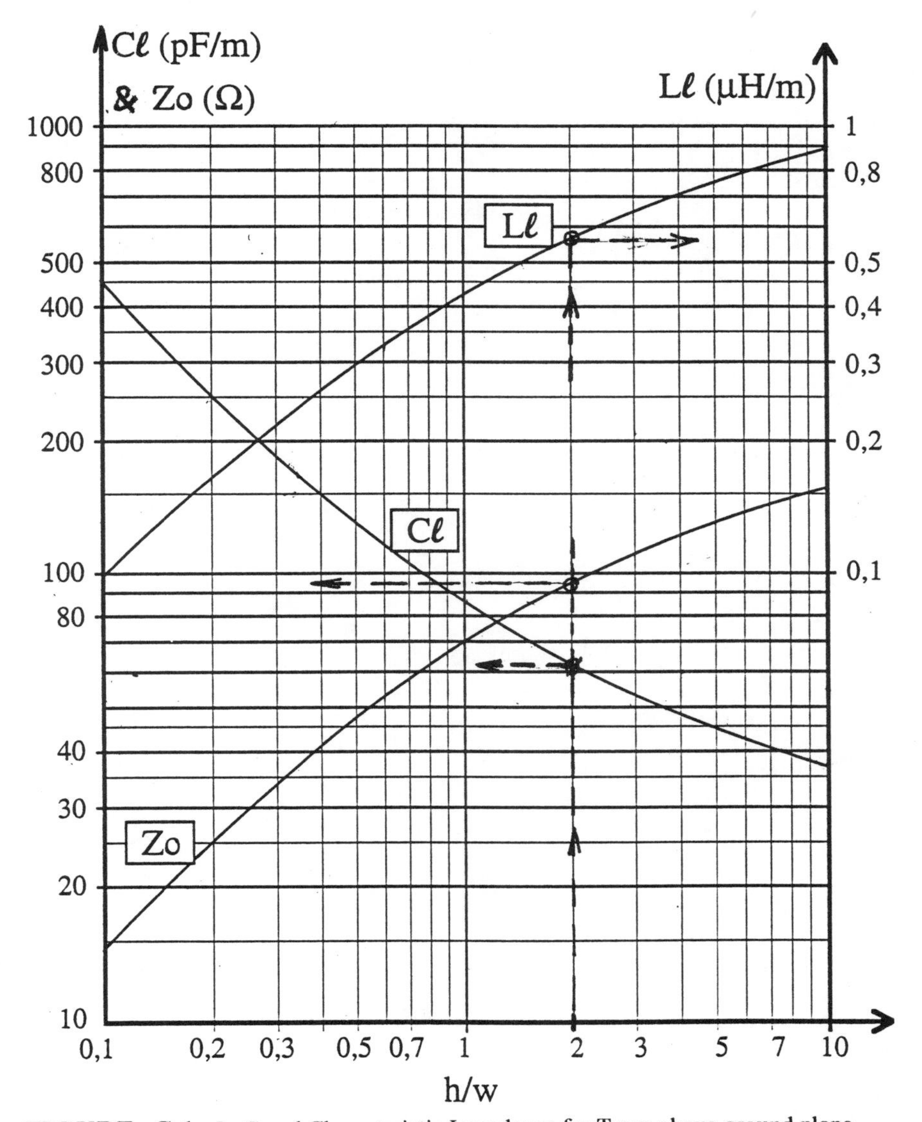

FIGURE C.1 L, C and Characteristic Impedance for Trace above ground plane
Example, as shown: w = 0.3, h = 0.6, h/w = 2.
L(right scale)= 0.55μH/m= 5.5nH/cm,C(left scale)=60pF/m=0.6pF/cm, Z0(left)=95Ω

Appendix D

A few equivalent circuits for component modeling via SPICE, MicroCap or similar simulation tools

D.1 Inductor (wound)

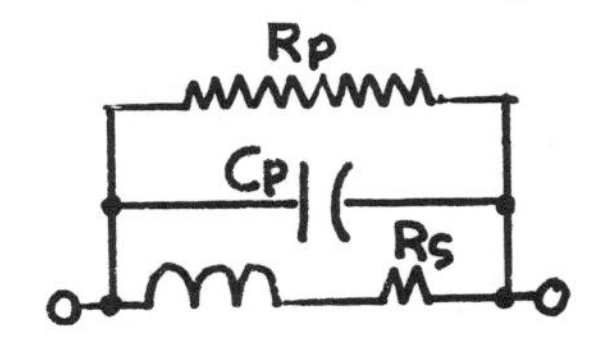

Example; typical values for a 50μH power line filter choke:
Cp = 30pF, Rs = 0.01Ω, Rp = 10kΩ

D.2 Ferrite bead

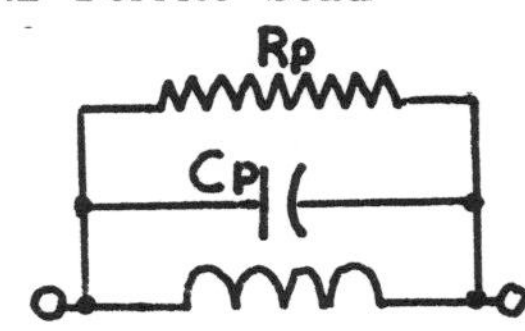

Lp, Rp, Cp depend on the material, number of turns and wire insulation diam.
Example; typical large bead, FAIRITE Mat.#43
L = 0.3μH (1 turn) x N2, R = 300Ω, Cp = 5pF

D.3 Capacitor

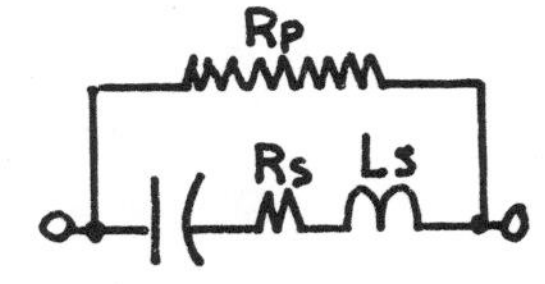

Example; 10 - 100nF, Ceramic, bypass type:
Rs = 0.2Ω, Ls = 6nH (discrete), 2nH (SMT)

D.4 Complete digital IC power supply circuit

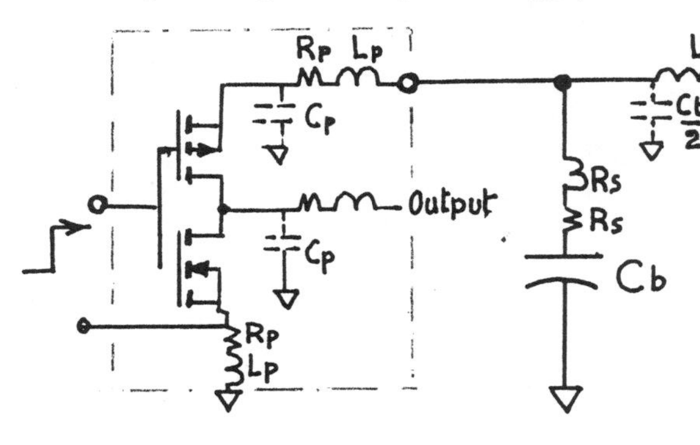

Lt, Rt, Ct = pcb trace.
Typ. values:
Lt = 10 nH/cm (no gnd plane)
Lt = 5 nH/cm (with gnd plane)
Rt = 0.01 Ω/cm
Ct =0.8 pF/cm

Rp, Lp, Cp = package leads.
Typ.values:

	surf. mount	DIP
Lp (nH)	2.5 - 5	5 - 15
Rp (Ω)	0.1	0.1
Cp (pF)	0.5 - 1.2	1 - 3

1. Keenan, K. 1983. *Digital Design for Specification Compliance.* Keenan Publ.
2. Costache, G., and R.L. Khan. *Finite Element Method Applied to Modeling Crosstalk on PC Boards.* IEEE EMC Transactions; February 1989.
3. Paul, C.R. and Bush,D. *Radiated EMI from Common-Mode Currents.* IEEE EMC Symposium (Atlanta), 1987.
4. Garn,H. *Proposal for new radiated emission test method..* EMC Sympos. Zurich,1991
5. Crawhall,R. *EMI potential of non-periodic signals.* IEEE/EMC Sympos., Anaheim, 1992

5A. Hill,L. *Video signal analysis for EMI Control.* IEEE/EMC Sympos. Chicago, 1994

6. Bush,D. *Spread Spectrum Clocks.* IEEE/EMC Sympos., Chicago 1994
7. Roy,O. *Influence of the resident sofware on radiated EMI.* Encyclopédie CEM. Dunod, Paris, 2000.
8. Mardiguian, M. *EMC in Components and Devices.* Gainesville, Virginia: Interference Control Technologies, Inc. 1987
9. Erwin, V. and K. Fisher. *Radiated EMI of Multiple IC Sources.* Proceedings of the 1985 IEEE EMC Symposium.
10. Johnson,H. *High speed Digital Design.* Prentice Hall, 1993.
11. Lutjens,W. *Reducing Electromagnetic Emissions of Micro-controllers.* COMPONIC Conference, Paris,1993
12. Perrin,J.C.*Increased EMI by CMOS Oscillators.* IEEE/EMCSympos. Santa Clara,1996
13. DeMoor,R. *Reducing EMI using PLL-based Oscillators.* IEEE/EMC Sympos. Santa Clara, 1996
14. Coenen,M. *EMC Workbench Faraday Cage.* Philips Journal of Research, 1994
15. Muccioli,J. *Radiation from MicroProcessors.* IEEE/EMC Sympos.,1990 (p.292) & 1997 (p.204).
16. Muccioli,J. *Radiation from MicroProcessors.* IEEE-EMC Transactions, Aug.99
17. Goulette, *Measurem. of Radiated EMI from ICs.* IEEE/EMC Sympos. Anaheim, 1992
18. Hockanson, *Finite Impedance of Reference Planes.* IEEE-EMC Transactions, Nov 1997.
19. Leferink, *Inductance Calculations.* IEEE/EMC Sympos,Santa Clara, 1996.
20. Val, C. *High Performance Surface-Mount VHSIC Packages .* Fourth Internatl Microelectronics Conference, Kobe, Japan, 1986.
21. Danker,B. *PCB Design Guidelines.* Philips Appliances Lab.
22. White, D.R.J. *EMI Control in the Design of PCBs and Backplanes.* Interference Control Technologies Inc. Gainesville, Virginia.1981
23. Charoy, A. PCB Design Seminars. (Various times and locations.)
24. Caniggia,S. *Reducing EMI in Multilayer Boards.* IEEE/EMC Sympos. Santa Clara, 1996
25. Heinke, R. *EMC Controlled PCB Design.* EMC Expo Conference. Interference Control Technologies, Inc.Gainesville, Virginia, 1989.
26. Fluke, J.C. 1991. *Controlling Conducted Emissions by Design.* New York: Van Nostrand Reinhold.

27. White, D.R.J. and Mardiguian,M. *Electromagnetic Shielding.* Interference Control Technologies, Inc.Gainesville, Virginia 1988
28. Cowdell, R. *Don't Experiment with Ferrite Beads.* Electronic Design, June 1976.
29. Gavenda, J.D. *Effectiveness of Toroid Choke in Reducing CM Current.* Proceedings of the 1989 IEEE EMC Symposium.
30. Ritenour, T.J. *Design to Control Common-Mode Current Emissions.* EMC design session, MIDCON 1982, Dallas, Texas.
31. Mayer, F. *Absorptive Low-Pass Cables.* IEEE/EMC Transactions, Febr. 1986.
32. Martin, A. *Introduct. to Surface Transfer Impedance.* EMC Technology.Jul 1982
33. Palmgreen, C. *Shielded Flat Cables for EMI/ESD Reduction.* 1981 IEEE/EMC Sympos. Also in EMC Technology, July 1982.
34. Federal Communications Commission (FCC) Parts 15 and 18. Washington, DC: U.S. Government Printing Office.
35. Military Standards 461-C,D and 462. Washington, DC: U.S. Department of Defense.
36. RTCA Standard DO160.Washington,DC:Radio Technical Commission for Aeronautics
37. Mardiguian,M. *EMI Troubleshooting Techniques.* Mc Graw Hill, N.Y. 1999
38. Chatterton,P., Houlden,M. EMC, *Theory to Practical Design..* Wiley
39. Bush,D. *Radiation from Printed Cct Board Oscillator.* IEEE/EMC Sympos.,1981
40. Walker,C. *Capacitance, Inductance and Crosstalk Analysis.* Artech House, 1990.

Index